Developments in Applied Phycology 9

Series editor
Michael A. Borowitzka, Algae R&D Centre, School of Veterinary and Life Sciences,
Murdoch University, Murdoch, WA, Australia

More information about this series at http://www.springer.com/series/7591

Anicia Q. Hurtado • Alan T. Critchley
Iain C. Neish
Editors

Tropical Seaweed Farming Trends, Problems and Opportunities

Focus on *Kappaphycus* and *Eucheuma* of Commerce

Springer

Editors
Anicia Q. Hurtado
Integrated Services for the Development of Aquaculture and Fisheries (ISDA) Inc.
Jaro, Philippines

Alan T. Critchley
The Evangeline Trail
Paradise, Nova Scotia, Canada

Iain C. Neish
PT Sea Six Energy Indonesia
Bali, Indonesia

Developments in Applied Phycology
ISBN 978-3-319-63497-5 ISBN 978-3-319-63498-2 (eBook)
DOI 10.1007/978-3-319-63498-2

Library of Congress Control Number: 2017952065

Printed on acid-free paper

This Springer imprint is published by Springer Nature
The registered company is Springer International Publishing AG
The registered company address is: Gewerbestrasse 11, 6330 Cham, Switzerland

K. alvarezii ***K. striatus*** ***E. denticulatum***

Hanging long-line technique, the most common technique of cultivating eucheumatoids

Foreword

According to recently published data, the seaweed hydrocolloid industry, comprising agar, alginate and carrageenan extracts, continues to grow in the order of 2–3% per year with the Asia-Pacific region increasingly dominating the raw material and manufacturing aspects of the industry.

Except for Japanese nori, the production of seaweed hydrocolloids consumes the largest amount of macroalgae annually, and carrageenan is the largest consumer of this group. The industry has been and is still undergoing structural changes largely led by Indonesia which is now the world's largest producer of agar- and carrageenan-bearing seaweeds. In addition, China is now the world's largest combined processor of seaweed hydrocolloids and is, to a considerable extent, reliant on imported raw materials. As noted in one of the chapters of this book, in 2016, about 250, 000 dry shades of *Kappaphycus* and *Eucheuma* (referred to together as the eucheumatoid seaweeds) entered value-chains as raw materials for single-stream processing that led to the production of carrageenan. The present work focuses fundamentally on these species in the industry of carrageenan extraction at the present time.

The developed topics cover areas ranging from the commercial development of eucheumatoid algae to biodiversity, biogeography, molecular genetics, ecophysiology, cultivation, micropropagation, diseases, the impact of climate change, harvesting and transport, biorefinery, applications of iota and kappa carrageenan variants and the development of value-chains.

Professor Leonel Pereira
Department of Life Sciences, IMAR-CMA
and MARE (Marine and Environmental Sciences, Centre)
University of Coimbra, Coimbra, Portugal

Contents

Contributors

Majid Khan Majahar Ali School of Mathematical Sciences, Universiti Sains Malaysia, Penang, Malaysia

Seaweed Research Unit (UPRL), Faculty of Science and Natural Resources, Universiti Malaysia Sabah, Kota Kinabalu, Sabah, USA

Erick Ask FMC Bio Polymer, Philadelphia, PA, USA

Rhodora V. Azanza The Marine Science Institute, University of the Philippines, Diliman, Quezon City, Philippines

Ross Campbell CyberColloids Ltd, Carrigaline Industrial Park, Carrigaline, Co. Cork, Ireland

Beatriz Castelar Fundação Instituto de Pesca do Estado do Rio de Janeiro (FIPERJ), Rio de Janeiro, Brazil

M.L. Cornish Acadian Seaplants Limited, J.S. Craigie Research Centre, Cornwallis, NS, Canada

Alan T. Critchley The Evangeline Trail, Paradise, Nova Scotia, Canada

Ik Kyo Chung Department of Oceanography, Pusan National University, Busan, Metro City, Republic of Korea

K. Eswaran Division of Marine Biotechnology & Ecology, CSIR-Central Salt & Marine Chemicals Research Institute, Bhavnagar, India

Ahmad Fudholi Solar Energy Research Institute (SERI), Universiti Kebangsaan Malaysia, Bangi, Selangor, Malaysia

Grevo S. Gerung Faculty of Fisheries and Marine Science, Sam Ratulangi University, Manado, Indonesia

Leila Hayashi Aquaculture Department, Universidade Federal de Santa Catarina (UFSC), Santa Catarina, Brazil

Integrated Services for the Development of Aquaculture and Fisheries (ISDA) Inc., Jaro, Philippines

Sarah Hotchkiss CyberColloids Ltd, Carrigaline Industrial Park, Carrigaline, Co. Cork, Ireland

Anicia Q. Hurtado Integrated Services for the Development of Aquaculture and Fisheries (ISDA) Inc., Jaro, Philippines

Danilo B. Largo Office of Research/Department of Biology, University of San Carlos, Talamban, Philippines

Phaik-Eem Lim Institute of Ocean and Earth Sciences (IOES), University of Malaya (UM), Kuala Lumpur, Malaysia

Institute of Biological Sciences, University of Malaya (UM), Kuala Lumpur, Malaysia

Rafael R. Loureiro Blue Marble Space Institute of Science, Seattle, WA, USA

Maria Rovilla J. Luhan Aquaculture Department, Southeast Asian Fisheries Development Center (SEAFDEC), Tigbauan, Iloilo, Philippines

Flower E. Msuya Institute of Marine Sciences, University of Dar es Salaam, Zanzibar, Tanzania

Mohana Sundaram Muthuvalu Department of Fundamental and Applied Sciences, Faculty Of Science and Information Technology, Universiti Teknologi PETRONAS, Tronoh Perak, Malaysia

Iain C. Neish PT Sea Six Energy Indonesia, Bali, Indonesia

José G. Ortiz-Tena Chemistry of Biogenic Resources, Technical University of Munich, Straubing, Germany

Siew-Moi Phang Institute of Ocean and Earth Sciences (IOES), University of Malaya (UM), Kuala Lumpur, Malaysia

Institute of Biological Sciences, University of Malaya (UM), Kuala Lumpur, Malaysia

C.R.K. Reddy Division of Marine Biotechnology and Ecology, CSIR-Central Salt and Marine Chemicals Research Institute, Bhavnagar, India

Renata P. Reis Instituto de Pesquisa Jardim Botânico do Rio de Janeiro (JBRJ), Rio de Janeiro, Brazil

Daniel Robledo Cinvestav Unidad Mérida, Departamento de Recursos del Mar, Mérida, Yucatán, Mexico

Mohd Hafidz Ruslan Solar Energy Research Institute (SERI), Universiti Kebangsaan Malaysia, Bangi, Selangor, Malaysia

Giselle P.B. Samonte ERT, Inc., Silver Spring, MD, USA

Alex Alves dos Santos Empresa de Pesquisa Agropecuária e Extensão Rural de Santa Catarina (EPAGRI), Centro de Desenvolvimento de Aquicultura e Pesca, Florianópolis, Santa Catarina, Brazil

Doris Schieder Chemistry of Biogenic Resources, Technical University of Munich, Straubing, Germany

Volker Sieber Chemistry of Biogenic Resources, Technical University of Munich, Straubing, Germany

Fraunhofer IGB, Straubing Branch BioCat, Straubing, Germany

Calvyn F.A. Sondak Faculty of Fisheries and Marine Science, Sam Ratulangi University, Manado, Indonesia

Jumat Sulaiman Mathematics with Economics Programme, Faculty of Science and Natural Resources, Universiti Malaysia Sabah, Kota Kinabalu, Sabah, Malaysia

Miguel Supelveda Ilha Grande, Rio de Janeiro, Brazil

Shrikumar Suryanarayan PT Sea Six Energy Indonesia, Bali, Indonesia

Ji Tan Department of Agricultural and Food Sciences, Universiti Tunku Abdul Rahman (UTAR), Kampar, Perak, Malaysia

Gloria Batista de Vega Director de Investigación y Desarrollo (I+D) de Gracilarias de Panamá S.A., and Facultad de Ciencias Naturales y Tecnología, Universidad de Panamá, Panamá

Suhaimi Md. Yasir Seaweed Research Unit (UPRL), Faculty of Science and Natural Resources, Universiti Malaysia Sabah, Kota Kinabalu, Sabah, Malaysia

Nair S. Yokoya Institute of Botany, Environmental Secretary of São Paulo State, São Paulo, Brazil

Wilson Thau Lym Yong Biotechnology Research Institute, Universiti Malaysia Sabah, Kota Kinabalu, Sabah, Malaysia

About the Editors

Anicia Q. Hurtado was a senior scientist at the Aquaculture Department, Southeast Asian Fisheries Development Center (SEAFDEC-AQD), Tigbauan, Iloilo, Philippines, for 20 years. She spearheaded the Seaweed Program of AQD during her entire career at the centre. She is at present the Chair of the Integrated Services for the Development of Aquaculture and Fisheries (ISDA Inc.), an organization of past and present scientists of SEAFDEC-AQD. She finished her Doctor of Agriculture (Phycology) at Kyoto University, Kyoto, Japan, as a Monbusho Scholar. She is an awardee of DA-BFAR Best Research Paper in Fisheries and Aquaculture (1991, 1992), Dr. Elvira O. Tan Memorial Award for Best Research Paper in Aquaculture in 2003 and Best Poster Presentation in 2007 at the International Seaweed Symposium, Kobe, Japan. She works mainly on the aquaculture of *Kappaphycus* as a consultant to international and local government and non-government agencies directly involved with seaweed farmers. At present, she is developing "new strains" of *Kappaphycus* using tissue culture techniques for possible sources of propagules for commercial farming. She works also on the mitigation of *Neosiphonia* infestation in *Kappaphycus* farms using seaweed extract from the brown seaweed *Ascophyllum nodosum*. She has published several papers on *Kappaphycus*, *Eucheuma* and *Gracilaria* as senior and co-author in peer-reviewed journals and book chapters; she has written manuals, flyers and posters as teaching and training materials for seaweed farmers. She is a trainer, resource person and lecturer to local and international training programmes and workshops particularly on ecophysiology, land-sea-based nurseries, seaweed aquaculture, crop management (disease and epiphytes), post-harvest management and research methodologies. She is a speaker in plenary and mini-symposium sessions in international congresses. She joined the editorial board of *Botanica Marina* for 6 years, and she is a regular reviewer of manuscripts submitted to the *Journal of Applied Phycology* and other international fisheries and aquaculture journals.

Alan T. Critchley grew up in Birmingham, UK. In 1981, he graduated with a PhD from Portsmouth University, in marine ecology, having been based at the university's Hayling Marine Laboratory studying the invasion of the brown seaweed *Sargassum muticum*. This was followed by a Royal Society European Postdoctoral Fellowship at the Delta Institute, Yerseke, Netherlands. He then moved to South Africa to teach in the Botany Department of the University of KwaZulu-Natal, Pietermaritzburg (7 years), and the University of the Witwatersrand, Johannesburg (10 years). He researched sub-tropical seaweeds to the east and cold-water, upwelling-influenced species to the west and developed a much keener interest in seaweed benefits and their applications. He also collaborated with Professor M. Ohno of Kochi University, Japan, to co-edit volumes of *Seaweed Cultivation and Marine Ranching* and *Seaweed Resources of the World*. In 1998, he moved to the University of

Namibia, Windhoek, to become a research professor at the Multidisciplinary Research and Consultancy Centre; seaweeds and their uses were at the centre of a number of novel projects there.

In 2001, he made the transition from academia to the commercial world. The call of carrageenan-bearing seaweeds took him to Normandy, France and a processing facility, then operated by Degussa Texturant Systems, to be in charge of their New Raw Materials Laboratory. It was at that time that his passion for the *Eucheuma* seaweeds, their biology, production, utilization and socio-economic importance, was fostered. It was here that he was most fortunate to begin collaborative research with Dr. Anicia Q. Hurtado.

In 2005, he moved to Nova Scotia, Canada, to work with a local seaweed processing company. This role involved the management of multidisciplinary, collaborative projects based on commercial seaweeds (browns and reds). Ensuring sustainable production techniques for seaweed resources and the scientific validation of biostimulant and bioactive properties of their extracts was a major focus.

Carrageenophyte seaweeds continue as a research focus. It was through these activities that Dr Iain Neish and Alan were introduced. A common goal led Anne, Iain and Alan to co-edit this current collation of the broad spectrum of science associated with the *Eucheuma* seaweeds, their production, processing, applications and most importantly socio-economic contributions to coastal communities. For this book, they turned to the expert teams included in each chapter to bring their vision to reality; the expert teams did a marvellous job!

He is currently pursuing sustainable utilization of resources, including encouraging research to stimulate 100% multiple-stream utilization of a variety of seaweed biomasses for their applications in a myriad of current and future potential uses, from nutrient-dense food to biologically active properties, which can collectively provide greatly beneficial goods and services to terrestrial and marine ecosystems, humans, animals, plants and microbes.

Iain C. Neish is a Canadian marine biologist and businessman who has worked with seaweeds in aquaculture systems since 1965. He followed the lead of his father, Arthur C. Neish, who played a role in the development of seaweed cultivation in Nova Scotia, Canada. Since 1977, most of Iain's career involved seaweed farm development and factory installations in the Philippines, Malaysia and Indonesia. He has also worked in the Americas, Africa and India. More than 25 years of this work was conducted while Iain was working as consultant, then employee of Marine Colloids (later FMC Corporation and then DuPont) until 1997. During that time, he was project manager for seaweed farm development and for the construction of the world's first factory that made semi-refined carrageenan (SRC) in Cebu City, Philippines. From 2003 to 2015, Iain undertook projects with various international organizations including IFC, GTZ, USAID, AusAID FAO, ILO and UNIDO, and he also undertook seaweed-related projects with several private companies. He is currently engaged with innovative seaweed business ventures in Indonesia in his role as a director of PT Sea Six Energy Indonesia. He lives with his family in Bali.

Reflections on the Commercial Development of Eucheumatoid Seaweed Farming

1

Iain C. Neish, Miguel Sepulveda, Anicia Q. Hurtado, and Alan T. Critchley

Abstract

The development of eucheumatoid seaweed agronomy is an outstanding example of widespread aquaculture that evolved from simple methods refined mainly by farmers in the sea. Innovations were stimulated, observed, recorded, modified and disseminated and driven toward commercial implementation by managerial, technical and scientific inputs from private- and public-sector organizations. The impetus for development was strong market demand from hydrocolloid manufacturers who desperately needed cultivated raw material sources by the 1960s. Such sources were required to augment, replace or complement limited supplies of seaweeds from wild stocks. Biomass selected from wild populations provided cultivars that formed the base for commercial seaweed farming. Cultivars were disseminated amongst farming locations through informal channels that often left the provenance of those "seedstocks" as a source of speculation, rather than with a basis in fact. Since the early 1960s, eucheumatoid seaweed production spread to several jurisdictions around the world, however, production volumes virtually leveled off to approximately 250,000 MT year^{-1} by about 2007 as the available seaweed supplies became adequate for the, by then, low-growth carrageenan markets. In what were effectively zero-sum carrageenan markets, production of raw materials came to be dominated by Indonesia and the Philippines. Indonesia attained the larger market share due to reduced production from the Philippines. During more than four decades of the development of eucheumatoid, seaweed agronomy, speculation flowed freely while the scientific testing of theories and hypotheses attracted little financial support. Innovation was minimal, not only in the field of seaweed agronomy, but also with respect to product and market development. It was mainly only within the last decade that commercial innovations commenced beyond the rheological applications of carrageenan and legacy agronomy production systems. These developments were initiated in regions as far-flung as India, Indonesia and Brazil. Such developments are seen as essential drivers for the next phase of commercial development and next-generation applications of the eucheumatoid seaweeds.

The varied chapters of this book fill in details of the historical narrative presented in the present chapter and they also describe some of the innovations that are driving step-changes in the industry as it evolves towards a promising, sustainable future.

I.C. Neish (✉)
PT Sea Six Energy Indonesia, Bali, Indonesia
e-mail: ineish@sea6energy.com

M. Sepulveda
Ilha Grande, Rio de Janeiro, Brazil

A.Q. Hurtado
Integrated Services for the Development of Aquaculture and Fisheries (ISDA) Inc., McArthur Highway, Tabuc Suba Jaro, 5000 Iloilo City, Philippines

A.T. Critchley
The Evangeline Trail, Paradise, Nova Scotia, Canada

A.Q. Hurtado et al. (eds.), *Tropical Seaweed Farming Trends, Problems and Opportunities*, Developments in Applied Phycology 9, DOI 10.1007/978-3-319-63498-2_1

1.1 Overview of the Development Time-Line for Eucheumatoid Seaweed Farming

In writing the present chapter, the authors must echo the *caveat* of Bixler and Porse (2011) and Porse and Rudolph (2017) who stated (prefaced from both): "*The information in these papers has been prepared by the authors primarily from their own knowledge of the seaweed hydrocolloid industry, and we take full responsibilities for its contents.*" The present reflections cover commercial developments where hard data were mostly commercially-confidential and the publicly-released numbers were often closer to being rhetoric than they were to being "true data".

A major source of the reflections in this chapter were from the authors' attendance at the seaweed gatherings that became increasingly numerous as the industry developed. An outstanding example was the triennial International Seaweed Symposia (ISS) of the International Seaweed Association (first attended by the corresponding author in 1965 at Halifax, Nova Scotia, Canada), where actors from all industry segments met for technical sessions and encounters over drinks and meals; for example at dinners of the 'Liars Club' where Harris J. (Pete) Bixler convened industry 'old-hands' and their descendants at every ISS over the past few decades. To a limited extent this role has also been adopted by the meetings of the International Society of Applied Phycology (ISAP). Many of the presentations on developments in carrageenan applications have been taken on by the "Wrexham Meeting".

Figure 1.1 traces key events in commercial farming development for eucheumatoid seaweeds from its crude beginnings to the time of writing. It indicates the pattern of global production based on estimates presented in Section 1.6, as eucheumatoid dry weight production reached levels in the order of 250,000 t year^{-1}. More details of the indicated events are presented in the following chapter sections; the over-arching trends were:

1.1.1 The major 'alpha' and 'beta' suppliers, Indonesia and the Philippines, were commercial sources of indigenous eucheumatoid seaweeds even before the development of farming and some wild harvest continued to enter commerce until within the past decade (Neish, personal observations).

1.1.2 By the time of writing, Indonesia had become an 'alpha' source capable of cost-effectively supplying all global eucheumatoid seaweed requirements of the carrageenan industry and much more besides.

1.1.3 By approximately 2007, the Philippines became a 'beta' source, with most domestic eucheumatoid seaweed production being processed by Philippines-based, carrageenan processors, who also imported substantial tonnages of raw materials from Indonesia.

1.1.4 Commercial farming of eucheumatoid seaweeds was trialed in about forty jurisdictions (Ask et al. 2003; Neish 2005). In many (if not most) jurisdictions, the eucheumatoid seedstocks used for farming were not indigenous. Farming was technically successful in many of those jurisdictions, but none developed production and markets anywhere near to the scales of Indonesia or the Philippines.

1.1.5 By the time of writing, carrageenan manufacturers were concentrated in China, Europe, the USA, the Philippines and Indonesia. Furthermore, these processors had little incentive to push development from scattered sources, with no apparent sustainable cost advantages over Indonesia and the Philippines.

1.1.6 As eucheumatoids begin to find markets beyond the mundane applications of carrageenan in processed foods, particularly during the past decade (see Chap. 12 of this book), innovative technologies began to develop in India, Indonesia and Brazil. Such innovations now provide opportunities for globally-dispersed development far beyond "just carrageenan".

By 2017 eucheumatoid seaweed-producing jurisdictions could be placed into four classifications, which have been identified by Greek alphabet labels in the spirit of notional nomenclature as applied to various types of extractable carrageenan (viz. iota, kappa and lambda). Value-chain implications for these categories are discussed fully in Chap. 12 and they are outlined here as follows:

The 'alpha' sourcewas Indonesia, which had risen to be capable of supplying the entire global requirement for cultivated eucheumatoid seaweeds, at prices that were hard to beat.

The 'beta' source was the Philippines, which could supply a full range of types eucheumatoid seaweed raw materials in industrial quantities, sufficient to supply much of the demands from the Philippines' domestic processors but was, nevertheless, a net importer of dried seaweeds which were required in order to enable processors to meet the needs of global customers. Limits to Philippines' seaweed production were largely attributable to the stochastic and catastrophic impacts of typhoons in the northerly regions and also armed conflicts in the southerly regions of the country.

The 'gamma' sources were jurisdictions that were farming one or more species of eucheumatoid seaweeds but had a total production that was relatively "minor" (i.e. less than 5% of those amounts which were available from the Philippines and Indonesia). Prominent examples were Malaysia, Tanzania, Brazil and India.

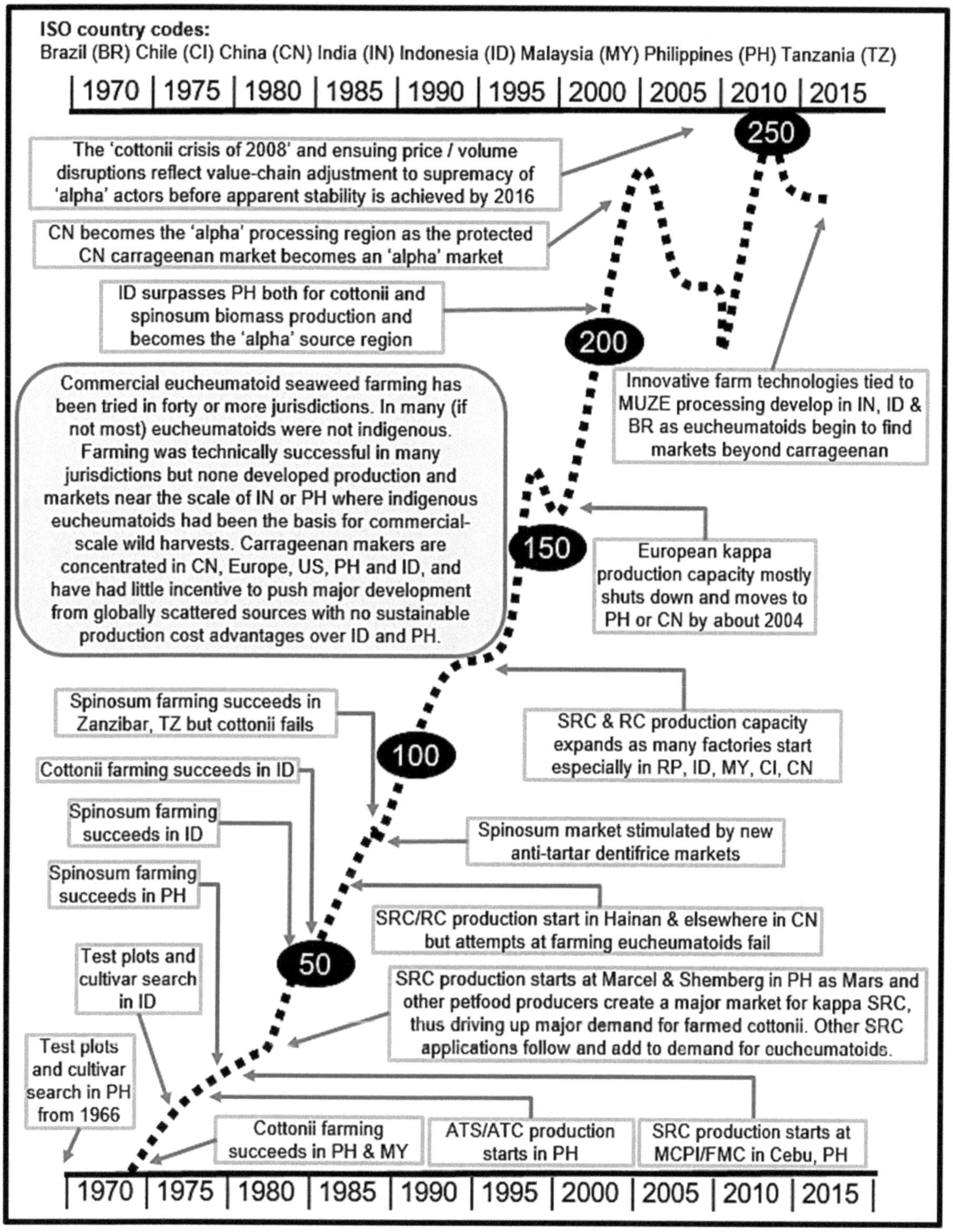

Fig. 1.1 Schematic overview of commercial farm development for eucheumatoid seaweeds. The *dotted line* indicates the pattern of global production based on the cottonii estimates, shown in Fig. 1.6.1, as the volume reached and began to plateau at about 250,000 MT year^{-1} (Note: "cottonii = *Kappaphycus alvarezii* and its various strains; "spinosum" = *Eucheuma denticulatum*; MCPI, FMC; SRC = semi-refined carrageenan; RC = refined carrageenan.)

The 'delta' sources were jurisdictions where farming had been successfully undertaken but robust commercial production of industrial quantities had not yet become established. Issues surrounding lower production levels included governance problems or weak value-chain links to global markets.

In addition to the four source categories described above, there were failed attempts to develop sources of raw materials and the history of such attempts also contributed to useful fore-ground knowledge (see Chap. 4 of this book). In some cases, failures in development of new sources of raw materials was due to systemic failures of crops to grow. In other cases, approvals and regulatory conditions required for the introduction of 'alien' cultivars had not been obtained, and/or such approvals had been denied.

1.2 The Urgent Need for Seaweed Farming by the Globalized Carrageenan Industry

During the decades before eucheumatoid seaweed farming first reached commercial proportions in around about 1974, the carrageenan industry was dependent on natural harvests of a wide range of tropical and cold water species from all over the world, including modest tonnages of eucheumatoid seaweeds from the Philippines and Indonesia. This created several problems, including:

1.2.1 Natural phenomena such as harvest pressures and trading factors caused the quantities, price and mixture of available seaweed species which varied unpredictably, geographically, seasonally and from year to year.
1.2.2 By the early 1970s, major wild stocks were being harvested at (or beyond) their sustainable yield limits.
1.2.3 The prevalence of raked and/or beach-cast raw materials caused deliveries to contain large quantities of extraneous algae and other debris as contaminants.

In a sense, these problems also created opportunities for the three, then dominant carrageenan processors of the early 1970s, namely Marine Colloids Inc. (later absorbed by FMC; then by DuPont/Dow Chemical in 2017), Copenhagen Pectin (eventually absorbed by Huber as CP Kelco) and Sanofi (ultimately absorbed by Cargill). Existing extraction processes were well-adapted to the 'cleanup' of poor raw materials and the energy required by those processes was still relatively cheap or cost-effective. The industry's extensive product lines; its technical expertise and its versatile production plant designs enabled the processing, blending, marketing and selling of varied products manufactured from a wide variety of seaweed mixes. Finally, the dominant market-share of the three, main established processors enabled them to buy and process all species of carrageenophyte available. This gave them dominance in the market place.

Before the eucheumatoid seaweeds came to be cultivated, there had already been trade in raw, dried seaweeds from wild stocks from Indonesia and the Philippines that amounted to a few thousand dry MTs per year and which served amongst many raw materials for the production of different types of carrageenans and their mixes. During the earliest days of evaluation and utilization of tropical, carrageenophyte seaweeds they were collectively referred to as: '*Eucheuma*' seaweeds or seaplants because the leading seaweed taxonomists of the time, considered them to all fall within that single genus. Seaweeds that were sources of iota carrageenan were nominated as '*Eucheuma spinosum*' and therefore acquired 'spinosum' as a name of the trade. This taxonomic designation was later amended to *Eucheuma denticulatum* (Burman) Collins *et* Hervey, however, the 'spinosum' name persisted within the trade.

Of lesser importance to the trade, at least in early days, was the kappa-carrageenan-bearing mix of seaweeds referred to taxonomically as '*Eucheuma cottonii*' and with 'cottonii' being used and persisting as its trade name. Taxonomic work by Professor Maxwell S. Doty and others, since the 1960s, ultimately resulted in the kappa-carrageenan-bearing eucheumatoid seaweeds being grouped under the common genus '*Kappaphycus*'. The predominant farmed species of the 1970s was designated as *Kappaphycus alvarezii* (Doty) Doty ex P.C. Silva in honor of Vicente B. Alvarez, the founding General Manager of Marine Colloids Philippines, Inc. Another species that has been identified prominently among the commonly cultivated varieties is *Kappaphycus striatus* (F. Schmitz) Doty ex P. C. Silva. Both of these *Kappaphycus* species continue to retain 'cottonii' as a name of the trade, although *K. striatus* is sometimes designated as 'sacol' after the island near Zamboanga, the Philippines, where early cultivars of the species were thought to have originated.

Materials gathered from wild populations in Indonesia and the Philippines in the early 1960s were purchased by carrageenan processors primarily due to the high value attached to iota carrageenan for specialty applications. Total production at that time only amounted to a few thousands of MTs per annum. It is interesting to note that the 'spinosum' and 'cottonii' types tended to be mixed, not only at their sources but also as they passed through the supply chain. As a result, processors tended to place a premium price on the far less abundant cottonii in order to encourage its separation from spinosum. It was known that cottonii contained kappa carrageenan, but it was of a type distinctly different from the kappa carrageenan extracts made from Irish Moss (*Chondrus crispus*) supplies that were then the mainstay of industry seaweed supplies (this extraction was known to be a mix of kappa and lambda carrageenans; among others). 'Cottonii kappa' was not yet an object of serious industry interest.

By the mid-1960s it was clear to the carrageenan industry that it could not diversify and grow based solely on the exploitation of wild seaweeds harvested from temperate-zone seashores. Investors in the marine hydrocolloid companies wanted growth and their managements perceived that the limits of natural seaweed supplies were being reached and therefore presented limitations to potential. It was concluded that cultivation would provide a way out of this predicament. Through liaison with researchers and entrepreneurs in various parts of the world, companies such as Marine Colloids Inc., Copenhagen Pectin and Sanofi stimulated and/or funded cultivation development projects in Canada, Mexico, the United States, Pacific Oceana, Indonesia, Malaysia and the Philippines, amongst other places on a global scale.

1.3 Early Development of Tropical Seaweed Agronomy

By 1965, finding ways to commercially farm carrageenophytes at an industrial scale became a high priority, especially for the world's leading, major carrageenan manufacturer: Marine Colloids, Inc. (MCI). Efforts to develop techniques for farming *Chondrus crispus* were already underway in collaboration with the National Research Council of Canada in Halifax Nova Scotia and the Nova Scotia Research Foundation. Stimulated especially by the 'holy grail' of abundant spinosum, for the manufacture of iota carrageenan, the then President of MCI: James (Jim) Moss, forged an alliance with Prof. Maxwell S. (Max) Doty at the University of Hawaii in Manoa (Doty, personal communications). At the time, Doty had been active in the seaweed world of the ASEAN region for several years and serendipitously amongst his graduate students he had three who would become pioneers in developing eucheumatoid seaweed farming, namely: Vicente B. (Vic) Alvarez (Filipino) in the business sector, Gavino C. Trono, Jr. (Filipino) and Keto Mshigeni (Tanzanian) in the academic sector.

Vic Alvarez was appointed as Managing Director of a newly formed company called Marine Colloids, Philippines Inc. (MCPI) and, with support from MCI, the parent company of MCPI, he commenced attempts to farm spinosum in the Philippines by 1966. Funding came mainly from Marine Colloids, USAID and SEAGRANT. As something of an after-thought, cottonii was included, with spinosum in farm development attempts but spinosum was the target of the most intense efforts. The development teams of Alvarez established test-plots and cultivar-search programs at several locations throughout the Philippines (Fig. 1.2) in collaboration with the Bureau of Fisheries and Aquatic Resources (BFAR, formerly called the Philippine Fisheries Commission) teams led by Dr. Inocencio A. Ronquillo and with University of the Philippines teams led by Dr. Gavino C. Trono, Jr.

Intensive surveys and collection of data were conducted from 1967–1970 to assess the best cultivation sites and seek out further suitable cultivars. Wild crops of spinosum in Panagatan Cay, Antique had continually produced more seaweed than any other area of the Philippines at that time (Parker 1974), so fronds from these wild stocks served as 'seed' (seedlings or seedstocks) for the domestication of cultivars in Panagatan Is., Caluya, Antique and Ilin Is. Occidental Mindoro, Philippines, which began in 1969. However, after a few years, the experimental cultivation sites in Caluya, Antique and Occidental Mindoro were abandoned due to their frequent typhoons, excessive grazing by herbivorous fish (siganids) and management issues (Alvarez, personal communications).

Cultivation trials were made in Tapaan Is., Siasi, Sulu in 1971 and these proved to be the most promising amongst all of the test sites, with an average growth rate of 1.5–5.5% day^{-1} (Doty 1973; Parker 1974). By 1972, experiments were also conducted at Sacol Island, Zamboanga, using fronds of indigenous, local species in order to develop further information on their agronomic value (Doty 1973). Unfortunately, 'peace and order' problems were endemic to the area. Meanwhile the Marine Colloids teams continued to seek more peaceful areas, deemed more suitable for cultivation within the Sulu Archipelago. Farm development efforts ultimately migrated to the vicinity of Sitangkai, Tawi-Tawi which is one of the most southerly islands of the Philippines. The region is home to thousands of shallow reef areas with sandy and coralline sea floors, clear waters, moderate to strong water movement and thousands of coastal dwellers seeking the means of a sustainable livelihood. Those people were to become the commercial, eucheumatoid seaweed farmers who have formed the core of the industrial work force ever since.

Characteristics of the Sitangkai region were ideal for eucheumatoid seaweed cultivation, as asserted by Doty (1973). Later surveys by Barraca and Neish in 1978 supported the assertion that regions of the islands of Sitangkai, Sibutu and Tumindao were "near-ideal" for seaweed farming; these areas were developed by Marine Colloids and have remained the major producing region of the Philippines to the present day.

For about five years little progress had been made towards commercial spinosum farming and skepticism prevailed in

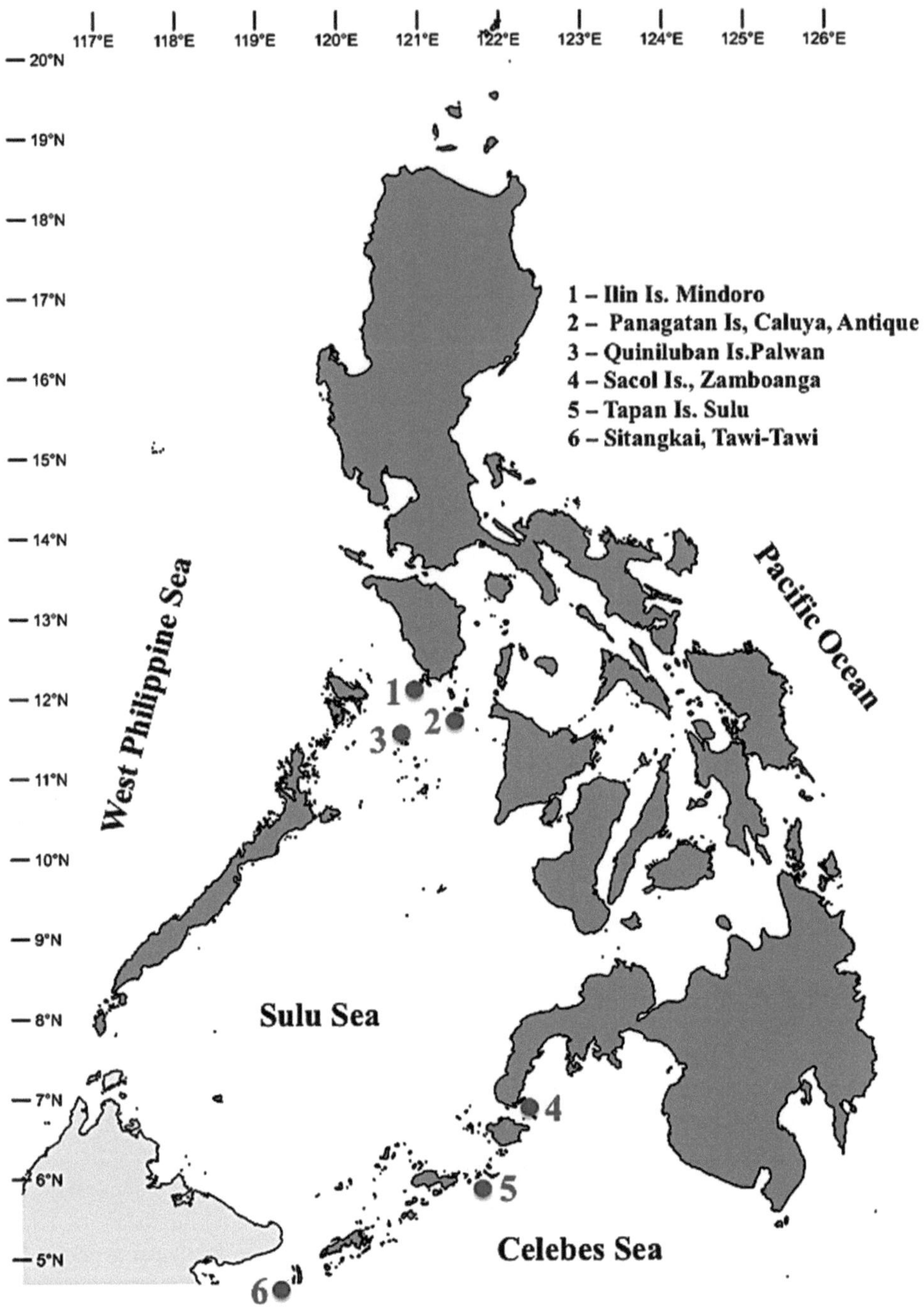

Fig. 1.2 Locations of earliest *Kappaphycus* and *Eucheuma* farming efforts in the Philippines

the industry. Development programs were in danger of cancellation when Louis E. Deveau of Marine Colloids took over as the manager of Vic Alvarez at MCPI. He found that as of 1971, fast growing, vegetative strains of *Kappaphycus* spp. had been grown by off-bottom nets and mono-lines in Tawi-Tawi. Deveau discovered that impressively large fronds could be produced (Fig. 1.3) and he strongly supported continuation of the farm development efforts made by Marine Colloids. Cultured cottonii then began to trickle into the market starting 1971–1973 and contributed to the annual harvests of about 400–500 MTs. By 1974, the impact of exponential vegetative growth finally manifested itself. The cottonii harvest leaped to 12,000 MT (Deveau, personal communications). At that time, that amount of biomass was three to four times what the market could absorb. The rapid and unexpected 'explosion' of farmed cottonii production came as a surprise to virtually everyone in the carrageenan industry and led to profound changes in it. For the first time, a major raw material was available in what seemed to be 'unlimited' quantities.

The turning point in the success of farming development and transition to well-developed seaweed agronomy was

Fig. 1.3 A large frond of *Kappaphycus alvarezii* from the early 1970s. L. Deveau is shown far right, with members of an MCPI farm development team in the Sitangkai region of Tawi-Tawi, Philippines (Photo: L. Deveau; first published in Hurtado et al. 2015)

arguably the point at which the persistently fast-growing, vegetative cultivars were finally propagated to provide a 'critical mass' of seaweed biomass. Fronds of the holotype *Eucheuma alvarezii* (Hurtado et al. 2016) were collected near a channel, west of Karindingan Island on Creagh Reef, South Semporna, Sabah, Malaysia (Doty 1985) and the teams of Vic Alvarez propagated several varieties of cultivar to the point where they were 'domesticated' and widely farmed in the Philippines (see Chaps. 2 and 4 in this book). A seaweed farmer named Mr. Tambalang was credited by Doty and Alvarez with first developing *Eucheuma alvarezii* var. tambalang near Omadal Is., Tawi-Tawi. The Tambalang cultivar is thought to have been the origin of most *K. alvarezii* which is farmed around the world to the present day, although many other cultivars of uncertain origin are also found in the mix and additional cultivars may still be recruited from time to time. A variety of morphotypes of various eucheumatoid seaweeds have also appeared throughout the regions where they are farmed, but lack of provenance information and the morphological plasticity typical of eucheumatoid seaweeds, makes it impossible to differentiate amongst them visually with any degree of certainty. Molecular taxonomy techniques are therefore best used to discriminate amongst the apparent myriad of cultivars (Hurtado et al. 2016).

Within the carrageenan industry, there was some disappointment that it was cottonii that first came to be farmed successfully. By 1977, however, Copenhagen Pectin (CP) and its local subsidiary Genu Philippines had established spinosum farming on Danajon Reef, Northern Bohol, the Philippines thanks to teams led by Hans Porse and managed by Jose R. Lim, Tirso Lirasan, Saturnino Soria Jr., Silver Cabanero and others (see Porse and Rudolph 2017). By 1979 the CP farm system and the MCPI farm system managed by Maximo A. Ricohermoso were amongst several enterprises supplying commercial quantities of spinosum to processors around the world. However, even then, it proved to be a difficult crop to grow. Spinosum cultivars of that time mainly showed sustainable growth in the far south island of Sibutu and on Danajon and Hingutanan reefs between Bohol and Leyte in the Central Visayas.

1.4 Successful Farming Leads to Disruptive SRC Technology

As supplies of raw materials for processing swamped demand in 1974, cottonii prices dropped below their production costs. Established processors did not want huge inventories and decided to buy only for their current needs. Processors took the position that they were "not in the seaweed business; they were in the carrageenan business". Therefore, established processors tended to buy cottonii at levels similar to those of prior years and sustained sources that had supplied them in the previous years, whilst also maintaining a price of about 1.10 PHP kg^{-1} (about 0.16 USD kg^{-1}, at that time).

This created an opportunity for traders; notably including Shemberg Marketing Corp. and Marcel Trading Corp. in the Philippines. These traders purchased excess cottonii stocks, on speculation, for prices as low as 15 centavos per kilogram (about 0.02 USD kg^{-1}). They rapidly bought and warehoused all available stocks and proceeded to dominate the market for the supply of cottonii raw seaweed to carrageenan manufacturers through 1976. They proceeded to become major seaweed suppliers to carrageenan processors before becoming carrageenan producers themselves in the early 1980s. Meanwhile, they were joined in the business by numerous traders as a 'seaweed gold-rush' developed.

By 1977, the 1974 cottonii production had been sold off. This was an incentive for established processors to put considerable efforts into reviving farm production. At MCPI,

Maximo A. Ricohermoso had become General Manager, Ruben T. Barraca replaced Vic Alvarez as farm development manager in the Philippines and Iain C. Neish was given responsibility and assigned budget funds in order to manage seaweed resource development for MCI, not only in the Philippines but also all other jurisdictions around the world. Meanwhile, Hans Porse led teams from CP who undertook similar functions. By 1978, the supply and demand struck a balance and the way was paved for the Philippines' dry cottonii production to reach its present level of about 60–80,000 dry MT per annum (Fig. 1.5; Baricuatro 2015; Bixler and Porse 2011; Porse and Ladenburg 2015).

Traders who had bought excess cottonii in 1974 immediately commenced a vigorous selling campaign. They knew little of the structure of the carrageenan industry itself and offered carrageenophyte seaweeds to any conceivable buyer. Eventually European, Japanese, Filipino and Chinese agar producers took notice. Agarophytes were by then in short supply and it was quickly discovered that kappa carrageenan could be processed to make KCl-precipitated carrageenan, in some agar factories. It was also at that time that Litex of Denmark came in to the picture. The red alga, *Furcellaria* that had been their main source of raw material, was becoming scarce from wild stock harvests and they discovered that cottonii could be processed in the Litex factory; they became a major buyer and by 1984 they were absorbed into FMC (later to be absorbed into DuPont/Dow Chemical, as of 2017).

Japanese agar producers were intent on increasing the gum concentration in their raw materials and were buying alkali-treated raw materials called 'colagar'. It was found that *Kappaphycus*, like *Gracilaria*, could be cooked in high concentrations of KOH then washed, dried and made into an attractive feedstock for extraction plants. Such processing continues until the present day and is increasingly being used by major carrageenan manufacturers to pre-process (to reduce volumes for transportation and also reduce a costly step in the process and associated, chemical waste disposal) thereby preparing cottonii for export to their factories, especially in China. This alkali-treated seaweed (ATS) came to be milled by some companies to make Alkali-Treated Chips (ATC). When milled to a powder form, ATC came to be called: Semi-Refined Carrageenan (SRC) which is also known as Processed *Eucheuma* Seaweed (PES) or as E407a (as the European Union utilization coding).

Immediately following the appearance of ATS and ATC in markets around 1975, Japanese/Filipino/Chinese joint-ventures sprang up in the Philippines in order to produce these 'upgraded' raw materials. By 1978, it was known by carrageenan makers and by the pet food industry that ATS and ATC could be milled to become SRC; thus supplying the world with a cheap, low-energy form of carrageenan, which was only made possible because sustainable, cottonii farming was capable of supplying seemingly unlimited amounts of consistently high-quality, raw material.

For the major carrageenan extraction plants of the day, farmed cottonii had become a mixed blessing. Their plentiful supply of a carrageenan-bearing seaweed had indeed become a commercial reality but, at the same time, and most importantly, they were losing control of their raw materials supply chains. SRC was a low-technology product that would enable small processors to proliferate and to 'ruin' their control over carrageenan markets. The major carrageenan makers had to decide whether to embrace SRC; try to 'kill' SRC; or to ignore SRC altogether and hope it would "go away". Ultimately the decision was forced in 1978 by the pet food divisions of the Mars group of companies. They decided to use SRC in their products and expressed their intention to even put new manufacturers into the SRC business, if necessary. Suppliers such as MCI were informed that they must develop and sell SRC, or expect to lose the Mars business (Neish, personal observations). Furthermore, SRC-based, ice cream stabilizers from USA blending houses also appeared on the scene at that time. This coincided with a time of great stress for carrageenan-extract producers. Global energy costs had pushed the price of carrageenan extracts upwards through the 1970s and they were becoming a high-cost product, relative to other ingredient solutions competing on a price/rheological performance basis.

It was in approximately 1978 that MCI was in the process of becoming a wholly-owned division of FMC Corporation and in-coming FMC managers decided to enter the SRC business (then called AMF for Alkali-Modified [seaweed] Flour). The other two major carrageenan players of the day, opted to oppose the use of SRC, especially in human food. Mars therefore recruited Shemberg Marketing and Marcel Trading as SRC suppliers, so by the early 1980s the carrageenan market underwent a step-change both in product volumes and product types. Proliferation of SRC producers had begun (see Chap. 12 in this book). Pet food applications for SRC were quickly followed by applications in air-freshener gels, dairy products, water-gels and meat-packing. Demands from these markets became a driver for the major expansion of cottonii farming to the point where, as of 2016, cottonii accounted for about 80% of the total carrageenan-bearing, seaweed raw materials worldwide, in a total carrageenan market volume in the order of 60,000 +/− 10,000 MT (after Bixler and Porse 2011 and Porse and Rudolph 2017).

1.5 Shifting Value-Chain Governance Models

Panlibuton et al. (2007) first applied the heuristic models of Gereffi et al. (2005) to eucheumatoid seaweed value-chain governance in 2007 and such models remain highly pertinent

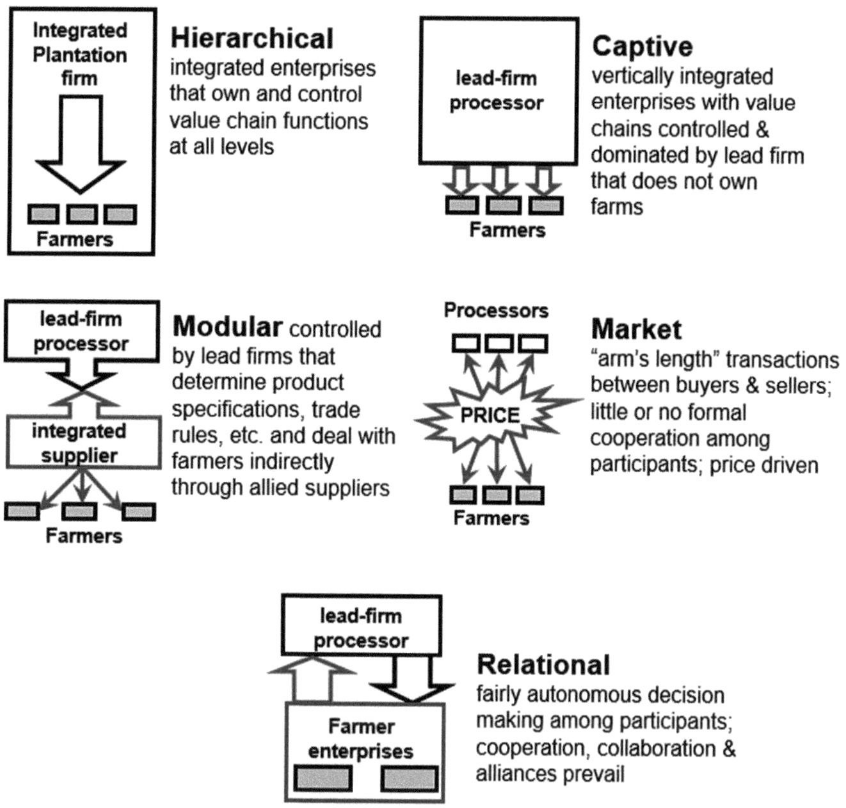

Fig. 1.4 Governance models that can be applied to eucheumatoid seaweed value-chains (After heuristics of Gereffi et al. 2005; Panlibuton et al. 2007)

to any analysis of commercial farming development (Fig. 1.4). As value-chain actors during the period in question; as value-chain analysts (e.g. Neish 2013a); and with insights communicated by other authors (Panlibuton et al. 2007; Porse and Ladenburg 2015; Porse and Rudolph 2017) the present authors postulated that the course of eucheumatoid seaweed value-chain governance in the Philippines and elsewhere followed the trajectory described below:

1.5.1 *Captive orhierarchical governance* was typical during the early development of eucheumatoid seaweed farming systems in the Philippines, Malaysia and Indonesia, from the late 1960s until the mid-1980s. During the initial development phase, there were many small sellers and just a few major buyers, so the market was an oligopsony (i.e. many sellers with few buyers). The carrageenan industry was dominated by a few innovative Small-to-Medium Enterprises (SMEs) amongst which Marine Colloids and Copenhagen Pectin were the most prominent. These companies funded the development of eucheumatoid seaweed cultivation and undertook necessary research and development (R&D) programs that were linked to academia and various government organizations. From about 1974, until 1986, the Philippines enjoyed a virtual monopoly in seaweed supplies. Carrageenan processors collaborated with local entrepreneurs to develop farms, through direct investment. Benefits were realized because they had a strong market position and built robust strategic alliances. Industry standards were mediated through MARINALG (the hydrocolloids industry association) but were enforced by each, individual processor. The copious availability of cultivated *Kappaphycus* made it possible by 1980 to introduce the 'disruptive' technology known as SRC production. The original process technology was copied as new industry players entered the SRC business and recruited former employees, consultants and equipment suppliers of previously established manufacturers.

1.5.2 *Modular governance* became increasingly common as the original carrageenan SMEs were absorbed and became divisions of larger companies. By the early 1980s the carrageenan business enjoyed rapid growth that was driven primarily by sales of SRC. Major traders became SRC processors and industrial innovation stagnated as 'R&D' became reduced to: 'copy and follow' or "me too" initiatives. Farm development was undertaken through supplier alliances and was driven by price manipulation. Development was mainly self-funded by farmers or was trader-funded (Neish 2013). Within modular governance systems, Indonesia and Tanzania developed as significant eucheumatoid seaweed sources by the late 1980s. Standards were still driven by MARINALG but weakened as processors proliferated, consolidated and failed. Since the mid-1980s the number of SRC and kappa-carrageenan producers has increased dramatically, especially in Asia (see Chap. 12 in this book). Proliferation of kappa-carrageenan capacity in Asia was partly driven by the fact that this hydrocolloid and agar were both "gelling gums" that could be produced in the same factories and in some applications could replace one another. The proliferation of new entrants into carrageenan value-chains resulted in successive shifts in their dynamics, from captive, or hierarchical governance in the mid-1980s, to modular governance by the mid-1990s. Captive and hierarchical governance had virtually disappeared by the mid-1990s. By 2017, modular value-chains still comprised a small proportion of the trade in carrageenan, especially for the longest established processors, but market governance dominated the market place.

1.5.3 *Market governance* became dominant in carrageenan value-chains by the turn of the century when much, if not most seaweed, was being sold in "spot-markets". The systems of standards which once enabled buyers to trace and control the quality of the raw materials purchased broke down. As supply sources began to develop in less-accessible, island locations, multiple levels of trading proliferated. Although they added little value, agents and officials were in a position to collect rents and gain trading advantage through their control of funds flow, possession of superior information and access to politically troubled regions, including the southern Philippines. Competition for reliable *Kappaphycus* sources intensified as market demands could not be met by the then available seaweed supplies. Farm development was driven by price manipulation and continued to be self-funded by farmers (i.e. the poorest members of the chain of supply; Neish 2013). Industry standards were generally not applied as buyers enforced their own standards through purchase orders. The use of letters-of-credit (L/C) virtually ceased as buyers settled payment for short-term purchase orders, usually on an FOB basis. Many buyers employed "hold-back" systems in order to guard against delivery of poor quality raw material (i.e. a percentage of the agreed sale price was with-held until the quality of a consignment was assessed when off-loaded at the processing center; Neish 2013).

1.5.4 *Relational governance* is seldom encountered at the seaweed-end of the carrageenan value-chains. As the twenty-first century commenced, it became clear to many in business, government and aid organizations that something was broken in seaweed-to-hydrocolloid value-chains. It was equally clear that developing diverse, transparent, relational value-chains could be utilized to drive the further growth of the industry and could also provide livelihoods to millions of coastal people who were living under the poverty line. Interventions were therefore initiated by several organizations including the International Finance Corporation Program for Eastern Indonesia Small and Medium Enterprise Assistance (International Finance Corporation Program for Eastern Indonesia Small and Medium Enterprise Assistance (IFC-PENSA) of the World Bank (2003–2008) which set up its Seaplant Network Initiative in Indonesia in order to address this issue. Since then, several private enterprises and agencies of government, non-governmental organizations and aid providers commenced support for relational, value-chain development. At the time of writing, some private-sector players were pursuing development through relational value-chains and there were fewer, but larger, sellers in the few cases where farmer enterprises had aggregated. A major incentive for the development of relational governance was to be found in examples where farm development and processing were moving toward **Mu**lti-stream **Z**ero-**E**ffluent (MUZE) processing systems (see Chap. 12 in this book).

1.6 Indonesia Becomes the Leading Producer of Eucheumatoid Seaweeds

Attempts to farm *Eucheuma* in Indonesia commenced as early as 1967 when Soerjodinoto and Hariadi Adnan undertook planting trials at Thousand Islands in a project which terminated by 1970 with Soerjodinoto's tragic death by a motorcycle accident. Commercial eucheumatoid seaweed farming was slow to take hold in Indonesia, but when it finally succeeded, it became firmly established and flourished (Fig. 1.5).

By the early 1970s, the three major carrageenan producers of the time united within a joint-venture, seaweed-source-development alliance known as ORDA (Ocean Resources Development Associates) with their offices at the Goodwood Hotel in Singapore. ORDA activities included both exploration for new sources of wild seaweeds and the development of seaweed farms. In the Philippines, cultivation took precedence but in the vast, under-explored archipelago of Indonesia, exploration was prioritized and that resulted in the expeditions of 1973–1975, which were led by Hans Porse. These came to be known as the 'True Blue' survey, actually named after the motor-sailing vessel chartered for the project. Unfortunately, major seaweed resources were not discovered during this survey and cottonii cultivation had succeeded in the Philippines by 1974, therefore efforts in

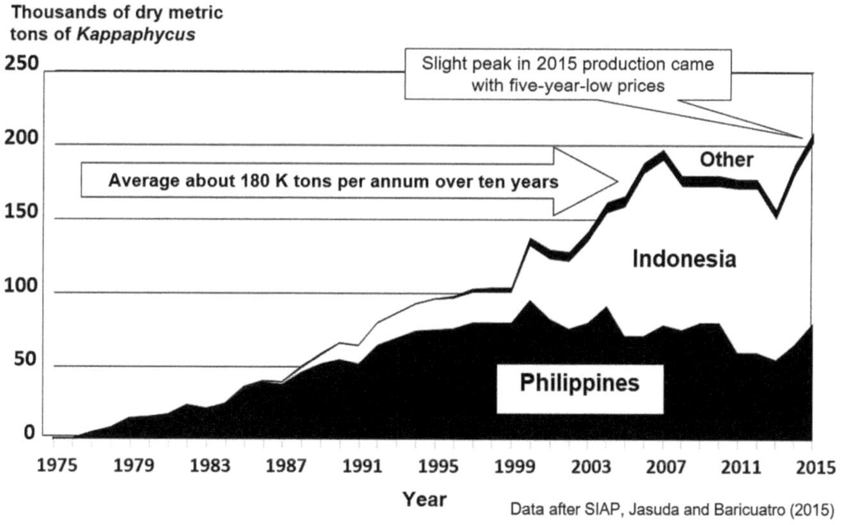

Fig. 1.5 Industry consensus for the estimated, total production of *Kappaphycus* spp. as raw, dried seaweed since commencement of commercial cultivation. These numbers were compiled by www.Jasuda.net from informal carrageenan industry sources and trade data analyses. See Baricuatro (2015) for similar figures. Similar estimates for *Eucheuma* were not available, but it followed similar trends for those years where data were available and production ran at approximately 20% of *Kappaphycus* volumes (Bixler and Porse 2011; Porse and Ladenburg 2015; Porse and Rudolph 2017)

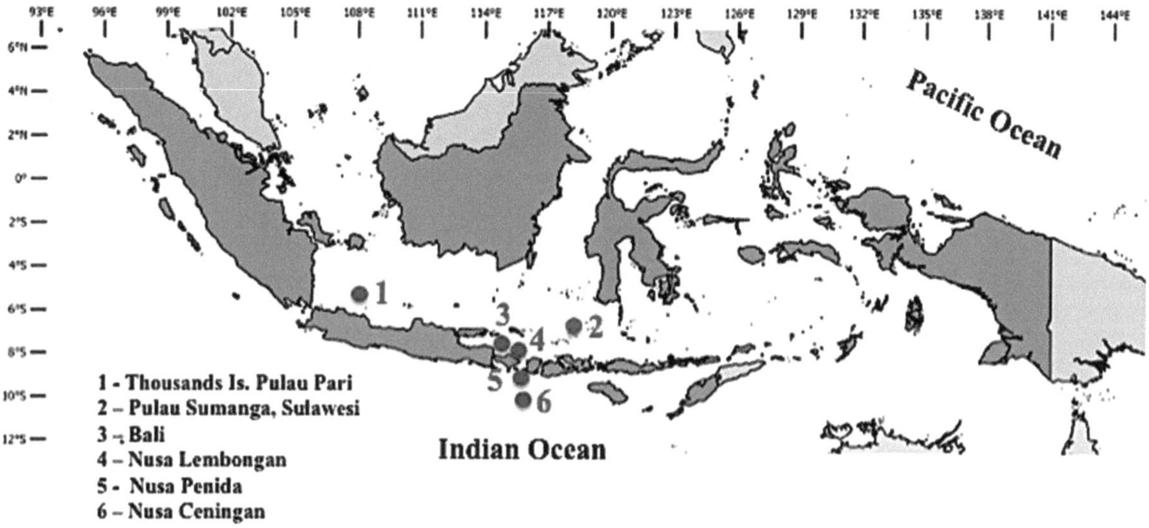

Fig. 1.6 Locations of the earliest *Kappaphycus* and *Eucheuma* farming in Indonesia

Indonesia also shifted towards farming development. Cultivation projects were launched by Hans Porse and Hariadi Adnan at Pulau Samaringa, Sulawesi in 1975–1977 and, under the auspices of the Indonesian Institute of Sciences (LIPI), represented by Hasan Mubarak (Porse and Rudolph 2017). In 1978, the project moved to Geger Beach at Nusa Dua, Bali, where 6 kg of fresh spinosum biomass from Kendari, Central Sulawesi produced the first commercial quantities which were shipped from family-operated farms. In 1984, CP assisted Bambang Tjiptorahadi and Made Simbik in establishing combined, spinosum and cottonii farms on Nusa Lembongan Island, near to Sanur Beach, Bali. Cottonii farming was introduced for the first time in Indonesia, based on 6 kg of *K. alvarezii* seedlings from Bohol, the Philippines (Porse and Rudolph 2017). Success at Nusa Lembongan was quickly followed by farm development at nearby Nusa Penida and Nusa Ceningan and also in the Nusa Dua area (Fig. 1.6), where the team distributed planting materials (stakes and ropes) and seaweed cultivars to independent farmers.

During early 1986, Hans Porse of CP convened a meeting in Bali with Iain Neish and Parker Laite from FMC (which had acquired MCI), Jean-Paul Braud from Sanofi and Max Doty from the University of Hawaii, under the auspices of ORDA to discuss collaborative efforts required for the further development eucheumatoid seaweed farming throughout Indonesia. At that time, the Philippines had an effective monopoly on the production of eucheumatoid seaweeds (Fig. 1.5) and major traders from there had formed a virtual cartel, hence the managers of processing companies were keen to develop Indonesia as a major, second-source, not only for *Eucheuma* but also *Kappaphycus*. The French company Sanofi, continued to emphasize development of seaweed farming in regions of the world with French influence, such as in Africa, Asia and in islands of the Pacific Ocean, however, CP and MCI decided to focus on Indonesia. The CP team continued to expand its farming activities under the leadership of Hans Porse and Hariadi Adnan, while FMC assigned Iain C. Neish to Indonesia and he teamed up with Made Simbik in order to develop farming-extension teams, under CV Duta Teknik. By 1986, Vic Alvarez and Ruben Barraca had both been side-lined by MCPI management in the Philippines so they also joined with the Indonesian farm development efforts. By the early 1990s, Erick Ask and Farley Baricuatro of FMC also commenced involvement with Indonesian seaweed farming developments and continue until the time of writing.

A reflective, 'bottom-up', iterative approach to seaweed farm development occurred in Indonesia, in concert with decentralization policies of the Indonesian government and traditional 'adat' forms of village government (Neish 2013). The strategy of CP and FMC was to develop farming systems which exercised modular governance (Fig. 1.4). Within such systems, the industrial manufacturers of carrageenan formed strategic business alliances with entrepreneurial enterprises within several regions of Indonesia. Care was taken to avoid territorial conflicts and all development activities were undertaken transparently, with the full knowledge and cooperation of various Indonesian government organizations that had responsibilities bearing on seaweed farming, including the Ministry of Marine Affairs and Fisheries (known as KKP in Indonesian) and the Agency for Assessment and Application of Technology (known as BPPT in Indonesian).

The system for development was that training-bases and cultivar nurseries were to be maintained by FMC and CP in Bali and staff from the collaborating companies could receive training there, as well as be accompanied to their development sites by FMC or CP field extension teams. These teams also brought eucheumatoid seaweed cultivars to their farm sites. Carrageenan companies would open irrevocable letters of credit (L/C) with collaborating suppliers for orders of substantial, industrial quantities of seaweed orders (typically about 1000 MT of dry seaweed in the case of FMC, for which the corresponding author was buyer). On this basis suppliers/developers could obtain some bridge financing from banks with their L/C as collateral. For the most part, however, it was aspiring suppliers who provided their own capital for farm development. These enterprises were not mere 'middlemen'; they were, and remain, key drivers of value-chain development (see Chap. 12 in this book).

A major reason for the primacy of seascape issues in seaweed farm development was that seascape communities held and continue to hold tenure rights to both the land and the sea regions where farming takes place. In Indonesia, coastal villages effectively control adjacent near-shore waters. This jurisdiction is delegated from the Kabupaten (Regency) level which has control over near-shore waters to the 4-nm (nautical miles) point; provinces control from 4–12 nm; and the national government has responsibilities from 12–200 nm. Law No. 32/2004 on Decentralization was enacted in order to clarify the authority granted to each level of government. It required that regional regulations comply with national law. It further emphasized that regional governments must act in partnership with the national government. Law 32/2004 gave local governments the authority over: "exploration, exploitation, conservation and management of marine resources"; control over administration and spatial planning; and, the enforcement of laws issued by the regions, or delegated by the central government.

Sea farming areas were initially not owned by anyone except the village at large, with ownership privileges earned through occupancy. According to respondents in recent surveys undertaken by the corresponding author (unpublished) the point related to tenure in the sea was highly significant for seaweed farmers and was even used as part of dowries handed down through marriage. Tenure privileges could be secured either by individual farmers, or by groups. When there were farming modules set in an area, farmers or groups were free to install them at that site. If the area already had farming modules, they had to ask the owners of the installed modules for permission to plant. The owners of the farming modules could sell or rent an area if he, or she, did not want to use it anymore. Sometimes, there was a problem among farmers if an area was empty, but someone else claimed the area. Such disputes were discussed at meetings (musyawarah) where it was decided who the rightful owner was. Due to the nature of Indonesian tenure systems, any enterprises involved with seaweed-sourcing projects had to ultimately deal with the incorporated seashore communities, as the ultimate controllers of the farming sites that developed as socio-economic, production seascapes/landscapes (SEPS/L). These are commonly referred to by the Japanese terms 'satoumi' and 'satoyama' (see Chap. 12 in this book for a full discussion).

During the late 1980s and early 1990s, literally dozens of eucheumatoid farming projects developed, all over Indonesia, and most of them eventually failed (Neish, personal observations). However, several succeeded and indeed some of the most successful farm developers also became carrageenan producers by the late 1990s (these are listed in Chap. 12 of this book). By the late 1990s, the total eucheumatoid seaweed production volume was in the order of 20,000 dry MTs, per annum (e.g. Fig. 1.5) and SRC factories were being established in Indonesia; in several instances with the Mars companies as 'anchor-buyers'. After the turn of the century, production in Indonesia began to increase relative to the production in the Philippines and by about 2007, Indonesia became the leading, global producer of farmed, eucheumatoid seaweeds. One consequence of industrial growth was that the modular governance systems broke down and were replaced by market governance systems where price was the main driver of transactions; business alliances broke down and quality controls weakened. Such trends were associated with the dominance that Chinese carrageenan processors achieved, after the turn of the century (Chap. 12 in this book).

1.7 Eucheumatoid Seaweed Production Levels Off

It must be emphasized that accurate seaweed production numbers are by far the exception and not the rule in producing jurisdictions. For the most part, 'hard' data are proprietary to value-chain actors, including the farmers, traders and processors. Zanzibar is one of the few raw material source areas where accurate data were available. This was because all production was weighed before export through the single port of Malindi and those data were made available by official sources. In the Philippines and Indonesia, shipping channels were complex and 'hard data' were difficult to obtain from official sources. In the Philippines, however, members the Seaweed Industry Association of the Philippines (SIAP) have pooled their commercial data in support of common causes (e.g. SIAP 2017). Industry numbers from SIAP indicated that the total Philippine production of all types of dried seaweeds, averaged almost 88,000 MT per annum, during the 5 year period from 2011–2015. Official government figures for 2015 indicated that production of live seaweeds was in the order of 1,566,161 MT, this was a year when SIAP reported 101,900 MT dwt production. That would indicate a wet-to-dry ratio of approximately 15.4 to 1.0 which is more than 1.5 times the level of wet: dry ratio experienced by the industry.

Although precise numbers were unavailable, general trends were evident from consensus estimates of production volumes and the tracking of price trends. Industry estimates of global *Kappaphycus* production indicated that, over the past decade, total production hovered in the range of 180,000 +/− 20,000 MT year^{-1} (Fig. 1.5; Baricuatro 2015; Porse and Ladenburg 2015; Porse and Rudolph 2017). This is roughly equivalent to 1.3–1.7 million MT fwt, if one assumes a wet-to-dry ratio of 8.5:1. Such numbers are consistent with industry production estimates, however they are not consistent with FAO (2014) estimates which were based on numbers as supplied by government sources. According to industry estimates (Fig. 1.5; Baricuatro 2015; Porse and Ladenburg 2015; Porse and Rudolph 2017) purchases of Indonesian *Kappaphycus*, during the period 2005–2015 were essentially flat. Based on Jasuda.net sources, production appeared to average about 104,000 dry MT; the minimum annual production was approximately 92,000, with a maximum production of about 122,000 MT dwt. Meanwhile, Indonesian *Eucheuma* exports rose from about 10,000 to 30,000 MT dwt, with approximately one third of that total being sold to the Chinese sea vegetable market. This would have been equivalent to a maximum, annual eucheumatoid seaweed production of approximately 250,000 MT dwt. Porse and Ladenburg (2015) estimated a slightly more conservative 215,000 MT dwt. Those levels of annual production would translate to live seaweed volumes of no more than 2.5 M MT fwt, assuming a very conservative 10:1 wet: dry ratio.

Despite the fact that *Kappaphycus* market volumes remained essentially flat from 2005–2015, there were considerable price fluctuations (Fig. 1.7). This coincided with a period when there was keen competition amongst processors, as they competed for market share in what were effectively zero-sum, carrageenan markets. During that period Chinese processors gained enough market share to capture at least half of the market (Porse and Rudolph 2017).

By about 2014, farm-gate prices showed a clear declining trend over the 2 years from May 2014–April 2015 (Fig. 1.8). Over that time, prices paid to farmers declined in an almost linear fashion, from just over \$1200 USD MT^{-1} dwt to just under \$550 USD MT^{-1} dwt. For most of 2016, however, prices were almost flat. It appeared that the cottonii supply and demand was balanced, with farm-gate prices at about \$550 USD MT^{-1}.

Although minor quantities of eucheumatoid seaweeds were sold as sea vegetables, as noted above, the only major market reported for them was that as a raw material for (iota) carrageenan production. Bixler and Porse (2011) reported global carrageenan sales volume of about 50,000 MT for 2009. From 1999–2009 they reported that the overall compounded, annual carrageenan volume growth rate was in the order of 2–3%. This growth rate was confirmed by Porse and Ladenburg (2015) and Porse and Rudolph (2017) for the period 2009–2015, in addition they reported a global carrageenan sales volume of about 57,500 MT. The authors therefore postulated that the lack of growth in seaweed production volumes reflected a lack in the growth of sales volumes for carrageenan.

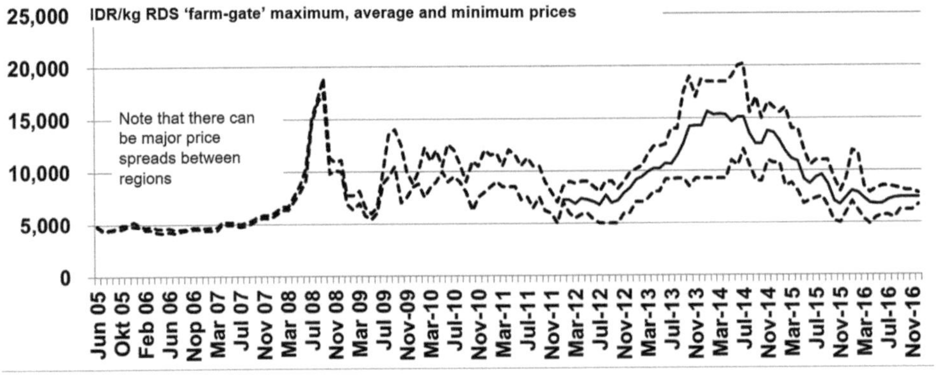

Fig. 1.7 *Kappaphycus* prices 2005–2016 indicate an approximate supply/demand balance after 10 turbulent years. Data are from PT Jaringan Sumber Daya (Jasuda). Jasuda does bi-monthly price polls of collectors and traders at transport hubs to establish what are effectively farm-gate prices. The polls include 22 trading-hub locations throughout all of the major seaweed producing regions of Indonesia

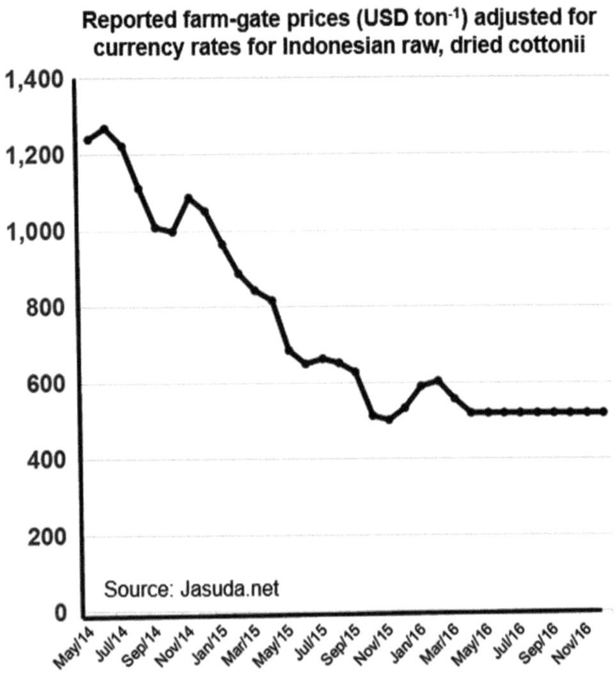

Fig. 1.8 Dried cottonii 'farm-gate' prices, adjusted for foreign exchange fluctuations, from May 2014 to November 2016. Prices declined steadily until early 2016 when they leveled to \$500–600 USD MT^{-1} (Source: www.jasuda.net)

Chapter 13 of this book cites Cybercolloids' information and analyses that give hope for continuing robust carrageenan markets and that there is still scope for market-building developments. Growing demand for processed foods is a global trend as increasingly affluent consumers seek more diverse and luxurious product offerings. The rheological properties of the carrageenans differentiate them from most other hydrocolloids and make them especially suitable for various niche markets. As such, carrageenans could still be important in growing global markets – if sufficient innovation takes place and is applied in the market place.

1.8 Eucheumatoid Seaweed Farming Spreads Around the World

With respect to carrageenan-bearing raw materials, according to estimates for the 2015 (MT dwt) by Porse and Ladenburg (2015), the Philippines accounted for about 60,000 MT (35%) of global *Kappaphycus* production and Indonesia accounted for about 110,000 MT (65%). These authors estimated that the Philippines accounted for about 6,000 MT (13%) of global *Eucheuma* production; Indonesia accounted for about 25,000 MT (56%); and Zanzibar, Tanzania accounted for about 14,000 MT (31%). No other jurisdiction produced enough seaweed to be a significant contributor. Nevertheless, there have been persistent efforts since the early 1970s to expand production beyond the major producing countries.

Eucheumatoid seaweeds are robust and easily transported, if they are kept moist with seawater and held within a temperature range which does not harm them. Under such conditions, propagules can live for several days. Biomass in the order of tens of grams can be propagated to yield thousands of MTs of material that form the basis of regional industries.

As asserted in Section 1.1 above, eucheumatoid cultivars have been dispersed by human activity to the point where marine out-plantings have occurred in 40 or more jurisdictions (Ask et al. 2003; Neish 2005). In many (if not most) jurisdictions the eucheumatoids cultivated were not indigenous. Farming was technically successful in many jurisdictions, but none developed production and markets near the scale of Indonesia or the Philippines, where indigenous eucheumatoids had been the basis for commercial-scale, wild harvests. Carrageenan manufacturers were concentrated in China, Europe, the USA, the Philippines and Indonesia and had little incentive to push major development in globally scattered sources, with no sustainable production cost advantages over Indonesia and the Philippines. During the past decade, innovative technologies were developed in India, Indonesia and Brazil as eucheumatoids began to find markets beyond those simply for carrageenan extraction (see Chap. 12 of this book). Such innovations provided opportunities for globally dispersed development beyond carrageenan as a hydrocolloid. Indeed, it may be due to the fact that these countries are 'outliers', that India and Brazil have taken the lead in innovation beyond legacy technologies tied to carrageenan production.

Aside from Indonesia, the Philippines, East Malaysia and Tanzania, most potential eucheumatoid farming jurisdictions were not home to indigenous populations of those species which became 'domesticated' as commercial cultivars through the 1980s. A few regions were home to indigenous species that could possibly be domesticated, but many had no known indigenous, eucheumatoid seaweed populations. The dispersal of eucheumatoid seaweeds for the purposes of cultivation was however not without controversy. There were reported cases in Hawaii, India and Kiribati where *Kappaphycus* and/or *Eucheuma* have been labelled as 'alien and invasive' species (Conklin and Smith 2005; Chandrasekaran et al. 2008). This generated instances of opposition to the farming of non-indigenous seaweed species, however, after the initial publications there seemed to be no further reports of such invasions continuing to spread widely. Protocols for seaweed introductions and quarantine procedures have been proposed to minimize the possible negative impacts of introductions (Hurtado et al. 2016; Sulu et al. 2003).

Processes for the legal introduction of non-indigenous seaweed cultivars into production seascapes have been a substantial hurdle for commercial farm development in jurisdictions where they are regarded as 'alien' or 'invasive' species and it was only through persistent, determined efforts over many years, that marine agronomists around the world have been able to plant eucheumatoid seaweeds and prove that they can thrive beyond their regions of origin. The following sub-sections of the chapter reflects on the on-going results of those determined efforts.

1.8.1 Sabah in East Malaysia

Commercial seaweed farming in Malaysia has been centered near the Semporna region and Darvel Bay, immediately adjacent to the Philippine waters of Tawi-Tawi. Local issues and difficulties in connecting with value-chains were preventing Malaysia from being a major global producer (Eranza et al. 2015 and Chap. 4 of this book). Production from the region has fluctuated depending on immigration issues (i.e. the necessary people to cultivate the farms) and value-chain linkages, Virtually all production has been of *K. alvarezii* and its varieties. Production volumes seem to have varied between about 2,000–4,000 MT dwt year^{-1} according to informants of the authors in Sabah. By the time of writing, a SRC factory established in Semporna was reported to have ceased operations and farmers were having trouble finding markets for their crops (Kasim, personal communications).

Eucheumatoid seaweeds are indeed indigenous to Sabah, Malaysia and farming commenced there even as it was developing in adjacent regions of the Philippines. Sabah is the only Malaysian state to have produced commercial volumes of eucheumatoid seaweeds (Sade et al. 2006). By 1977, Vic Alvarez was leading development efforts in association with Basarun bin Kasim; first with Marine Colloids Inc. and then with Malaysian Development Plan funding, under a project managed by Prof. Maxwell Doty, through a company known as: Aquatic Resources Limited. Most of the seaweed farmers in Sabah have their origins as migrants, known locally as 'Suluk' (mainly ethnic Tausug), originating from the southern Philippines and Bajau people with indeterminate national origins (Eranza et al. 2015). It came to be that many people who were "un-documented" became involved in seaweed farming in Sabah and that created value-chain issues for local traders and for ATC/SRC producers since the uncertain immigration status of farmers impacted on their ability to operate farms in a predictable way.

Since 1980, management of seaweed farming in Sabah fell within the purview of the Department of Fisheries, Sabah (DOFS). Various government-led initiatives operated in parallel with private sector ATC/SRC endeavors, such as Prompt Access Sdn. Bhd., Omnigel, Tawau Carrageenan and others that have both come and gone. A Seaweed Industrial Zone (SIA) was declared by the National Aquaculture Program in 2000 and the Seaweed Commercial Project Approach (SCPA) and New Innovation Technology and Environment (NITE) strategy were declared in 2010 within the Government Transformation Program (GTP) in efforts to strengthen R&D and commercialization of technologies (Azhnar 2016). Malaysian development from an academic and government perspective has been reviewed in Chap. 4 of this book and was also reviewed by Ask et al. 2003; Hurtado et al. 2014; Hurtado et al. 2016.

1.8.2 China

Commercial-scale farming of eucheumatoid seaweeds in China was limited to the Hainan region at the time of writing. During the mid-to-late 1980s cultivation trials were carried out in Hainan Island using *Kappaphycus*cultivars imported from the Philippines (Wu et al. 1989). Adequate growth occurred during part of the year, but production tended to be seasonal. *Kappaphycus* farming in Hainan was reported to support a local carrageenan industry (Chap. 4 of this book), however, production figures were unavailable and Chinese production volumes were thought to be a small fraction of the volumes imported from Indonesia or the Philippines.

Wild harvest and locally-farmed cultivars of *Betaphycus* spp. (a.k.a. *Eucheuma gelatinae*) were the basis for local production of 'seaweed noodles' as seen by the corresponding author during an FAO mission to Hainan in 1988 (unpublished).

1.8.3 Cambodia, Myanmar and Vietnam

Since the 1980s, there have been numerous introductions of *Kappaphycus* cultivars from the Philippines or Malaysia to Cambodia, Myanmar and Vietnam and details concerning several such introductions can be read in Chap. 4 of this book. Commercial-scale production has been achieved, at least for limited periods, by several projects operating amongst these three countries, but data pertaining to production volumes were unavailable to the authors at the time of writing and production volumes were thought to be a small fraction of volumes from Indonesia and the Philippines. Several projects showed initial promise, but failed for various reasons and in the case of Cambodia, production of eucheumatoid seaweeds has not been reported since 2006 (Lang 2015 and Chap. 4 of this book).

1.8.4 India

Eucheumatoid seaweed production in India has been limited to a *Kappaphycus alvarezii* cultivar stock that originated in the Philippines, then passed to India from Japanese tissue cultures in 1984. Production grew from 21 MT dwt in 2001 to a peak production of 1,490 MT dwt in 2013, according to Mantri et al. (2017) but after that there appeared to be a catastrophic failure of seaweed production. This was referred to as a 'mass mortality' by Mantri et al. (2017), however, industry sources cite value-chain failure as a contributing cause as well. Specifically, industry sources reported that Indian seaweed exports became commercially impractical due to restrictions imposed under provisions of the Biological Diversity Act – 2002. 2016 was a difficult year for eucheumatoid seaweed production in India so the reported production was only about 195 MTs (see Chap. 4 of this book). The struggle to increase production was continuing at the time of writing (Vadassery, personal communications).

Per Mantri et al. (2017), coastal out-planting experiments were carried out during 1989–1996 at Port Okha. Experimental trials were extended to the Mandapam region of Tamilnadu from 1995 until late 2000 when a subsidiary of PepsiCo India Holdings Ltd. showed interest in seaweed cultivation with encouragement from the pet food-producing segment of Mars. Farm development projects focused on the Palk Bay and Gulf of Mannar regions of coastal Tamil Nadu. By August 2008, cottonii farm development was spun-off from PepsiCo to Aquagri Processing Pvt. Ltd., a company formed by entrepreneurs led by former PepsiCo executive, Mr. Abhiram Seth, in an agreement that engaged with the Indian Government Council of Scientific and Industrial Research, Central Salt and Marine Chemicals Research Institute (CSIR-CSMCRI). Commercial cultivation activities were well established in the southern state of Tamil Nadu by that time and there were periodic, though less successful, attempts to establish cottonii farming in other regions of India including: Gujarat, Andhra Pradesh, and Maharashtra (Selvavinayagam and Dharmar 2017).

CSIR-CSMCRI was the implementing organization for the introduction of eucheumatoid seaweeds to coastal farming efforts in India and the time-line for development of their projects were presented in detail in Mantri et al. (2017). Starting in 1984 with small amounts of *K. alvarezii*biomass, in laboratory cultures, scientists and technicians of CSIR-CSMCRI methodically introduced biomass into ocean farming systems using cautious and environmentally sensitive procedures. Nevertheless, issues were raised, for example by Chandrasekaran et al. (2008), who reported a bio-invasion by *K. alvarezii* on to corals in the Gulf of Mannar and cautioned against allowing commercial farming in bio-diverse habitats. Industry took such precautions seriously and adopted agronomy procedures geared to minimizing or eliminating damage to sensitive marine habitats. Emphasis was on floating systems that created new habitats near to the sea surface, over substrata that were not amenable to the re-attachment of loose *Kappaphycus* fronds.

Selvavinayagam and Dharmar (2017) summarized government approvals for farming of *K. alvarezii* (a.k.a. *K.cottonii*) in the sea. These included The National Academy of Agricultural Sciences Policy paper No.22 of 2003; and Government of Tamil Nadu permission for farming in Palk Bay, per G.O. MS. No.229 E&F (EC3) Department of 20 December, 2005. Tamil Nadu authorities reportedly had no objections to near-shore, seaweed cultivation, since it did not fall under coastal regulation zone rules.

Commercial, cottonii farm development in India has involved a variety of agronomy systems including off-bottom, longline, net-bag, tube-net and floating bamboo raft systems. Methods adopted for commercial cultivation were developed with a view towards minimizing or eliminating any damage to coral communities and other sensitive habitats. As a result, off-bottom farming fell out of favor. Preferred methods have varied amongst locations depending on the local, prevailing conditions, but tube-net systems, of the type initially developed in Mexico in the early 1990s (Zertuche-Gonzalez et al. 1999, 2001), were proved to be successful in India (e.g. Vadassery et al. 2016) and net-bag systems were found to be almost twice as productive as floating bamboo raft systems, at least for seed-stock maintenance during north-east, monsoon conditions (Selvavinayagam and

Dharmar 2017). Such methods may eventually enable year-around cultivation, along the south-east coast of India, where turbulent seas have tended to make seaweed farming a seasonal occupation.

Selvavinayagam and Dharmar (2017) described the distinctly seasonal aspect of *K. alvarezii* farming in the Ramanathapuram district of Tamil Nadu which has been the reported source of about 75% of Indian eucheumatoid production. The authors stated that about 3–4 crop cycles of about 45 days each, tended to occur from late April until early October, however, crop seasonality in India has proven to be variable from year to year and has been difficult to predict (Vadassery, personal communications). The seasonality of seaweed farming activities was tied to the presence of heavy waves and other environmental factors associated with the north-east monsoon in association with seasonal rains that also prevented drying of crops. During the approximate six month period when substantial harvests were not obtained, farmers strived to maintain a biomass reserve and protect their farm infrastructure from damage or destruction. During the first quarter of the year, biomass was propagated to levels sufficient for the initiation of the necessary crop cycles.

The socio-economic impacts of seaweed farming in India were assessed by Krishnan and Narayanakumar (2013), Selvavinayagam and Dharmar (2017) and Mantri et al. (2017). These studies concluded that the participatory approach to farm development and the application of contract farming models had enabled further expansion of what began as a livelihood option, to develop into institutionalized systems which supported socio-economic transformations of seaweed farming villages in India. This is especially the case because seaweeds farmed in Tamil Nadu have not been limited to be sold solely as a raw material for carrageenan production. The cultivated biomass was instead sold as live seaweed biomass for the production of agricultural biostimulants by Aquagri Processing Pvt. Ltd., Sea6 Energy Pte. Ltd. and M/s Prasmo Agri, who were the only three seaweed buyers known to be operating in India at the time of writing.

The collapse of eucheumatoid seaweed production in India, subsequent to 2013, appeared to be related both to crop failures and to the impact of regulatory actions that induced value-chain failure. Crop failures seemed to be related to excessive rises in near-shore, water temperatures and excessive grazing by siganid fish (Vadassery, personal communications). Biomass reserves sequestered by farmers in cool pockets near to the sea floor were mostly consumed by grazers, so that biomass intended for restocking farms was almost wiped out. At the time of writing, farm operators were still struggling to increase the necessary biomass reserves (Vadassery, personal communications).

Impacts of the Indian Biodiversity Act resulted from a requirement that international customers wanting to buy Indian seaweeds must directly obtain permission from the Indian government, before they could buy any seaweed biomass. This therefore by-passed the roles of the Indian producers and exporters. For a commodity product such as seaweed, which was readily available from international sources, this was almost a death-blow to Indian seaweed farmers. The only saving grace was that Sea6 Energy, Aquagri and Prasmo Agri were able to continue buying live seaweeds from farmers in order to locally manufacture agricultural, food ingredient and well-being products. However, these products were also fairly early in their development-cycle such that the Indian manufacturers could only absorb a few hundreds of MTs of dry seaweed equivalent. Their operations were also hampered because they had lost the option of generating cash-flow and building buffer-stocks through the selling and exportation of dried seaweeds.

Despite those positive factors which favored expansion of Indian eucheumatoid seaweed cultivation, it seemed that, at the time of writing, production was vastly below potential levels and expansion was not occurring. Reasons for this appeared to be a combination of factors that were discussed in the various sources cited above. Issues and potential actions per cited studies included:

1.8.4.1 Eucheumatoid seaweed farming in India has been based entirely on a limited set of cultivars of *Kappaphycus*. If there is a path toward diversification of this germ-plasm the industry could benefit from following that path. If introducing 'alien' cultivars is a problem, perhaps clones could be developed using the native eucheumatoid biomass as reported to be in Indian waters (Mantri et al. 2017).

1.8.4.2 Although India has a coastline of about 7000 km, there appeared to be few regions that included 'Goldilocks zones' for eucheumatoid seaweed farming (Mantri et al. 2017). Some potentially suitable regions such as the Gulf of Mannar Biosphere Reserve were out of bounds for seaweed farming. Industry expansion required continued searching for suitable sites and pilot-trial farming, along the lines suggested by Selvavinayagam and Dharmar (2017).

1.8.4.3 According to the cited eucheumatoid seaweed literature, farm production was sharply seasonal; growing seasons were limited to half of the year or less; there were recurring issues with the availability of biomass for annual replanting. Innovations cited by Selvavinayagam and Dharmar (2017) indicated that shifting toward some form of net bags might help to solve these problems.

1.8.4.4 A strong case can be made that floating, seaweed farm habitats, in open waters, add to habitat diversity and do not reduce biodiversity, especially if such farms are installed in deeper waters (e.g. over 10 m in depth). It is problematic to argue that exporting seaweed crops from such habitats reduces biodiversity along Indian seashores. However, a very strong case can be made to support the contention that socio-economic benefits are a direct result and of benefit to coastal communities (Krishnan and Narayanakumar 2013; Mantri et al. 2017; Selvavinayagam and Dharmar 2017). In support of those communities, appeals could be made to the appropriate authorities to reduce restrictions which currently seem to be blocking the exportation of Indian dried seaweeds.

1.8.4.5 The installed capacity of Indian producers remained small, at the time of writing. Although small amounts of carrageenan were produced in India, the main thrust of Indian processors was for the production of crop-care and livestock-well-being products that were at the start of their market expansion curve. Although such products have enormous potential, market development is a protracted and expensive undertaking that requires years of trials and testing in order to enter regulated markets. On a positive note, the market for eucheumatoid seaweeds can be expected to grow rapidly once markets for the products of MUZE-processing reach their logarithmic growth phase.

1.8.5 Sri Lanka

According to their website index page, Sri Lanka Hayleys Aquagri was producing 20 MT of cottonii per month, by early 2014 and aspired to grow up to 100 MT per month by 2015, with further expansion thereafter (http://hayleysagriculture.com, April, 2017). According to these reports, 50–60 farm families yielded an actual production on the order of 120 MT, for the period 2014–2016 (Chap. 4 of this book and Shanmugam et al. 2017) and the Sri Lanka Government was actively promoting expansion of the industry.

In 2008, the Sewalanka Foundation, in collaboration with Panamanian partners, undertook a pilot assessment for *Kappaphycus alvarezii* farming, using cage-culture on Sri Lanka's east coast. It was reported that there were grazing, drying and marketing issues and also a perceived need for an ecological impact assessment, all of which delayed commercial development (see Chap. 4 of this book). *Kappaphycus alvarezii* seedstock was also imported to Sri Lanka from Malaysia in 2011, following published procedures for the introduction and quarantine; commercial farming had commenced by mid-2012 (Shanmugam et al. 2017). The average monthly production of dry cottonii was reported to be about 1.5–2.0 MT mo^{-1}, per farm family of three people and production was sustainable. Since 2014, Hayleys Aquagri promoted cultivation using an out-grower system that supplied biomass; planting materials for bamboo rafts and off-bottom systems; drying racks, training; and marketing assistance. Focus was in the western part of Sri Lanka, in the Gulf of Mannar, opposite the most productive regions of India. Expansion to other regions of the country appeared to be promising, such that the government of Sri Lanka was vigorously supporting the concept of industry expansion. Although eucheumatoid farming in Sri Lanka appeared to be limited to a cultivar of *K. alvarezii*, as in the case of India, industry expansion involved an established Indian company (Aquagri) that was known for adding value, not only though carrageenan manufacture, but also through production of agricultural biostimulants. Industry development should therefore not be limited by any constraints of the global market for carrageenan.

1.8.6 Tanzania and Other East African/ Western India Ocean Regions

An overview of eucheumatoid seaweed developments in Kenya, Madagascar and Mozambique is presented in Chap. 4 of this book, although species and cultivars are not specified beyond a reference to the genus *Eucheuma* being farmed in Madagascar (Randriambola and Rafalimanana 2005). According to 2015 Kenya Coastal Development Project-KCDP, annual production was about 600 MTs and the crop was sold in Tanzania (Chap. 4 in this book).

Seaweed comprises more than 90% of Zanzibar's total exports of marine products. At about 25%, it is the third largest contributor to Zanzibar's GDP, after tourism and cloves. Zanzibar and other regions of Tanzania have produced farmed eucheumatoid seaweeds, at commercial levels, since about 1989. Development of that industry was described by Lirasan and Twide (1993), Porse and Rudolph (2017) and in Chap. 4 of this book. The subject was reviewed in some detail in Msuya (2013), Neish (2013b) and Neish and Msuya (2013 and 2015). The information material that follows was sourced from those reviews.

Tanzania has been a commercial source of dried *Kappaphycus* seaweeds, but repeated crop failures were reported. *Kappaphycus*biomass was introduced from the Philippines and, as yet, Tanzania has not emerged as a substantial source of cottonii or sacol seaweeds. Spinosum has flourished in the two main islands of Zanzibar (Pemba and Unguja) and it comprises most of the seaweed production of Tanzania. More than 15,000 MT dwt were exported during the peak year of 2012, production fell to as low as 11,000 MT dwt in subsequent years (Neish and Msuya 2015). This represented about one third of the global production of spinosum with Indonesia and the Philippines accounting for most of the remainder. Spinosum from Zanzibar was sold to companies that produced refined iota carrageenan to be utilized as a dentifrice stabilizer and applications in high-end, processed food products.

During the studies cited above, the Zanzibar seaweed industry was found to have the following salient features:

1.8.6.1 Virtually all Zanzibari seaweed production was weighed and recorded by authorities as it cleared through the port of Malindi. Arguably the Zanzibar production is the best-reported data of any eucheumatoid seaweed source in the world.

1.8.6.2 Virtually all production was of raw, dried *Eucheuma*.

1.8.6.3 Most, if not all Zanzibar dried seaweed was sold as raw material for making iota carrageenan; there was a market of limited size that had exhibited low growth for the past two decades (Bixler and Porse 2010; Porse and Ladenburg 2015; Porse and Rudolph 2017).

1.8.6.4 Most, if not all major exporters and buyers were known and leading exporters also acted as processors/packers of dried seaweeds; system developers; and system financiers. An outstanding example was Murtaza (Morty) Fazal's C-Weed Corporation, a company that was amongst the pioneers of commercial Tanzanian seaweed production which accounted for 45–60% of Zanzibari seaweed exports during the past 5 years (Fazal, personal communications).

1.8.6.5 One prominent feature of Zanzibar farming was the key role of contract-farming in development and ongoing operations. Companies such as C-Weed made advances to farmers for their farm inputs and boats, or other equipment and they also maintained posted prices for the raw materials. They also offered farmer training and contributed to socio-economic and coastal community development.

1.8.6.6 There was substantial communication both between and amongst buyers and sellers with some buyer-seller relationships comprising decades-long alliances.

1.8.6.7 As the world's third largest producer of spinosum, Zanzibar faced extremely strong competition with much larger seaweed source areas in the Philippines and Indonesia, both of which had the advantage of producing, marketing and selling *Kappaphycus* seaweed and other seaweeds in addition to spinosum.

1.8.6.8 According to industry sources, diminishing seaweed quality was an issue in Tanzania, as in other spinosum source areas. Such quality diminution has been attributed, among other causes, to "… the ongoing, general seaweed deterioration, experienced in cultivated species …" (e.g. Porse and Rudolph 2017). In Indonesia and the Philippines several initiatives were underway, at the time of writing, to improve the qualities of spinosum crops and other eucheumatoid seaweeds. Similar initiatives seemed to be necessary for Tanzania as well.

By the time of writing, the Zanzibar seaweed industry had developed to a point where major market expansion seemed unlikely unless the industry went well beyond the practice of mainly selling RDS to an oligopsony of global, carrageenan producers. Expansion of Zanzibar seaweed markets depended on the development and installation of crop diversification; improvements in farm productivity; and development of new products for novel and additional markets. Emerging technologies give reason to believe that new processes and innovative products can substantially increase the size and quality of Zanzibar seaweed production and could indeed build participation in value-added markets. The basis of these technologies is advanced methods in seaweed agronomy combined with biotechnology-based process methods which commence with fresh, live seaweed that is then processed near to the major farming sites (Neish 2013b).

1.8.7 Island Jurisdictions in the Pacific Ocean

Kappaphycus farming was introduced to island jurisdictions of the Pacific Ocean as far back as 1976 when cultivars were brought from the Philippines to Fiji (Solly and Booth 1977). Since then there were several other introductions, either from Southeast Asian sources, or within the region and many of these have been reported in Chap. 4 of this book, where there are accounts pertaining to the Cook Islands, Fiji, Kiribati, the Marshall Islands, Papua New Guinea and the Solomon Islands. In addition to those efforts, there was a project conducted by Max Doty in Ponape where Basarun bin Kasim was able to grow *Kappaphycus* successfully during the 1980s (Doty and Kasim, personal communications). In addition, farm development was attempted by Antoine Teitelbaum and others in the French-influenced Pacific Islands. Commercial production has waxed and waned several times, in several jurisdictions, during the past four decades, but none had achieved sustained, commercial production, by the time of writing, with the possible exceptions of Kiribati and the Solomon Islands.

As of 2017, although a major market-share was not yet gained by Pacific Island jurisdictions, development efforts in the regions led to some advances in agronomy-related technologies and training. For example, Sulu et al. (2003) advanced the literature regarding cultivar introductions and quarantine procedures and training manuals by Teitelbaum (2003) and Tiroba (2013) and were shining examples of the use of effective training and extension materials.

Production numbers were not available to the authors for most jurisdictions, however the Solomon Islands reported their 2014 seaweed production figures as 1,520 MT, with a value of about 721,000 USD to the local seaweed farmers

(Algae World News 2015, as reported in Chap. 4). At the time of writing, the Central Pacific Producers Ltd., a government-supported company in Kiribati, had about 130 MT of unsold, dried cottonii in warehouses and reported that Fanning and some other islands were supplying about 15 MT per month, but that markets for the materials were still being sought (Nicholas Paul, personal communications).

The accumulated experience reported in Chap. 4 and also the personal experience of the corresponding author, make it clear that *Kappaphycus* can be farmed commercially in several Pacific Ocean jurisdictions, where it had already been introduced. The varied authors concluded, that based on available information, that value- chain factors and not failures of agronomy accounted for the fact that eucheumatoid seaweed supplies from Pacific Island jurisdictions have not yet achieved a significant market share relative to production in Indonesia and the Philippines. No Pacific Island jurisdictions appeared to have sustainable, production cost-advantages relative to the major sources and all had cost-adding logistical issues which were mainly tied to the cost and frequency of available, commercial shipping options for exportation of dried goods.

The authors here postulate that as long as their raw materials are sold only into carrageenan markets, seaweeds from the Pacific Islands will remain as 'gamma' or 'delta' suppliers, with little prospect for further successes unless major, new regional carrageenan markets develop (see Chap. 12 of this book). On the other hand, development of a MUZE-approach to processing could lead to significant opportunities because agricultural bio-stimulant products and other products from a MUZE processing chain could indeed find local and regional markets. Opportunities seemed especially attractive for Papua New Guinea and the Solomon Islands, both of which are jurisdictions within the Coral Triangle.

1.8.8 Mexico, the Caribbean and Central America

Eucheumatoid seaweeds have been successfully farmed at numerous locations throughout Mexico, the Caribbean and Central America but, as of 2017, robust, large-scale commercial production had not been established.

Emphasis was placed on indigenous and introduced species of *Eucheuma* and introduced cultivars of *Kappaphycus*. The technical capacity to farm eucheumatoids was proven at several locations, but in many jurisdictions there were issues pertaining to industry sustainability, especially with regard to environmental impacts and concerns over 'alien invasion' by a non-indigenous seaweed (e.g. Robledo et al. 2013). For example, such concerns were expressed in the case of Cuba, where tests were conducted to evaluate the potential ecological risks of an introduced species. A summary of the research results was published by Areces et al. (2014) and as a result of recommendations, *Kappaphycus* was officially banned from cultivation in Cuba.

Chapter 4 of this book includes narratives of the experiences of eucheumatoid seaweed farm development in several Caribbean Islands and also Belize, Columbia, Ecuador, Mexico, Panama and Venezuela. In this region, *Eucheuma isiforme* has received much attention since it is a locally abundant, indigenous species (Cheney 1988) that can serve as biomass for production of both iota carrageenan and use as a sea vegetable. Various red algae have been harvested for the preparation of traditional drinks and puddings in the region (Smith 1997) and they were indeed cultivated for such purposes in the Caribbean, since at least the first half of the nineteenth century (Smith 1992). No domestication was known for indigenous species other than *E. isiforme* but eucheumatoid seaweeds were also introduced in to several countries including: Mexico (Muñoz et al. 2004), Panama (Batista de Vega 2009) and several Caribbean Islands (Smith and Rincones 2006). As reported in Chap. 4 of this book, processing facilities for sea vegetable products were established in several jurisdictions including Jamaica, Barbados, Grenada, Antigua and Barbuda, Trinidad and Tobago, St. Vincent and the Grenadines, St. Lucia and Dominica, as a result a variety of packaged and bottled products became available. The supply of commercial species from wild stocks declined over time so the region's largest processor in Jamaica resorted to importing an average of 13 MT of *K. alvarezii* per year, from the Philippines in order to meet demands (Smith and Rincones 2006).

Test plots conducted using introduced *Kappaphycus* spp. and *Eucheuma* spp.cultivars met with varying degrees of success at several venues in the region, as reported in Chap. 4 of this book, much of this work was described in various publications of Rincones and Smith which are cited above. Raul Rincones (personal communications) has observed flourishing growth of *K. alvarezii* on long-line and tube-net test plots in Columbia, Ecuador, Panama and Venezuela. In Venezuela high growth rates were reported, but problems arose related to bio-invasion and coral reef-bleaching issues, as a result farm developments ceased (Smith and Rincones 2006; Barrios et al. 2007). In Columbia, farm development also stopped for similar reasons. In Panama *Kappaphycus* farming has evolved to the point where by 2015, Panama Sea Farms (PSF) were able to make trial shipments of dried cottonii to Malaysia for the production of SRC (see Chap. 4).

In Ecuador, *Kappaphycus* was introduced in 2011 by the Brazilian company Seaweedconsulting by Miguel Sepulveda (co-author of the present chapter), in association with the Ecuadorian companies Ecuaalgas S.A. and Lobelia S.A. The seaweeds were cultivated experimentally in *Litopenaeus vannamei* shrimp farms. The results were very positive and indicated great potential for *Kappaphycus* in polyculture systems. Two shrimp farms were cultivating *K. alvarezii* with this system, at the time of writing, and the Ecuador government was also encouraging the cultivation of *K. alvarezii* in the sea for artisanal fishermen, through a local fishermen's cooperative known as Federacion Nacional de Cooperativas Pesqueras del Ecuador (FENACOPEC).

Commercial and socio-economic aspects of eucheumatoid production in Mexico were reviewed in Chap. 4 of this book and in Valderrama et al. (2015). Aside from test plots using introduced cultivars, farm development initiatives focused on using the indigenous*Eucheuma isiforme* varieties, which were native to the Gulf of Mexico and the Caribbean. This species seemed amenable to commercial cultivation in some areas, although production was highly seasonal, with peak harvests during the cold season of November and December, at least off the coast of Yucatan. (Freile-Pelegrin and Robledo 2006). Robledo and Freile-Pelegrín reported that commercial exploitation of *E. isiforme* off the Yucatan Coast, cannot be successfully implemented without a management plan, or proven cultivation practices and there have been no official initiatives to promote seaweed farming in Mexico. Although seaweed farming had not yet established as a major economic activity in Mexico, at the time of writing, scientific and technical work conducted in Mexico had impacts on global developments, notably with the initial development of tube-net farming technology by Jose Zertuche-Gonzalez and his collaborators in the early 1900s (Zertuche-Gonzalez et al. 1999, 2001).

At the time of writing for Mexico, the Caribbean region and Central America, the prognosis for eucheumatoid seaweed farming seemed to be that, pending necessary environmental impact and quarantine clearances, commercial farming could be feasible in several jurisdictions. However, it seemed to be unlikely that dried seaweeds could compete in international markets for carrageenan raw materials, for many of the same reasons cited for island jurisdictions in the Pacific Ocean. Similarly, development of a MUZE-approach to processing could lead to significant opportunities because agricultural biostimulants and other products could find local and regional markets. There are enormous plantation and livestock-rearing enterprises in the region that could probably benefit and become more profitable once they apply seaweed-based biostimulants to their crops and feed supplements to their livestock.

1.8.9 Brazil

Algasbras Biorrefinaria Ltda. (http://www.carragenabrasil.com.br) reported on their website that they produced dry tonnages of *Kappaphycus* amounting to about 121, 285 and 326 MT for the years 2014–2016, respectively, a total of 734.2 MT overall. At the time of writing, Algasbras and Seaweed Consulting SA planned the 2017/2018 cultivation season for *Kappaphycus* with a production target of 500 MT of fresh seaweed per month; rising to 1,000 MT mo^{-1}. These operations were to be based in the region of Ilha Grande Bay, south of Rio de Janeiro.

Chapter 4 of this book presents a history of the substantial academic research and development that have been undertaken with eucheumatoid seaweeds in Brazil. The following reflections give a commercial perspective from Miguel Sepulveda (www.seaweedconsulting.com) whose company was actively producing farmed seaweeds in Brazil at the time of writing (Goes and Feder-Martins 2015; Sepulveda 2016) (Fig. 1.9).

The first introduction of *K. alvarezii* in Brazil was by Édison de Paula, from the Universidade de São Paulo (USP), who experimentally introduced a clone (which originated from southern Japan, in Ubatuba Bay), on the São Paulo coast in 1995 (Paula et al. 2002). The introduction of a non-indigenous species was a response to the lack of native species that were economically available and viable for mariculture. Since then, this species has proven to be an excellent choice because it was easy to handle and multiply; it had a high daily growth rate; it was easy to market, at attractive prices to countries that imported large volumes of this species; and, it generated direct jobs in rural, coastal areas that increased family incomes.

After a few years, in 1998, a Venezuelan clone of *K. alvarezii* was introduced experimentally into Ilha Grande Bay, on the south coast of the State of Rio de Janeiro by Miguel Sepulveda, with the objective of testing its viability for cultivation both on a pilot and commercial scale. This initiative also introduced a prototype, culture structure known as the 'Float Raft System' (FRS). The Sete Ondas Biomar Company (now extinct) was founded in order to expand crop production in Ilha Grande Bay and it had an important role in leveraging farm activities which produced about 600 MT fwt in the region of Marambaia (Rio de Janeiro). These initiatives were supported by the Ministry of Fisheries and Aquaculture (MPA), IBAMA (Brazilian Institute of the Environment) and researchers from several institutions, which further stimulated the regulation of the activity. In 2008, after several environmental studies by universities and institutes, the Normative Instruction – IN No. 185 (IBAMA) was issued. This regulation allowed for the cultivation of *Kappaphycus alvarezii* in the coastal range between the Bay of Sepetiba (RJ) and the Ilha Bela (SP) region in Brazil.

On the south coast, in the littoral zone of Santa Catarina, the potential of *Kappaphycus* mariculture had already been shown to be promising by researchers from the Federal University of Santa Catarina and the EPAGRI Research Institute. A license was issued for new farm developments, at the time of writing. The studies carried out by these institutions showed that the coast of the State of Santa Catarina had

Fig. 1.9 (**a**) Young farmers working with their *Kappaphycus alvarezii* farm near to Ilha Grande, Brazil. (**b**) Co-author Miguel Sepulveda with *K. alvarezii* cuttings. (**c**) *K. alvarezii* crop drying in Brazil (Miguel Sepulveda photos)

areas with potential to produce approximately 730 MT dwt, eucheumatoid seaweed per year. In addition, there were encouraging technical indicators of productivity for the production of *K. alvarezii* in polyculture systems with some of the mollusks also cultivated in this region, which further increased the potential for successful mariculture of multiple species (also see Chap. 4).

Along the north coast of the State of São Paulo, some producers were still struggling to expand their farming activities, at the time of writing, due to the lack of an Environmental Management Plan within the Environmental Protection Area (EPA), although researchers of the Fisheries Institute were in the process of addressing this problem with the Brazilian Institute of the Environment (IBAMA).

In 2013, research at the Federal University of Pernambuco, as part of their Postgraduate Program in Oceanography, aimed to evaluate the cultivation of *Kappaphycus* introduced to the coast of Paraíba, northeast Brazil, in the early 2000s. The species was cultivated there by local, artisanal fishermen, raising questions as to whether or not there was a risk of negative environmental impacts to the local ecosystem. Results of the study (unpublished), showed that there was no establishment of *K. alvarezii* on the coast of Paraíba, hence the invasion potential of this species was considered low for that region, although it would be important to continue environmental monitoring actions. The issue remains controversial, but strong indications were that there was an enormous potential for the cultivation of *Kappaphycus* on the northeastern coast of Brazil and that studies should continue in order to clarify the nature and magnitude of the potential environmental impacts and scale of future commercial cultivation of allowed.

In Brazil, it was found that protected coves, with strong wave surges, offered favorable conditions, including adequate light levels; surface seawater temperatures above 20°C, at an average depth of 0.50 m and salinity above 20 ppt for the successful cultivation of *K. alvarezii* at a commercial scale. The cultivation of this seaweed did not require high technology and could be initiated with a relatively modest investment, which allowed for expansion of the activity in relatively, economically-challenged, rural communities. In addition, it is worth noting that, in general, the producers were able to start farms by purchasing seedling biomass once, after which further propagation was vegetative, with no further need to buy additional 'seedlings' for replanting. However, it remains to be seen if repeated clonal production results in any quality or seedstock issues as have been experienced in other jurisdictions (see other sections in this chapter and Chap. 4).

In Brazil, the cultivation structure known as the 'Float Raft System' (FRS), typically consisted of a set of PVC pipes, 100 mm diameter, 3 m length, acting as floats which were connected to one another by 8 mm, polypropylene ropes. The dimensions of the raft structures were 150 m × 3 m (450 m^2) and each was anchored to the sea bottom using cement blocks (Fig. 1.8.1). On average, one RFS could produce seven, live MTs over a 50 day crop cycle. During the summer months yield could reach eight MTs per unit, depending on the cultivation site and other factors such as algal density, surface seawater temperature, salinity, availability of light and pressures of herbivory. These values cited were crude estimates since during the harvest, 20% of the cultivated biomass was retained for replanting. The cost of materials for each FRS of 450 m^2 was approximately 1,100 USD ha^{-1} equivalent. Considering an average capacity of 15 rafts (+/− 7,000 m^2), requiring the navigation and transportation of materials by boats and canoes in the cultivation area, it was estimated that four people could manage and operate a one hectare system, hence 15 FRS would provide jobs for 28 people.

Traditionally, the most widely used cultivation or propagation technique for the cultivation of *Kappaphycus* required cutting 100 g pieces of seaweed, then tying cuttings to lines to be placed in parallel rows, spaced about 20 cm apart. Such lines were fixed to stakes driven into the sea floor, or suspended from floating rafts. These methods are known, respectively, as the off-bottom and long-line methods and the latter is limited to regions of relatively shallow water where farmers can handle the crop more easily. In Brazil, another cultivation technique known as Tubular Networks (TN) was introduced by the businessman Alexandre Feder of the Algasbras Biorrefinaria Ltda. (Rio de Janeiro) in 2005, after a visit made by him to some production sites in the Philippines, where this TN method had been introduced, after the techniques developed by Zertuche-Gonzalez et al. (1999, 2001) in Mexico. The TN was found to be simple and easy to use, the method involved placing seaweed cuttings of about 100 g in a tubular net, or sock similar to those used for the cultivation of mussels. Typically the nets were about 10 m long and filling the open tube was accomplished with the aid of a 75 mm PVC tube as a hopper. After inoculation, the nets were stretched on to the FRS modules, close to the sea surface. At the time of writing, this was the most frequently used method in Brazil, due to the speed and efficiency of planting and also the facilitation to improve harvesting.

In Brazil, the scenario at the time of writing for *Kappaphycus* cultivation in the range approved by the Brazilian Institute of the Environment (IBAMA), was still considered to be incipient, even after 8 years of operations. However, investments were being made by private entrepreneurs for a number of years and could be considered to have a quite promising future. As a result, there were two commercial farming operations in the Paraty region and three in Ilha Grande Bay, at the time of writing. These operations were suppliers to Algasbras, which was located in the municipality of Itaguaí, Rio de Janeiro. Algasbras has played an important role in the development of Brazilian seaweed farming in recent years. By 2015, this company had facilities dedicated to the processing of seaweed in order to extract carrageenan. According to the Director of the company, Alexandre Feder, in 2015, Algasbras bought and processed about 400 MT live of *Kappaphycus*, paying 140 USD MT^{-1} fwt. Algasbras provided logistical support for transport of the seaweed biomass to the factory. All of the production was processed to obtain kappa carrageenan, a product which has commercial value on in the national market place. Meanwhile, Brazil had imported about 2,000 MT of carrageenan annually, valued at $22 M USD. Therefore, there was a favorable scenario conducive to the expansion of seaweed farming activities.

For 2017/2018, Algasbras, in partnership with Seaweed Consulting SA, were planning to cultivate 500 MTs fresh weight, per month of *Kappaphycus* in the region of Ilha Grande Bay, south of Rio de Janeiro. Production was expected to reach 1,000 MT fwt mo^{-1}, in order to supply the local Algasbras factory. Ilha Grande Bay, has been the target of investors and NGOs to foster development of cultivation and encourage young farmers in the region to not only supply Algasbras, but also to export dried seaweed to countries such as Chile and Argentina, which have processing plants such as Gelymar and Soriano SA, respectively. Those companies import more than 100 MT dwt of *Kappaphycus* mo^{-1} at international prices.

According to the experience shared by several institutions and experts in Brazil, over the past 30 years, for a commercial seaweed farming program to succeed in Brazil, the following aspects must be taken into account:

1.8.9.1 Government enabling support is required, including the granting of licenses necessary for farm development;
1.8.9.2 Existence of a safe and reliable markets;
1.8.9.3 Availability of the economic resources necessary for the support of the program in order to reach commercial volumes;
1.8.9.4 Design, management and competent execution by the project leader. This point includes the linkage of technical and professional personnel to the necessary field work, as well as adequate identification of barriers which preclude entry of farmers to the commercial activity and the development of strategies to overcome these challenges;
1.8.9.5 Appropriate selection of suitable sites to establish seaweed farms, so as to enhance the chances of

success of the program in the community and to justify the investment of time, effort and resources.

1.8.9.6 Clear and precise identification of the main constraints that are shared by members of the coastal communities in order to join the project (i.e. investment capital for seedlings and cultivation materials, training and technical assistance, business organization, basic services, etc.)

1.8.9.7 Availability of sufficient seaweed biomass and adaptation to local ecological and environmental conditions, as well as cropping systems. In addition, the *Kappaphycus*cultivars should be selected on the basis of carrageenan with the highest quality and therefore commercial interest.

1.8.9.8 Technical assistance and a permanent business partner, from the installation of the cropping system, to the commercial phase, including subsequent monitoring to ensure reliable, responsible and sustainable production of raw materials which generate sufficient income in order to meet the socio-economic needs of the farmers.

1.8.9.9 Guarantee of marketing to producers with fair and competitive prices through long-term purchase and sale agreements.

1.8.9.10 Creation of a healthy and pleasant work environment with safety and hygiene conditions that allow for the long-term motivation of the farmers and their families.

1.8.9.11 The incorporation of women and young people, in particular, offering tools for their integration, since they are frequently at the margin of productive activities in most coastal communities of Brazil.

The development of commercial eucheumatoid seaweed farming in Brazil has been a notable example of how key factors for success have been brought together through sustained, persistent effort in order to develop systems with robust potential for future sustainable development. In so doing, these activities can deliver positive, socio-economic impacts in several regions of the Brazilian coast. An essential foundation for such development was the concerted effort by scientists in the private and public sectors, over the course of almost two decades, to introduce *Kappaphycus* from its origin in the Philippines to be legally sanctioned for cultivation in Brazil. Efforts were directed at establishing clear rules regarding environmental licensing and the assignment of areas in Union waters. That step was followed by the practical innovations of the farmers themselves that led to commercially viable agronomy systems (Chap. 4 of this book; Goes and Feder-Martins 2015; Sepulveda 2016).

Local development projects in several fishing communities on the south-east coast of Brazil offered opportunities for people to exit the national, social assistance programs such as the Bolsa Família Program. Besides being an attractive business option, seaweed farming was contributing to the reduction of poverty on the Brazilian coast through the generation of jobs, income and providing food, as diversified polyculture opportunities, beyond seaweed monoculture, were being developed. Aquaculture is therefore becoming a socio-economic foundation for traditional, coastal communities. In the governmental sphere of Brazil, the Aquaculture and Fisheries Secretariat was paving the way for the success of seaweed farming, with a series of actions being implemented in the zoning of areas suitable for the installation of seaweed farms.

In establishing commercial eucheumatoid seaweed farming, Brazilian innovators have been pioneers in developing next-generation, methods in seaweed agronomy such as the tube-net technology which originated from work in Mexico by Zertuche-Gonzalez et al. (1999, 2001). Although initial developments were incentivized by the lure of carrageenan-based markets, they also followed the path towards MUZE-processing as both Algasbras and Seaweed Consulting branched into product lines including human food, animal feed ingredients and biostimulants for plants. The massive potential for such products in the agricultural economies of South America ensured that the development of a Brazilian seaweed farming industry would not be limited by lack of innovation in developing new markets beyond carrageenan. It will be interesting to plot the future track and successes of these initiatives (Fig. 1.10).

1.9 Developing Technologies Fuel Hopes for Further Growth

The authors postulate that the future growth in eucheumatoid seaweed production will be driven by new products and technologies as described in Chap. 12, the value-chain chapter of this book. Events have shown that major expansion of eucheumatoid agronomy cannot be supported as long as the industry is primarily dependent on sales as raw material for carrageenan production.

Emerging, mechanized cultivation technology and multi-stream, zero-effluent (MUZE) processing technologies are now providing paths toward increased utilization of 100% of eucheumatoid seaweed biomass. Such technologies will enable production of a range of value-added, agricultural, chemical and biofuel products that can support future value-chain development. However, major developing markets, supplied by MUZE-production systems will require many-fold increases in the production volumes of raw materials that can be achieved, without creating inflated unit-costs.

Farm systems, current at the time of writing, were generally, small-holder operations, involving a large, repetitive 'drudge-labor' component that comprised most of the sea-

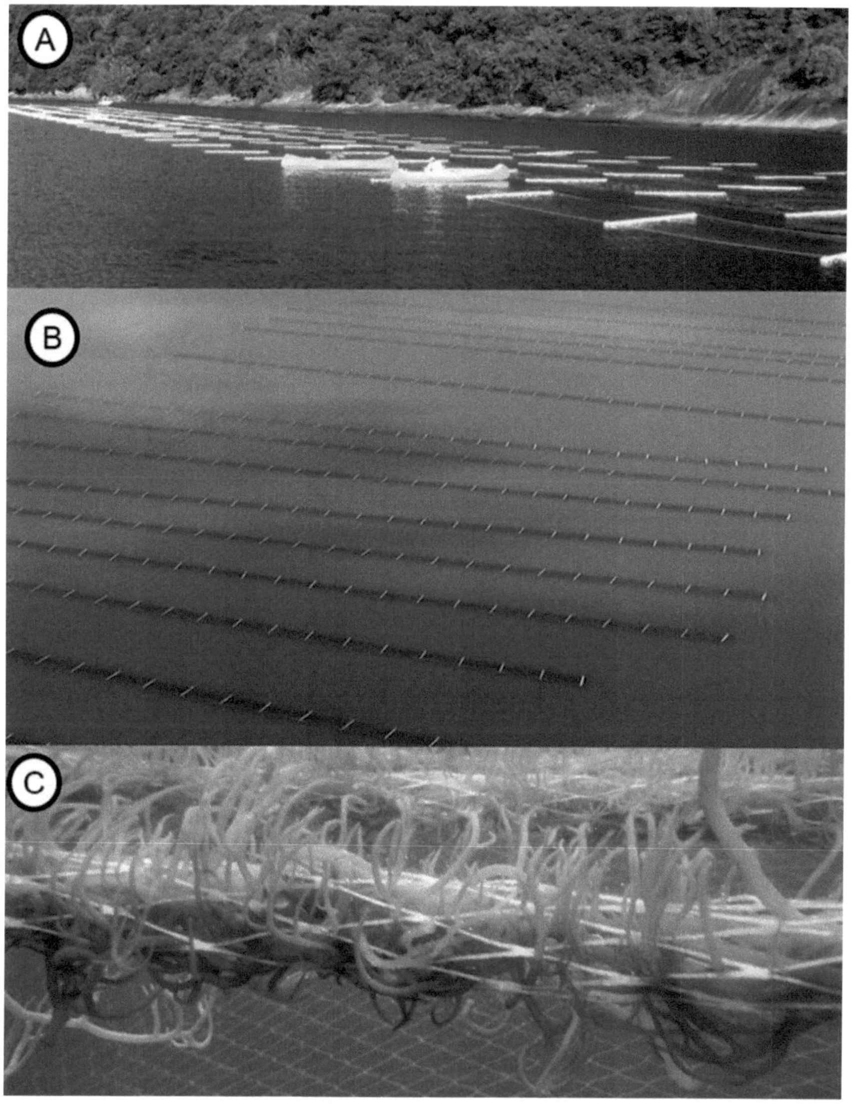

Fig. 1.10 (**a**) 'Float Raft System' (FRS) installed at a Seaweed Consulting SA farm in Ilha Grande Bay. (**b**) Aerial view of a multi-hectare array of FRS in Marambaia Bay, Brazil. (**c**) A newly inoculated tubular net containing *K. alvarezii* cuttings near to Ilha Grande, Brazil (Miguel Sepulveda photos)

weed production cost (Vadassery et al. 2016). As of 2017, methods of seaweed farming were therefore being developed to enable simple mechanization of repetitive, menial tasks that previously involved drudge-labor and hence enable farmers to increase their farm productivity, per unit-of-effort (Goes and Feder-Martins 2015; Sepulveda 2016; Vadassery et al. 2016). In such systems, *Eucheuma* seaweed biomass is inoculated into tubular nets, rather than being fastened manually to ropes (Fig. 1.8.2). Planting, crop-tending, harvesting and crop handling are to be mechanized using simple machinery that can be operated either on shore or at sea, thus eliminating most the most labor-intensive farm chores. Farming is to be done within contract systems, with relational governance and the systems managed such that the flow of fresh seaweed biomass to process facilities is sustainable, even, predictable and reliable on a daily basis. Further advances in agronomy systems also require that designs to operate in deeper and more turbulent waters than the current coastal based systems can withstand, thus expanding to areas of open ocean that are suitable for the support of seaweed cultivation.

During several decades of involvement in eucheumatoid seaweed development, the corresponding author repeatedly pitched the concept of MUZE-processing which was tied to out-grower farming systems, to the point where both he and the targets of such pitches reached 'concept fatigue'. The concept was repeatedly dismissed by counter-entrepreneurial comments such as: "If it is such a great idea why is nobody doing it?", or "We tried it, but it didn't work" or "It does not fit with our core business" or "we have always done it this way". The fact is that value-chains of the MUZE + outgrower type (see Chap. 12 of this book) were very difficult to set up, they required innovative technology and management systems and they also required a long process of relational capital/governance development. This was not a "get-rich-quick" opportunity. It was an opportunity that was barely beginning to be realized even as of 2017, despite decades of halting attempts at implementation. Persistent innovation, intrepid entrepreneurship and dogged determination are the key drivers which are indeed building major new supplies of raw materials and demands as eucheumatoid seaweed farming launches into a step-change phase of renewed, sustainable, global expansion.

Acknowledgements The authors have had the privilege of the experience of working with thousands of seaweed farmers and with numerous colleagues from both the private and the public sectors. We hope that this chapter does justice to their work and we apologize for any errors or omissions. Reflections such as these are necessarily somewhat subjective and are limited by the personal scope of authors such as ourselves who have been in the midst of many of the events written about here. We hope and trust that our reflections are complemented by the writings of reflections by colleagues from around the (seaweed) world!

References

Algae World News (2015) www.news.algaeworld.org

Areces MAJ, Alvarez Villanueva FC, Bernardi J, Cabrera R (2014) Ecological risk assessment of the introduction of exotic carrageenophytes in the tropical Western Atlantic. J Appl Phycol 26(5):2055–2063

Ask EI, Batibasaga A, Zertuche-González JA, de San M (2003) Three decades of *Kappaphycus alvarezii* (Rhodophyta) introduction to non-endemic locations. In: Chapman ARO, Anderson RJ, Vreeland VI, Davidson IR (Eds.) Proc Internat Seaweed Symp 17: 49–57

Azhnar K (2016) Seaweed development and production of seaweeds in Sabah, Malaysia. Paper presented at the Seaweed Workshop, University of Malaysia at Terengganu, 26 Feb 2016

Baricuatro F (2015) Challenges in increasing Indonesia seaweed exports. Presentation at the 7th meeting of the ASEAN Seaweed Industry Club (ASIC) Novotel Hotel, Makassar, Indonesia, 1–3 August 2015

Barraca RT, Neish IC (1978) A survey of *Eucheuma* farming practices, I. Tawi-Tawi. I The Sitangkai, Sibutu and Tumindao Region. Marine Colloids Inc. Report, 74pp

Barrios J, Bolaños J, L6pez R (2007) Blanqueamiento de arrecifes coralinos por la invasi6n de *Kappaphycus alvarezii* (Rhodophyta) en Isla Cubagua, Estado Nueva Esparta, Venezuela. Boletin del Instituto Oceanografico de Venezuela 46:147–152

Batista de Vega G (2009) Cultivo Ecosostenible de *Kappaphycus alvarezii* en Panamá. Tesis Ph Doctoral en el Departamento de Biología, Fisiología y Biotecnología Vegetal Marina, en La Universidad Las Palmas de Gran Canaria. España

Bixler HJ, Porse H (2011) A decade of change in the seaweed hydrocolloids industry. J Appl Phycol 23:321–335

Chandrasekaran S, Nagendran NA, Pandiaraja D, Krishnankutty N, Kamalakannan B (2008) Bioinvasion of *Kappaphycus alvarezii* on corals in the Gulf of Mannar, India Current Sci 94 (9) 10 May 2008

Cheney DP (1988) The genus *Eucheuma* J. Agardh in Florida and the Caribbean. In: Abbott IA (ed) Taxonomy of economic seaweeds with reference to some Pacific and Caribbean species, vol 2. California, California Sea Grant College Program, University of California, La Jolla, pp 209–219

Conklin EJ, Smith JE (2005) Increases in abundance and development of management options for the invasive red alga *Kappaphycus* in Kaneohe Bay, Hawaii. Biol Invasions 7:1029–1039

Doty MS (1973) Farming the red seaweed, *Eucheuma*, for carrageenans. Micronesica 9:59–73

Doty MS (1985) *Eucheuma alvarezii* sp. nov. (Gigartinales, Rhodophyta) from Malaysia. In: Abbott IA and Norris JN (eds), Taxonomy of Economic Seaweeds with special reference to some Pacific and Caribbean species. California Sea Grant Coll Prog, 33–45

Eranza DR, Bahron A, Alin J (2015) Sustaining seaweed farming in Malaysia. Internat J Bus Mgt 3:201–205

Eswaran K, Ghosh PK, Mairh OP (2002) Experimental field cultivation of *Kappaphycus alvarezii* (Doty) Doty ex Silva at Mandapam region. Seaweed Res Utilin 24:67–72

FAO (2014) The State of World Fisheries and Aquaculture Opportunities and Challenges 2014. E-ISBN 978-92-5-108276-8 (PDF). Publishing Policy and Support Branch Office of Knowledge Exchange, Research and Extension FAO

Freile-Pelegrín Y, Robledo D (2006) Carrageenan of *Eucheuma isiforme* (Solieriaceae, Rhodophyta) from Yucatan, Mexico. II. Seasonal variations in carrageenan and biochemical characteristics. Bot Mar 49:72–78

Gereffi G, Humphrey J, Sturgeon T (2005) The governance of global value chains. Rev Inter Political Econ 12:1 February 2005: 78–104

Goes H, Feder-Martins A (2015) Manual para cultivo de carraginofitas no Brasil. Online PDF manual by Algasbras Biorrefinaria Ltda. [April, 2017] (http://www.carragenabrasil.com.br)

Hurtado AQ, Gerung GS, Yasir S, Critchley AT (2014) Cultivation of tropical red seaweeds in the BIMP-EAGA region. J Appl Phycol 26:707–718

Hurtado AQ, Lim PE, Tan J, Phang SM, Neish IC, Critchley AT (2016) Biodiversity and biogeography of commercial tropical carrageenophytes in the southeast Asian region. In: Pereira L (ed) Carrageenans: sources and extraction methods, molecular structure, bioactive properties and health effects. Nova Science Publishers, Hauppauge, New York, pp 67–90

Krishnan M, Narayanakumar R (2013) Social and economic dimensions of carrageenan seaweed farming in India. *In:* D. Valderrama, J. Cai, N. Hishamunda, N. Ridler (eds) Social and economic dimensions of carrageenan seaweed farming, Fish Aqua Tech Paper 580. FAO, Rome, p 163–185

Kronen M (2013) Social and economic dimensions of carrageenan seaweed farming in the Solomon Islands. *In*: D. Valderrama, J. Cai, N. Hishamunda, N. Ridler. (eds) Social and economic dimensions of carrageenan seaweed farming. Fish Aqua Tech Paper 580. FAO, Rome, p 147–161

Lang O (2015) Current status of sustainable aquaculture in Cambodia. In: Romana-Eguia MRR, Parado-Estepa FD, Salayo ND, Lebata-Ramos MJH (Eds.), Resource Enhancement and Sustainable Aquaculture Practices in Southeast Asia: Challenges in Responsible Production of Aquatic Species: Proceedings of the International Workshop on Resource Enhancement and Sustainable Aquaculture Practices in Southeast Asia 2014 (RESA). Aquaculture Dept. Southeast Asian Fisheries Development Center Tigbauan, Iloilo, Philippines: pp. 27–40

Lirasan T, Twide P (1993) Farming *Eucheuma* in Zanzibar, Tanzania. Hydrobiologia 260/261:353–355

Mantri AV, Eswaran K, Shanmugam M, Ganesan M, Veeragurunathan V, Thiruppathi S, CRK R, Seth A (2017) An appraisal on commercial farming of *Kappaphycus alvarezii* in India: success in diversification of livelihood and prospects. J Appl Phycol 29:335–357

Msuya FE (2013) Social and economic dimensions of carrageenan seaweed farming in the United Republic of Tanzania. In: Valderrama D, Cai J, Hishamunda N, Ridler N (eds) Social and economic dimensions of carrageenan seaweed farming. Fish Aqua Tech Paper 580. FAO, Rome, pp 115–146

Muñoz J, Freile-Pelegrín Y, Robledo D (2004) Mariculture of *Kappaphycus alvarezii* (Rhodophyta, Solieriaceae) color strains in tropical waters of Yucatán, México. Aquaculture 239:161–177

Neish IC (2005) The *Eucheuma* Seaplant Handbook Volume I: Agronomics, Biology and Crop Systems. Seaplantnet Technical Monograph No. 0505- 10A. ISBN 979 99558 0 7

Neish IC (2013a) Social and economic dimensions of carrageenan seaweed farming in Indonesia. *In:* D. Valderrama, J. Cai, N. Hishamunda, N. Ridler (eds) Social and economic dimensions of carrageenan seaweed farming. Fish Aqua Tech Paper 580. FAO, Rome, p 61–89

Neish IC (2013b) Feasibility assessment for a Zanzibar MUZE seaweed processing facility (ZanMUZE). Report submitted for UNIDO project no 13083 "building seaweed processing capacities in Zanzibar and Pemba: creating value for the poor" Neish IC, Msuya FE (2013) Seaweed value vhain assessment of Zanzibar. Report submitted for UNIDO Project no 13083 "Building Seaweed Processing Capacities in Zanzibar and Pemba: Creating value for the poor"

Neish IC, Msuya FE (2015) An analysis of production, markets and market systems for farmed seaweeds from Zanzibar. Report submitted for FAO Project No. TCP/URT/3401

Panlibuton H, Porse H, Nadela E (2007) Seaweed/Carrageenan Value Chain Assessment. Final Report submitted to IFC/PEP Philippines and GTZ. 47 pp

Parker HS (1974) The culture of the red algal genus *Eucheuma* in the Philippines. Aquaculture 3:425–439

Paula EJ, Pereira RTL, Ohno M (2002) Growth rate of the carragenophyte *Kappaphycus alvarezii* (Rhodophyta, Gigartinales) introduced in subtropical waters of São Paulo state, Brazil. Phycol Res 50:1–9

Porse H, Ladenburg S (2015) Seaweed value chain programme: final report. Submitted to Smart-Fish Indonesia

Porse H, Rudolph B (2017) The seaweed hydrocolloid industry: 2016 updates, requirements, and outlook. J Appl Phycol doi: 10.1007/s10811-017-1144-0

Randriambola T, Rafalimanana T (2005) L'algoculture ou la culture d'algues rouges du genre *Euchema*. Premier atelier national sur l'aménagement de la pêcherie de céphalopodes, 28–29 juillet 2005. Toliara, Madagascar. 8 pp

Robledo D, Gasca-Leyva E, Fraga J (2013) Social and economic dimensions of carrageenan seaweed farming in Mexico. In: Valderrama D, Cai J, Hishamunda N, Ridler N (eds) Social and economic dimensions of carrageenan seaweed farming. Fish Aqua Tech Paper 580. FAO, Rome, pp 185–204

Sade A, Ali I, Ariff MRM (2006) The seaweed industry in Sabah, East Malaysia. Jati 11:97–107

Selvavinayagam KT, Dharmar K (2017) Selection of potential method for cultivation and seed stock maintenance of *Kappaphycus alvarezii* during the northeast monsoon in southeast coast of India. J Appl Phycol 29:359–370

Sepulveda M (2016) O potencial da maricultura: oferta mundial da macroalga *Kappaphycus alvarezii* cresce para attender demanda por carragena. Panorama da Aquicultura (February 2016)

Shanmugam M, Sivaram K, Rajeev E, Pahalawattaarachchi V, Chandraratne PN, Asoka JM, Seth A (2017) Successful establishment of commercial farming of carrageenophyte *Kappaphycus alvarezii* Doty (Doty) in Sri Lanka: Economics of farming and quality of dry seaweed. (In prep.)

SIAP (2017) Report from Technical Working Group Meeting, Seaweed/Carrageenan Industry, Department of Trade and Industry – BOI Bldg., Sen. Gil Puyat Ave. Makati City, April 7, 2017

Smith AH (1992) Seaweed cultivation in the West Indies. In: Mshigeni K, Bolton J, Critchley A, Kiangi G (eds) Sustainable seaweed resource development in sub-Saharan Africa. University of Namibia, Windhoek, pp 337–351

Smith AH (1997) Seamoss cultivation in the West Indies. CANARI Guidelines Series 23 pp

Smith AH, Rincones RE (2006) Seaweed resources of the Caribbean. In: Critchley AT, Ohno M, Largo DB (eds) World seaweed resources: an authoritative reference system DVD-ROM multimedia interactive software – UNESCO Publishing/ETI Bioinformatics, University of Amsterdam, ISBN 90 75000 80 4

Solly RK, Booth W (1977) Seaweed colloids from the South Pacific. South Pacific Commission, Ninth Regional Technical Meeting on Fisheries, Noumea, New Caledonia, 24-28th January 1977, SPC/Fisheries 9/WP10

Sulu R, Kumar L, Hay C, Pickering T (2003) *Kappaphycus* seaweed in the Pacific: review of introductions and field testing proposed quarantine protocols. ISSN 1683–7568

Teitelbaum A (2003) Farming seaweed in Kiribati: A practical guide for seaweed farmers. Compiled and designed for the Secretariat of the Pacific Community

Tiroba G (2013) Seaweed quality manual: a practical guide for seaweed farmers, buying agents, fisheries officers and exporters [ed. 2013] compiled and designed for the Secretariat of the Pacific Community

Vadassery N, Neish IC, Suryanarayan S (2016) Systems for achieving high productivity during cultivation of *Kappaphycus* and *Eucheuma*. Paper presented at the 22nd Int Seaweed Symp, Copenhagen, Denmark, 19–24 June 2016

Valderrama D, Cai J, Hishamunda N, Ridler N, Neish IC, Hurtado AQ, Msuya FE, Krishnan M, Narayanakumar R, Kronen M, Robledo D, Gasca-Leyva E, Fraga J (2015) The economics of *Kappaphycus* seaweed cultivation in developing countries: a comparative analysis of farming systems. Aquaculture Econ Manag 19(2):251–277

Wu C, Li J, Xia E, Peng Z, Tan S, Li J, Wen Z, Huang X, Cai Z, Chen G (1989) On the transplantation and cultivation of *Kappaphycus alvarezii* in China. Chin J Oceanol Limnol 7:327-334www.carragenabrasil.com.br

www.jasuda.net

www.seaweedconsulting.com

Zertuche Gonzalez JA, Garcia-Lepe G, Pacheco-Ruiz I, Chee A, Gendrop V, Guzman jm (2001) Open water *Chondrus crispus* Stackhouse cultivation. J Appl Phycol 13: 249–253

Zertuche-Gonzalez JA, Garcia-Lepe G, Pacheco-Ruiz I, Chee-Barragan A, Gendrop-Funes V (1999) A new approach to seaweed cultivation in Mexico. World Aquaculture 30(2)

2 Biodiversity, Biogeography and Molecular Genetics of the Commercially Important Genera *Kappaphycus* and *Eucheuma*

Ji Tan, Phaik-Eem Lim, Siew-Moi Phang, and Anicia Q. Hurtado

Abstract

Through the years, several strains and color morphotypes of *Kappaphycus* and *Eucheuma* have emerged in cultivation areas as a result of environmental and anthropogenic factors. This led to confusion in nomenclature of these seaweeds. However, the advent of molecular technology has brought notable changes to seaweed research over the past two decades. New molecular techniques have proved useful, particularly in the molecular identification and cataloging of economically important carrageenophytes, i.e. *Kappaphycus* and *Eucheuma* which were commercially introduced into many countries worldwide. The following chapter discusses the application and findings of molecular studies within the context of these red seaweeds, along with potential areas for future research.

2.1 Introduction

Kappaphycus and *Eucheuma* are two genera of red seaweeds which are economically important in the global market because they are valuable sources of natural phycolloids, in the form of *kappa* and *iota* carrageenan, respectively. Originally, *Chondrus crispus* was gathered from the wild for carrageenan processing, however the natural population was not able to sustain the ever increasing demands of the industry. Following years of research and development in field cultivation, carrageenan characterization and product development, *Kappaphycus* and *Eucheuma* were selected as the primary sources of raw materials for *iota* and *kappa* carrageenan extraction.

In addition to its use as a raw material for hydrocolloid, *Kappaphycus* has also been used experimentally in aquaculture feed for abalone (Capinpin and Corre 1996). The seaweed was also used with promising results as a sink or sponge for the extraction of organic and inorganic products from fish and invertebrate aquaculture in land based, Integrated Multi-trophic Aquaculture (IMTA) systems (Rodrigueza and Montano 2007; Hayashi et al. 2008; Namukose et al. 2016). Positive results provided the impetus for further research on other "alternative" uses of eucheumatoid biomass. These commercial trials are crucial to the further development and sustainability of the cultivation industry. Likewise, initial results on the use of *Kappaphycus* as a source of bioethanol (Khambhaty et al. 2011; Meinita et al. 2012; Fakhrudin et al. 2014) were encouraging.

Kappaphycus and *Eucheuma* were successfully farmed in the Philippines in 1971 after almost five years of field cultivation trials (Doty 1973; Doty and Alvarez 1981; Hurtado et al. 2014, 2015). Presently, *Kappaphycus* is commercially farmed over a greater area than *Eucheuma*, in several

J. Tan
Department of Agricultural and Food Sciences, Universiti Tunku Abdul Rahman (UTAR), 31900 Kampar, Perak, Malaysia

P.-E. Lim (✉)
Institute of Ocean and Earth Sciences (IOES), University of Malaya (UM), 50603 Kuala Lumpur, Malaysia
e-mail: phaikeem@um.edu.my

S.-M. Phang
Institute of Ocean and Earth Sciences (IOES), University of Malaya (UM), 50603 Kuala Lumpur, Malaysia

Institute of Biological Sciences, University of Malaya (UM), 50603 Kuala Lumpur, Malaysia

A.Q. Hurtado (✉)
Integrated Services for the Development of Aquaculture and Fisheries (ISDA) Inc., MacArthur Highway, Tabuc Suba, Jaro, 5000 Iloilo City, Philippines
e-mail: anicia.hurtado@gmail.com

A.Q. Hurtado et al. (eds.), *Tropical Seaweed Farming Trends, Problems and Opportunities*, Developments in Applied Phycology 9, DOI 10.1007/978-3-319-63498-2_2

countries above and below the 10° latitude (Kronen 2013; Krishnan and Narayanakumar 2013; Neish 2013; Hayashi et al. 2014; Msuya et al. 2014).

2.2 Biodiversity

The late Dr. MS Doty together with Drs. Abbott, Cheney and Norris worked diligently and contributed to the present nomenclature of *Kappaphycus* and *Eucheuma*. A series of workshops and conferences were conducted under the California Sea Grant Program, University of California, La Jolla, California which resulted in notable publications in 1985 and 1988. These publications involved thorough examinations of external and internal morphologies of tetrasporophyte, carposporophyte, male and female gametophytes of *Eucheuma* which contributed to the eventual separation of the genus *Kappaphycus* from *Eucheuma*. Prior to this taxonomic separation, Doty and Norris (1985) classified *Eucheuma* into four (4) sections (Table 2.1).

However, Doty (1988) classified *E. alvarezii* and *E. striatum* as *Kappaphycus alvarezii* and *K. striatum* (now *K. striatus*), respectively based on the following characteristics:

(a) fronds are of many forms but commonly cylindrical, bearing simple blunt or spiny protuberances
(b) protuberances are irregularly arranged or, in some cases, appearing in rows so segments become angular
(c) branching is irregular but sometimes irregularly pinnate or, by chance opposite of falsely dichotomous
(d) hyphal axial core usually present and vaguely cylindrical
(e) carrageenan extracted is kappa
(f) cystocarps are on main axes
(g) no laterals are associated with cystocarp

Kappaphycus alvarezii,K. striatus and *Eucheuma denticulatum* are endemic to the tropics, especially within the Coral Triangle, where the biodiversity of flora and fauna is very rich. Tanzania is also known to have natural populations of *E. denticulatum* (Sen 1991). *Kappaphycus* has been introduced to several countries, ranging from sub-tropical to tropical waters (Ask et al. 2003; Hurtado et al. 2015). At present, there are several morphologicalplasticities and color morphotypes (Fig. 2.1a–m) of *K. alvarezii,K. malesianus*, *K. striatus* and *E. denticulatum*. Some of these are possibly morphological responses to crop selection and environmental conditions in the commercial farms (Hurtado 2013; Hurtado et al. 2016). Details of the biodiversity and distribution of *Kappaphycus* and *Eucheuma* in Southeast Asia, where the holotypes of the two seaweeds were collected, were reported by Hurtado et al. (2016). Similarities and differences in their molecular systematics will be discussed in detail in the next section.

Table 2.1 Classification of *Eucheuma* (Doty and Norris 1985)

Section	Section	Section	Section
Eucheuma	Anaxiferae	Gelatiformia	Cottoniformia
(iota)	(iota)	(kappa)	(kappa)
	(Kraft 1972)		
E. denticulatum	*E. arnoldii*	*E. gelatinae*	*E. alvarezii*
E. isiforme			*E. striatum*
E. uncinatum			

The first stock of *Kappaphycus* and *Eucheuma* to be used for farming purposes was collected from the wild, which was later domesticated and formed the original stock (Doty 1973; Doty and Alvarez 1975, 1981). Vegetative cuttings are the only means to propagate the raw materials for carrageenan extraction (deemed to be called eucheumatoids or carrageenophytes); the methods have remained the same to the present day. Cultivars of *Kappaphycus* and *Eucheuma* have been transplanted to different areas of the Philippines where the favorable environment encouraged robust growth. Likewise, these seaweeds were introduced to other parts of the world. Due to several transplantations, in and out of the original area of growth in the Philippines, *Kappaphycus* and *Eucheuma* cultivars adopted several phenotypic or morphological plasticities – probably in response to each new cultivation environment. The following shows some photos of various morphotypes of *K. alvarezii* (Fig. 2.1a–m), *K. striatus* (Fig. 2.2a–r), *K. malesianus* (Fig. 2.3a and b) and *E. denticulatum* (Fig. 2.4a–d).

2.3 Development of Genetic Markers for *Kappaphycus* and *Eucheuma*

A genetic or molecular marker is a gene, or region of DNA, that can be used to identify and differentiate species. Although genetic markers serve various purposes, their application towards *Kappaphycus* and *Eucheuma* stemmed mainly from the need for reliable identification of a very confusing array of color and morphological variants. *Kappaphycus* and *Eucheuma* are morphologically plastic, in fact seaweeds of the same species may vary substantially in appearance. The accuracy of a molecular marker also allows for the identification of species in immature or damaged samples, e.g., broken thalli, carpospores etc. Thus, application of genetic marker technology allows an alternative, independent and reliable method for identification and therefore discrimination of these red seaweeds.

Genetic markers displaying notable levels of genetic variation were those most commonly used in the molecular studies of *Kappaphycus* and *Eucheuma* (Conklin et al. 2009; Fredericq et al. 1999; Freshwater et al. 1994; Tan et al. 2012,

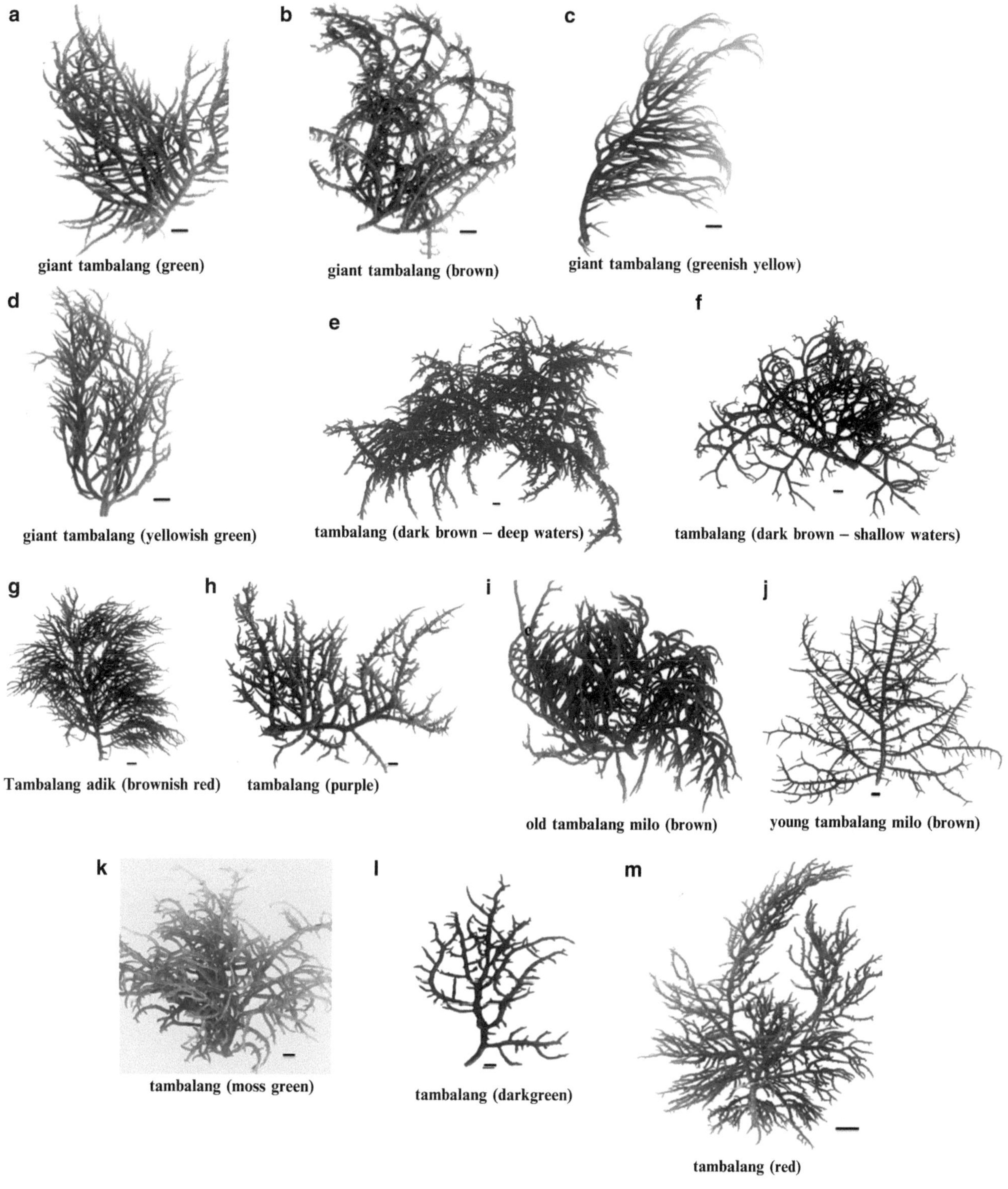

Fig. 2.1 (**a–m**) Several morphologicalplasticities and different color morphotypes of *K. alvarezii* (All photos courtesy of AQ Hurtado, bar = 1 cm)

2013, 2014; Zuccarello et al. 2006; Lim et al. 2014a; Zhao and He 2011). Other markers, i.e., random amplification of polymorphic DNA (RAPD) (Dang et al. 2008) and single locus markers (Maili et al. 2016) were also employed. Apart from molecular identification, genetic markers are also invaluable in: (i). elucidating the phylogenetic relationship between *Kappaphycus* and *Eucheuma*; (ii). identifying seaweed strains with favorable traits for cultivation and commercialization; (iii). mapping the biodiversity and genetic

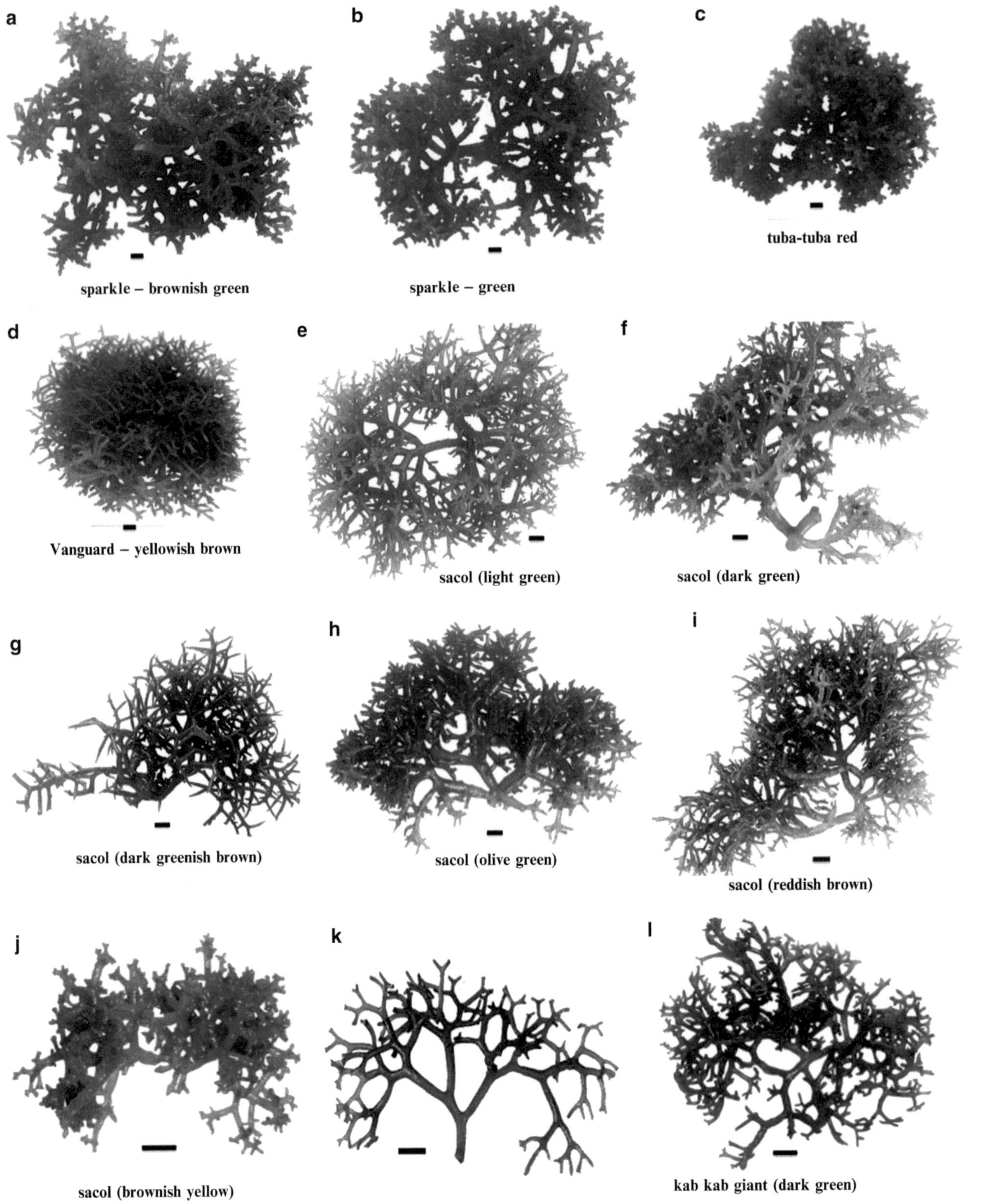

Fig. 2.2 (**a**–**r**) Several morphologicalplasticities and different color morphotypes of *K. striatus* (All photos courtesy of AQ Hurtado, bar = 1 cm)

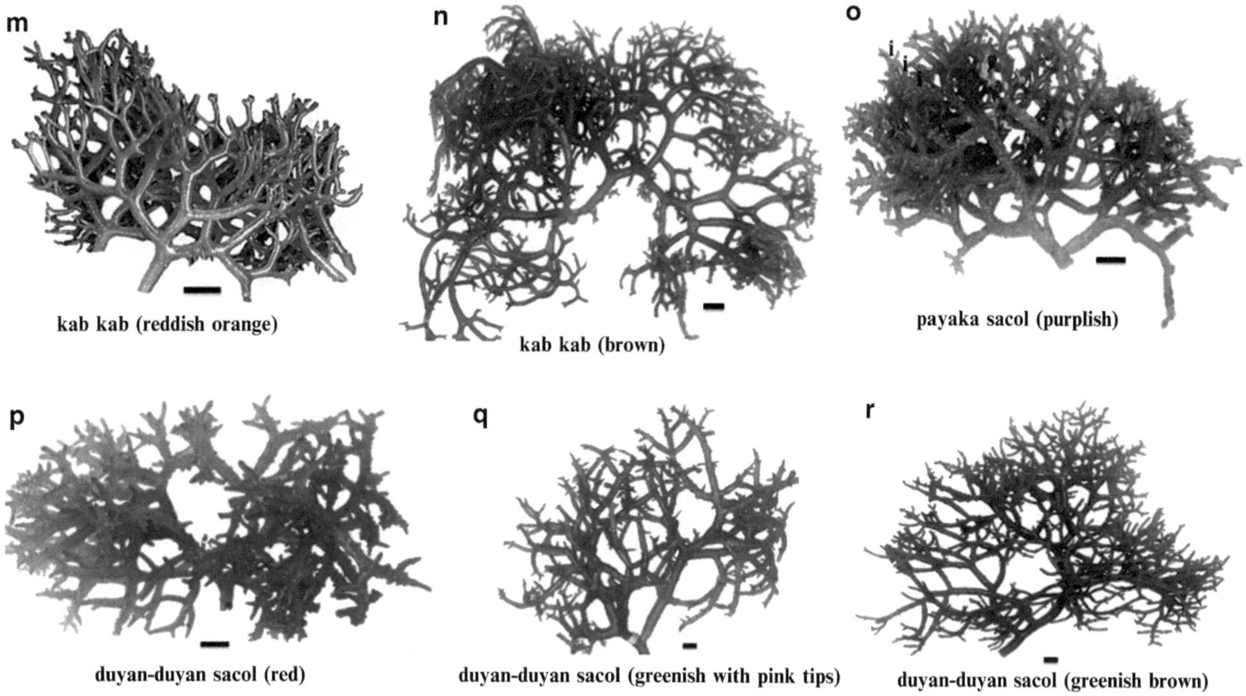

Fig. 2.2 (continued)

Fig. 2.3 (**a** and **b**) *K. malesianus* (Photo courtesy of AQ Hurtado and Tan J, bar = 1 cm)

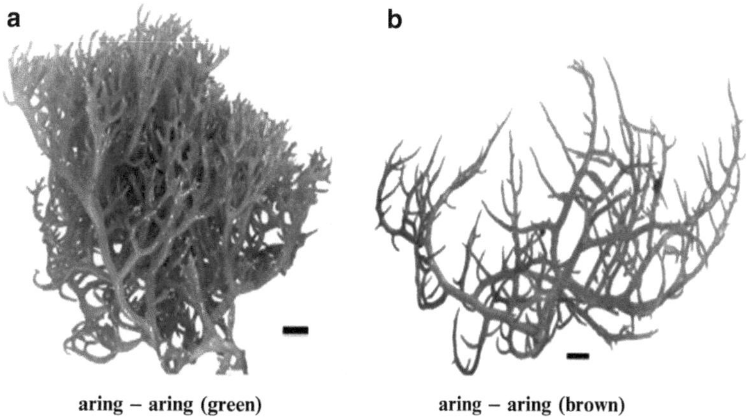

diversity of these seaweeds on a global basis; (iv). detecting bio-invasive species, etc.

The first extensive use of genetic markers for *Kappaphycus* and *Eucheuma* was conducted by Zuccarello and co-workers (2006), in which a mitochondrial-encoded *cox*2–3 spacer and a plastid-encoded RuBisCO spacer were used for phylogenetic reconstruction of these seaweeds. The study demonstrated the usefulness of molecular markers in addition to revealing valuable insights into the phylogeny of *Kappaphycus*, *Eucheuma* and *Betaphycus* seaweeds which had been collected from multiple sites around the world. The *cox*2–3 spacer has since become a frequently used genetic marker due to its formidable resolving capabilities (Conklin et al. 2009; Dumilag and Lluisma 2014; Tan et al. 2012, 2013, 2014; Lim et al. 2014a). Over the last decade, several new DNA markers, i.e., mitochondrial *cox*1 and *cox*2; nuclear internal transcribed spacer (ITS), as well as plastid *rbc*L, were designed or utilized for various purposes, including the detection of invasive *K. alvarezii* in Hawaii (Conklin et al. 2009), development of DNA barcodes (Tan et al. 2012), molecular identification and phylogenetics of species or varieties of *Kappaphycus* and *Eucheuma* (Araújo et al. 2013; Tan et al. 2013; Zhao and He 2011; Dumilag and Lluisma 2014), species descriptions (Ganzon-Fortes et al. 2012; Tan et al. 2014), as well as genetic studies (Lim et al. 2014a). The details of these primers are summarized in Table 2.2

The resolution of genetic markers depends on multiple factors, including the number of phylogenetically informa-

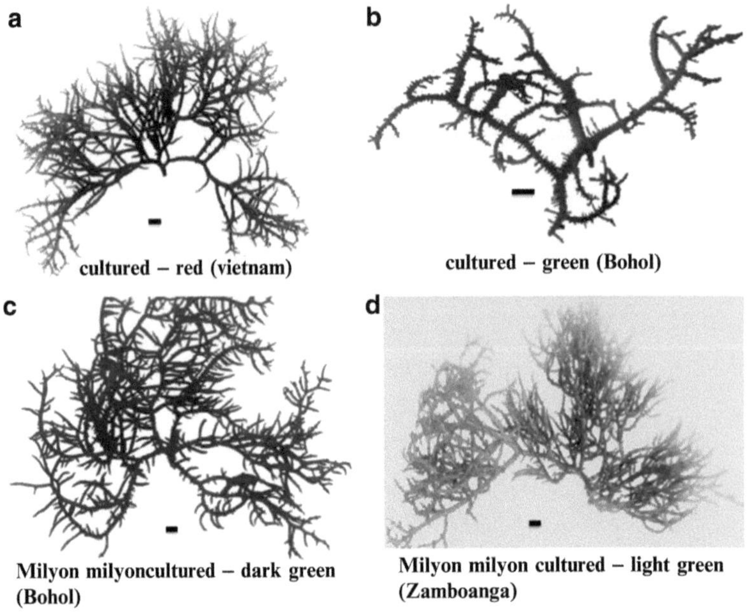

Fig. 2.4 (**a**–**d**) Different color morphotypes of *E. denticulatum* (All photos courtesy of AQ Hurtado, bar = 1 cm)

Table 2.2 Common genetic markers used in molecular studies of *Kappaphycus* and *Eucheuma*

Genetic marker	Primer	Primer sequence*	Annealing temperature (Tm)	References
*Cox*1	COXI43F	5′-TCAACAAATCATAAAGATATTGGWACT-3′	52 °C	Geraldino et al. (2006) and Yang et al. (2007)
	C622F	5′-CCTGTNTTAGCAGGWGCTATTACAATGC-3′		
	C880R	5′-ACAGTATACATATGATGNGCTCAAAC-3′		
	COXI1549R	5′-AGGCATTTCTTCAAANGTATGATA-3′		
*Cox*2	Cox2_for	5′-GTACCWTCTTTDRGRRKDAAATGTGATGC-3′	50 °C	Tan et al. (2012)
	Cox3_rev	5′-GGATCTACWAGATGRAAWGGATGTC-3′		
*Cox*2–3 spacer	Kcox2_F71	5′-TTCAAGATCCTGCAACTCC-3′	51 °C	Zuccarello et al. (1999a)
	Kcox2_R671	5′-ATTTCACTGCATTGGCCAT-3′		
*rbc*L	F-7	5′-AACTCTGTAGTAGAACGNACAAG-3′	50 °C	Freshwater and Rueness (1994) and Gavio and Fredericq (2002)
	F-577	5′-GTATATGAAGGTCTAAAAGGTGG-3′		
	R-753	5′-GCTCTTTCATACATATCTTCC-3′		
	R-rbcS start	5′-GTTCTTTGTGTTAATCTCAC-3′		
RuBisCO spacer	RuBisCO_for	5′-TGTGGACCTCTACAAACAGC-3′	55 °C	Zuccarello et al. (1999b)
	RuBisCO_rev	5′-CCCCATAGTTCCCAAT-3′		

* Ambiguous nucleotide codes are in accordance to IUPAC: K = G/T; R = A/G; W = A/T; N = A/T/C/G

tion characters, interspecific divergences and intraspecific variations. According to Tan and co-workers (2012), the mitochondrial-encoded *cox*2–3 spacer was the most suitable marker for molecular identification, basic molecular phylogeny and DNA-barcoding (discussed below) of *Kappaphycus* and *Eucheuma*, considering the large database of DNA data already available in GenBank. However, the *cox*2–3 spacer, along with *cox*2, were more suitable at delineating inter-generic and inter-specific relationships. Phylogenetic analysis of higher taxa requires a combination of both *cox*1 and *rbc*L which are genetically more conservative (Tan et al. 2012). In terms of phylogeny, a multi-gene approach, involving DNA from different organelles, is most likely to best represent the evolutionary pathways of these commercially important rhodophytes, regardless of the taxonomic rank being analyzed.

DNA-barcoding is the use of short, universal genetic markers in DNA in order to determine its identity, based on online databases. The technique was first introduced by Herbert and co-workers (2003a, b, 2004). Although conceptually similar to the use of genetic marker(s) for the molecular identification of organisms, DNA-barcoding emphasizes the preferable use of just one DNA barcode for rapid and reliable identification, particularly by non-experts. Thus, the assessment and selection of DNA-barcodes are generally more thorough and comprehensive, also taking into account the universality of the DNA-barcode, popularity, ease of amplification, length of the DNA barcode (short barcodes may contain less information, whereas longer ones may take too long to analyze), in addition to resolving power, as well as inter-specific divergences and intra-specific variations (Jinbo et al. 2011; Meier et al. 2008; Meyer and Paulay 2005; Tan et al. 2012). The mitochondrial *cox*2 and *cox*2–3 spacers were identified as suitable DNA-barcodes for *Kappaphycus* and *Eucheuma* (Tan et al. 2012), however the extent of their effectiveness will become more evident as an increasing number of species are sampled.

2.4 Molecular Identification of *Kappaphycus* and *Eucheuma*

The morphologically plastic nature of eucheumatoidseaweeds (e.g. *Betaphycus*, *Eucheuma* and *Kappaphycus*) was well known in the early 1980s (Doty 1985, 1988; Doty and Norris 1985), whereby different species of seaweeds may resemble one another, or conspecifics appearing different between individuals. The issue led to confusion in the identification of these seaweeds, which not only impeded taxonomic studies, but affected farmers as well as the entire carrageenan industry. The confusion was exacerbated by the use of colloquial and commercial names which differ amongst countries (Doty 1985; Doty and Norris 1985; Zuccarello et al. 2006). For example, the common trade name for *Kappaphycus alvarezii* (formerly known as *Eucheuma alvarezii*) is "cottoni" which is not to be confused with *Kappaphycus cottonii* (Weber-van Bosse) Doty ex P.C.Silva, a different species within the genus. Another example involves the use of the local term "*Aring-aring*" which refers to *K. malesianus* in Malaysia but *K. alvarezii* in the Philippines (Dumilag and Lluisma 2014; Tan et al. 2013, 2014).

Comprehensive work, mostly by MS Doty led to the eventual classification of eucheumatoid seaweeds into the genera used at present, viz. *Betaphycus*, *Eucheuma* and *Kappaphycus* (Doty 1985, 1988; Doty and Alvarez 1975; Doty and Norris 1985). The delineation of these genera was based on morphological and biochemical characteristics of these red seaweeds. Some distinctive morphological characters included: branching patterns, the presence of spines, cystocarp structure, as well as cell arrangement (microscopy required). The most notable biochemical differences between *Betaphycus*, *Eucheuma* and *Kappaphycus* is the type of carrageenan produced, namely *Beta* (β), *Iota* (ι) and *Kappa* (κ) carrageenan, respectively.

Although the morphological and biochemical features allow for the inter-generic distinctions to be made between-*Betaphycus*, *Eucheuma* and *Kappaphycus*, these features were less reliable in differentiation at the species level (Lim et al. 2014b; Tan et al. 2013; Zuccarello et al. 1999). Species identification has proven to be exceptionally challenging in Southeast Asia which harbors a rich diversity of eucheumatoids. For example, Tan and co-workers reported at least eight varieties of *Kappaphycus* that were recognized by the locals in Sabah, Malaysia (Lim et al. 2014b; Tan et al. 2013; Phang et al. 2010). Likewise, numerous varieties of *Kappaphycus* and *Eucheuma* are also cultivated in Indonesia and the Philippines (Lim et al. 2014a). Without a reliable means of identifying the farmed seaweeds, most farmers end up planting mixed populations of *Kappaphycus* and *Eucheuma*seaweeds. This translates to profit losses in the carrageenan industry in Malaysia as the harvested *Kappaphycus* and *Eucheuma*seaweeds produce different types of carrageenan which have to be separated prior to processing (Lim et al. 2014b; Tan et al. 2013; Phang et al. 2010). In other countries, concerns were mainly due to the uncontrolled spread of invasive seaweed strains arising from the commercially introduced *Kappaphycus alvarezii* (Araújo et al. 2013; Conklin et al. 2009; Dumilag and Lluisma 2014; Zhao and He 2011). These issues promoted the need for the application of molecular techniques in identification of *Betaphycus*, *Eucheuma* and *Kappaphycus*seaweeds (Araújo et al. 2013; Conklin et al. 2009; Dumilag and Lluisma 2014; Lim et al. 2014a; Tan et al. 2012, 2013, 2014; Zhao and He 2011; Zuccarello et al. 2006; Maili et al. 2016). The identity of a seaweed is usually determined via genetic comparison against DNA data within an online database such as GenBank (http://www.ncbi.nlm.nih.gov/genbank/). Despite the ease and convenience of this method, care has to be taken in verifying the authenticity and reliability of DNA sequences available within the repository. In addition, the confirmation of a seaweed species should always be supported with morphological examination.

In Malaysia, the application of the several mitochondrial and the plastid-based genetic markers has led to the clear identification of several varieties of seaweed. For example, the "*Aring-aring*" variety of *Kappaphycus,* which is morphological similar to *K. alvarezii* was shown to be a different species, now described as *K. malesianus* J. Tan, P. E. Lim et S.M. Phang (Lim et al. 2014b; Tan et al. 2013, 2014). Likewise, the commonly cultivated "*Cacing*" variety was shown to be *Eucheuma* instead of *Kappaphycus*. These findings had immediate practical implications in that they minimized the cultivation of mixed populations of *Kappaphycus* and *Eucheuma*seaweeds within the country, thereby decreasing processing issues and in so doing increasing the value of the biomass. Similar approaches were applied to farmed and wild (growing on the seabed far away from any seaweed farms) specimens elsewhere in a first attempt to identify the bio- and genetic diversity of these seaweeds.

2.5 Molecular Phylogenetics of *Kappaphycus* and *Eucheuma*

Molecular phylogeny is often performed alongside molecular identification as both studies involve the same DNA sequences or dataset. Unlike molecular identification which employs the use of genetic data for species identification, molecular phylogeny infers relationships between various species (or taxa) of organisms. As genotype is inherently linked to phenotype, so molecular-based phylogenetic results are often used to supplement morphologically-based phylogeny, or taxonomy. Often this analysis reveals evolutionary lineages as well as the genetic distribution of organisms from various regions of the world, which is particularly important in strain selection and the detection of bio-invasions (Conklin et al. 2009; Tan et al. 2012; Zuccarello et al. 2006).

Molecular phylogeny of seaweeds, at the family and genus level, was reported during the 1990s, mostly relying on the plastid-encoded *rbc*L gene (Fredericq et al. 1999; Freshwater et al. 1994). Freshwater and co-workers (1994) reported the paraphyly of *Eucheuma*, which was supported by subsequent studies employing different sets of molecular markers (Tan et al. 2013; Zuccarello et al. 2006). The phylogeny of *Betaphycus*, *Eucheuma* and *Kappaphycus* (Fig. 2.5) is based mainly on the mitochondrial *cox*2–3 spacer marker due to the large number of sequences available in GenBank.

The *cox*2–3 spacer phylogenetic tree (Fig. 2.5) has shown that *Kappaphycus* and *Eucheuma* are genetically different. However, the phylogenetic resolution of the genetic marker was not sufficient to infer the relationships between *Betaphycus*, *Eucheuma* and *Kappaphycus*. The paraphyly of *Eucheuma* was observed with the inclusion of *Eucheuma isiforme* specimens from America (Doty 1988; Doty and Norris 1985; Zuccarello et al. 2006).

The genus *Kappaphycus* was inferred to be monophyletic although the taxonomic position of the poorly studied *K. cot-*

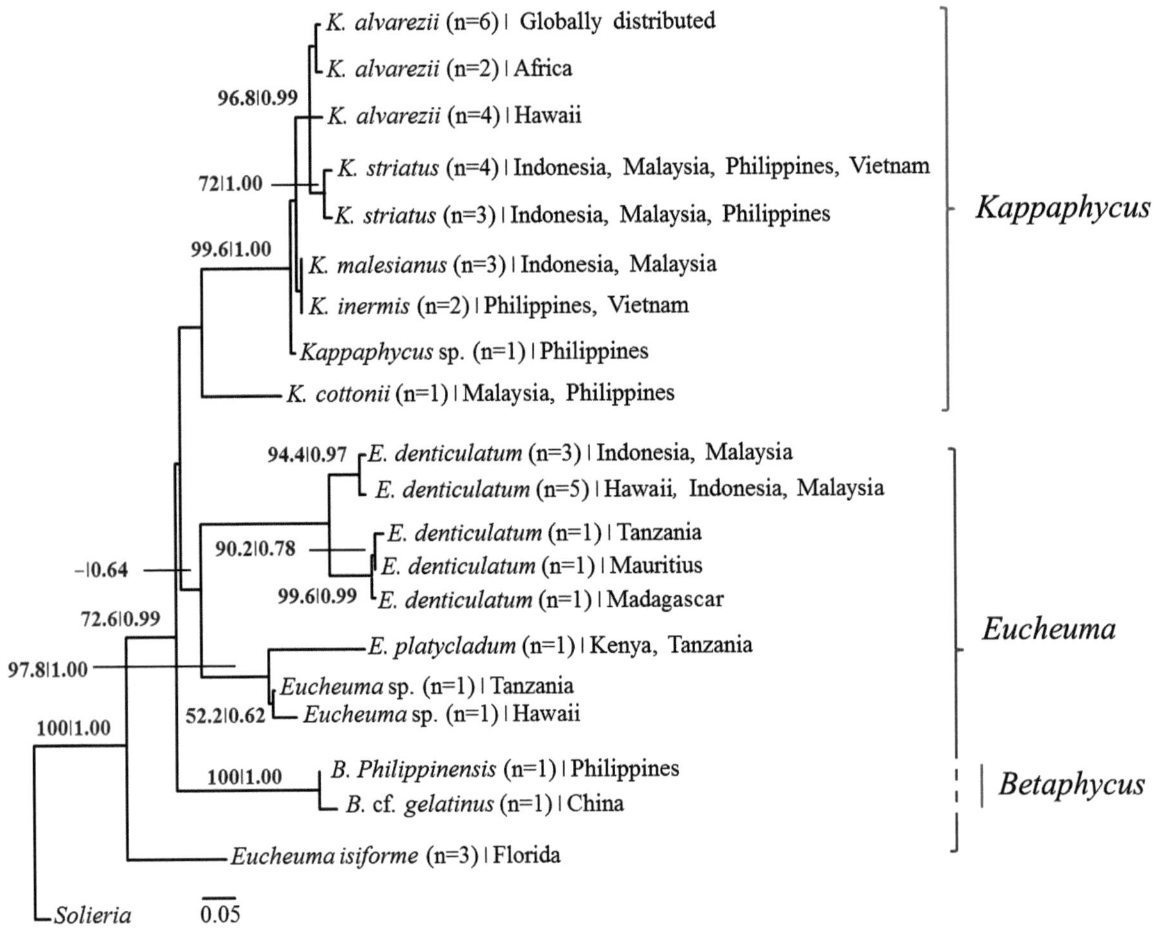

Fig. 2.5 Simplified phylogenetic tree showing the current relationships between *Betaphycus*, *Eucheuma* and *Kappaphycus* based on the mitochondrial-encoded *cox*2–3 spacer. Nodal values denote Maximum Likelihood bootstrap support and Bayesian posterior probabilities (Phylogenetic tree constructed based on DNA data adopted from Dumilag and Lluisma 2014; Tan et al. 2013; Zuccarello et al. 2006)

tonii is unresolved (Doty 1988; Doty and Norris 1985; Zuccarello et al. 2006). It is also worth noting that DNA records of *K. procrusteanus* (Kraft) Doty 1988, first reported from the Philippines, were unavailable (Tan et al. 2014). Based on genetic results, three different genotypes of *K. alvarezii* were reported (Tan et al. 2013; Zuccarello et al. 2006; Conklin et al. 2009): (i). A *K. alvarezii* strain that is globally distributed, and is likely the commercial strain initiated in the Philippines; (ii). A genotype unique to Africa; and (iii). a strain exclusive to the Hawaiian islands. These genotypes were not known to be morphologically different from one another. Based on the *cox*2–3 spacer, the globally distributed *K. alvarezii* specimens were genetically closer to those collected from Africa (Genetic Distance, GD, of <0.9%) than those from Hawaii (GD of <2.1%). Comparative growth and carrageenan studies between these genotypes may lead to the identification of potentially better cultivars.

Likewise, two different genotypes of *Kappaphycus striatus* were observed within Southeast Asia (GD < 1.2%; *cox*2–3 spacer), in which one of the genotypes is constituted of commercially cultivated strains, whereas the other involves specimens from the wild. On the other hand, *K. malesianus*, a newly described species (Tan et al. 2014) was inferred to form a sister clade to *K. inermis* from the Philippines, based on both mitochondrial and plastid-encoded genetic markers (Dumilag and Lluisma 2014). Sampling efforts by Lim and co-workers (2014a) revealed several new genotypes of *Kappaphycus* which may be potential new species worth further research.

Research on *Eucheuma*seaweeds has been generally less extensive than that of *Kappaphycus*, which may be attributed to: (i). sampling difficulty; (ii). relatively lower economic value and (iii). the vast number of species available, many of which may be synonymous. Three main genotypes of *E. denticulatum*, the most widely cultivated *Eucheuma* species, were reported (Lim et al. 2014a; Tan et al. 2013; Zuccarello et al. 2006): two genotypes which are cultivated in Southeast Asia (and countries where they have been introduced) and one genotype exclusive to Africa. A genetic distance of <9.1% was observed between the Southeast Asian and African genotypes of *Eucheuma*. The taxonomic positions of other *Eucheuma* species were poorly resolved and can only be inferred when more samples become available.

The genus *Betaphycus* is also poorly studied, although its taxonomic differentiation from *Eucheuma* and *Kappaphycus* was reported from earlier studies (Doty 1988; Doty and Norris 1985; Dumilag et al. 2014; Tan et al. 2013; Zuccarello et al. 2006). Dumilag and co-workers (2014) suggested that the type of cell wall polysaccharide may not be a reliable character in the taxonomy of *Betaphycus* due to the presence of hybrid carrageenan structures (Chopin et al. 1999). The authors also suggested the possible conspecificity of *B. philippinensis* and *B. speciosus* based on morphological and molecular analyses.

Phylogenetic studies of *Betaphycus*, *Eucheuma* and *Kappaphycus* within the past decade (Conklin et al. 2009; Dumilag et al. 2014; Dumilag and Lluisma 2014; Tan et al. 2012, 2013, 2014; Zuccarello et al. 2006) have revealed valuable insights into the taxonomic positioning of these species. Although the commonly used genetic markers, viz., *cox*1, *cox*2, *cox*2–3 spacer, *rbc*L were quite capable of species delineation, resolving inter-generic relationships would most likely require slower-evolving markers, or the use of concatenated datasets involving nuclear, plastid and mitochondrial genetic regions.

2.6 Genetic Diversity of *Kappaphycus* and *Eucheuma*

Documenting the genetic diversity of *Kappaphycus* and *Eucheuma* provides promising benefits to the seaweed industry, which include: marker-assisted selection of good seaweed strains for cultivation, DNA-barcoding, monitoring of invasive strains of *Kappaphycus* or *Eucheuma*, as well as studies on biodiversity and phylogeography. Lim et al. (2014a) reported for the first time, a collaborative effort between major carrageenan producers within Southeast Asia to document the genetic diversity of *Kappaphycus* and *Eucheuma* within the region. The study involved a large-scale sampling and study program for *Kappaphycus* and *Eucheuma* involving Malaysia, the Philippines, Indonesia and Vietnam. The study employed the use of the mitochondrial-encoded *cox*1 and *cox*2–3 spacers to generate haplotype (DNA polymorphism) networks of these seaweeds. Haplotype networks based on the concatenated *cox*1 and *cox*2–3 spacer markers of *Kappaphycus* and *Eucheuma* are summarized in Figs. 2.6 and 2.7, respectively.

In brief, the study revealed many different haplotypes within Southeast Asia for most species of *Kappaphycus*, as well as *Eucheuma denticulatum*. A total of seven known haplotypes was revealed for *K. alvarezii*, seven for *K. striatus*, four for *K. malesianus*, as well as two unknown *Kappaphycus* spp. from Vietnam and the Philippines. The main haplotype of *K. alvarezii* "KA" is cultivated throughout Southeast Asia and is likely to be distributed globally. Interestingly, three different major genotypes of *K. striatus* were observed in Southeast Asia, namely "KSA" which was present in Malaysia, Indonesia, Philippines and Vietnam; "KSB" which was observed in Malaysia and Indonesia, and "GUI2" which was exclusive to the Philippines. These genotypes are morphologically similar and may potentially be cryptic species. The major haplotype of *K. malesianus* "MY216" (including cultivars), showed 75 nucleotide differences from *K. alvarezii* "KA". The two unidentified *Kappaphycus* species recorded a nucleotide difference of four and 13 steps from

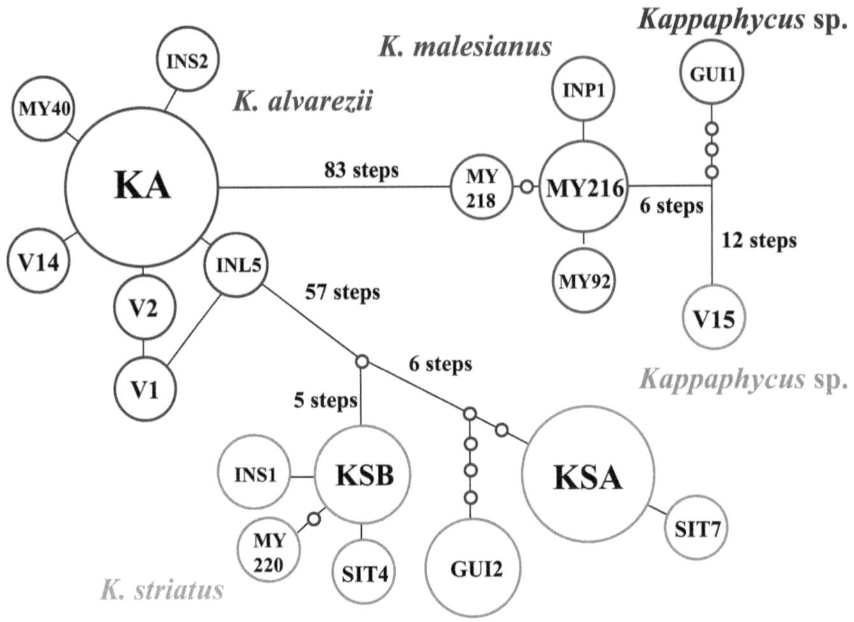

Fig. 2.6 Haplotype networks of *Kappaphycus* based on the concatenated *cox*1-*cox*2–3 spacer markers. Each *line* represents a point mutation (a step) whereas circles indicate hypothetical haplotypes. Common haplotypes are represented by *larger circles* (not drawn to scale) (Haplotype network adopted from Lim et al. 2014a, b)

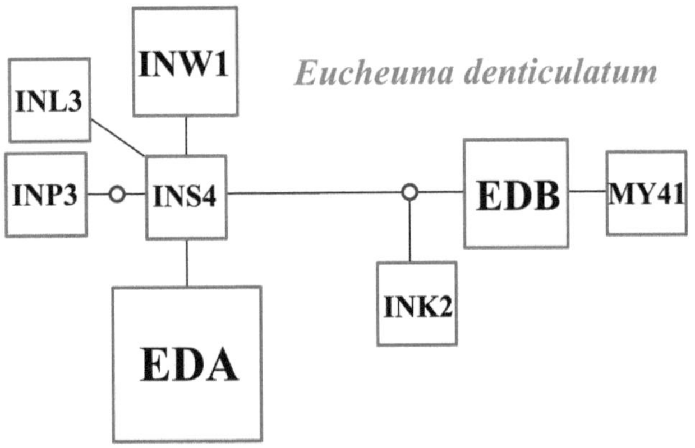

Fig. 2.7 Haplotype network of *Eucheuma* based on the concatenated cox1-cox2–3 spacer markers. Each *line* represents a point mutation (a step) whereas circles indicate hypothetical haplotypes. Common haplotypes are represented by *larger circles* (not drawn to scale) (Haplotype network adopted from Lim et al. 2014a)

"MY216", which inferred genetic relatedness to *K. malesianus*, although there was no geographical pattern associated with these seaweeds.

The scarcity of *Eucheuma* spp. (and *Betaphycus*) samples impeded the genetic documentation of these carrageenophytes. Only *E. denticulatum* was sampled in Lim et al. (2014a) which hinted at the rarity of other *Eucheuma* spp., at least within the vicinity of shores and farms in Southeast Asia. Two main haplotypes of *E. denticulatum* (which differed by ten nucleotides), namely "EDA" and "EDB" were sampled within the region. Based on the preliminary study, Lim and collaborators (2014a) believed that numerous additional haplotypes and new species of *Kappaphycus* and *Eucheuma* are present within the seas of Southeast Asia, particularly within the Coral Triangle. However, sampling the thousands of islands within this region will require concerted efforts, particularly in terms of funding and logistics between Southeast Asian countries in order to comprehensively document the genetic diversity of these commercially important rhodophytes. Similarly, genetic diversity studies could be performed for countries or continents with native populations of *Kappaphycus* or *Eucheuma*, e.g. Africa and Hawaii.

2.7 Introduction of *K. alvarezii* to Non-indigenous Areas

Non-indigenous means a species occurs outside its native range generally arising by human influences. It is synonymous to "non- native," "alien" and "exotic" (Lodge et al. 2006). Species that spread widely beyond the location of initial establishment, become locally abundant, or spread into natural areas, are referred to as invasive and this depends on time and spatial scale. Within the Areschougiaceae, 8 from the total of 96 species have been successful invaders (Williams and Smith 2007), including members of the genera *Kappaphycus*, *Eucheuma* and *Sarconema*, many of these were intentionally introduced for commercial carrageenan production (Zemke-White and Smith 2006). Amongst those red species which has been introduced to many places outside their original location of the Philippines are *Kappaphycus alvarezii*, *K. striatus* and *Eucheuma denticulatum* (Ask et al. 2003; Hurtado et al. 2016; Fig. 2.8).

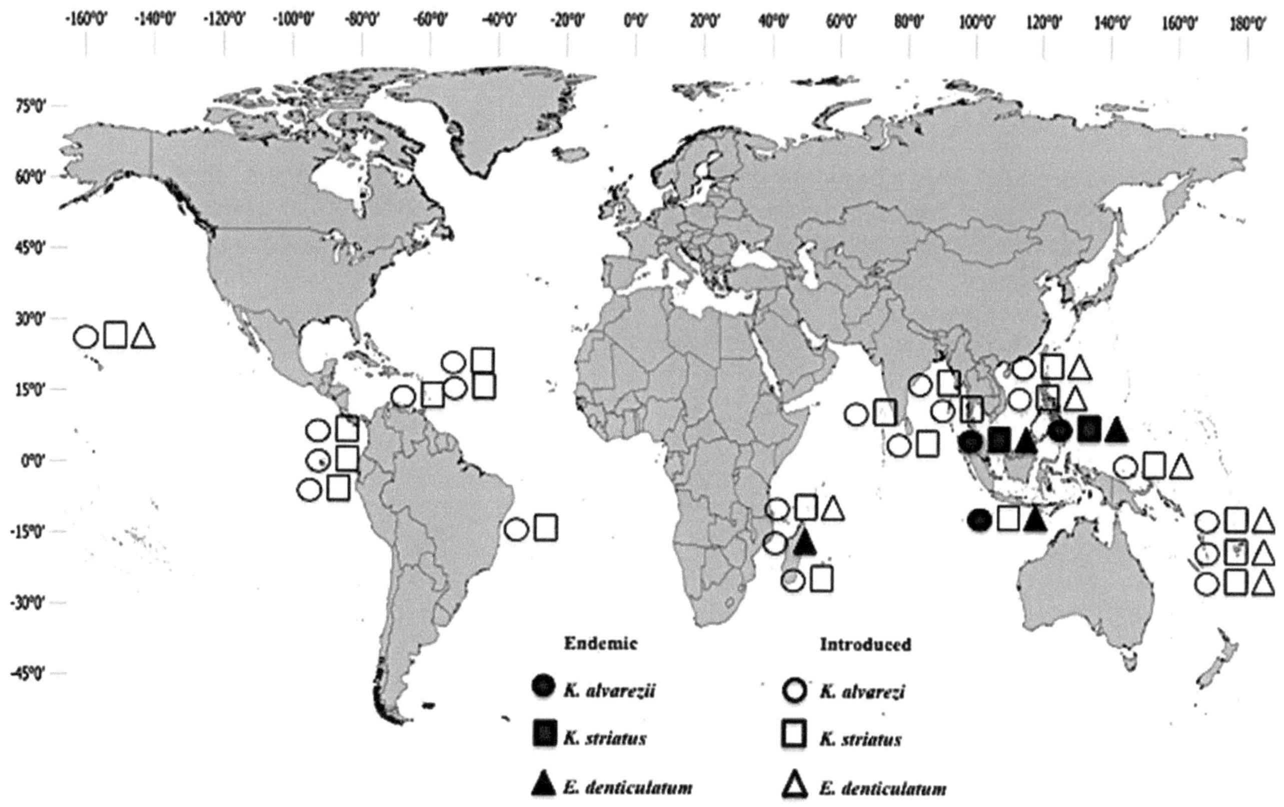

Fig. 2.8 Endemic and introduced *K. alvarezii,K. striatus* and *E. denticulatum*

Of the almost 30 countries where these seaweeds were introduced, only Hawaii and India reported invasion on corals. The introduction of *K. alvarezii, K. striatus* and *E. denticulatum* in areas which were not indigenous was always made with good intentions, i.e. for research purposes and ultimately for commercial cultivation, since these species bring economic gains to seaweed farmers as well as to the whole seaweed carrageenan industry (Porse and Rudolph 2017).

The only reason for the introduction of *Kappaphycus alvarezii, K. striatus* and *E. denticulatum* outside the Philippines, Indonesia and Malaysia was for their use as raw materials for aquaculture purposes. These seaweeds were intentionally introduced in 1974 to late 1976 into the fringing reef surrounding the Hawai'i Institute of Marine Biology (HIMB) at Coconut Island (Moku o Lo'e), Kane'ohe Bay, O'ahu, Hawaiian Islands for experimental research and commercial aquaculture projects (Doty 1978; Russell 1983). Though the pilot-scale evaluations were positive, even after several years, no commercial cultivation was initiated in the area. Sadly further to Prof Doty's death and subsequent lack of sustained attention, the experimental pilot farms were abandoned, the cultivars which had been grown there then very surprisingly gave rise to "alien species", which began to expand their range beyond the cultivation sites, as claimed by Woo (1999), Rodgers and Cox 1999; Smith et al. (2002), Conklin and Smith (2005). The remarkable observations were that once "benign strains", were able to reproduce via sporulation and further more attach (via rhizoids), something unknown (or unobserved) to that point in cultivation sites. These were definitely important lessons learned. The success of the non-indigenous eucheumatoid algae may have been the result of a variety of factors including adaptation to low and high-wave energy environments, extraordinary phenotypic plasticity, chemical or physical defense from herbivory and diverse physiological characteristics that lead to rapid growth rates from vegetative propagation (Borowitzka 1981; Duffy and Hay 1990; Holmlund et al. 1990; Hay et al. 1994; Bolser and Hay 1996; Hay 1997; Paul 1997). Despite the absence of reproductive plants in Ka¯ne'ohe Bay, O'ahu, vegetative propagation from small fragments allowed these seaweeds to spread within the vicinity to a limited extent, as no growth of *Kappaphycus* was found outside the Bay, probably due to its heavy weight which limits long dispersal (Woo 1999).

Management strategies were implemented to reduce the spread of the invasive *Kappaphycus* in Hawaii. Manual removal of the biomass of *Kappaphycus* was time consuming for large-scale removals, though a modified dredging is under test. Populating the area with *Tripneustes gratilla* as bio-control in reducing the biomass of *Kappaphycus* was promising (Conklin and Smith 2005), but the low palatability of the algae to native herbivores seemed not to show positive effect as claimed further by the authors.

A similar example leading to the notoriety of cultivated eucheumatoids was later provided by an example from another side of the world. *Kappaphycus alvarezii* (a Philippine-derived macroalga) was introduced by PepsiCo into the Gulf of Mannar (GoM) Marine Biosphere Reserve, South India for mariculture in 2000. However, at about the same time as matters in Hawaii were becoming evident, the ecological threat from this invasive alga to coral species in GoM was first reported by Pereira and Verlecar (2005) followed by Tewari et al. in 2006 and later picked-up by the Daily Spreadsheet (Anon. 2006; Vijayalakshmi 2007). These reports were affirmed by the report of Chandrasekaran et al. (2008) who made qualitative observations and quantitative analysis on the bio-invasion of *K. alvarezii* on corals in the Kurusadai Island of the GoM. The authors observed that *K. alvarezii* had successfully "invaded" and established by attachment on both dead and live corals, specifically *Acropora nobilis* and *A. formosa* in Kurusadai Island; death of the corals followed due to shadowing and smothering effects. The invasive populations occurred as either mono-specific beds or mixed with other marine communities on live and dead corals, coral rubble and pavement. Furthermore, the authors observed an extraordinary phenotypic plasticity in terms of color and shape of the thallus, i.e. thickness of its major axis, morphological features and frequency of primary and secondary branching. In terms of the cover of coral by *K. alvarezii* (sq. cm/0.5 sq. m) between two sites studied, there were significant differences ($P < 0.05$), i.e. a higher cover of *K. alvarezii* was measured towards the open sea; but there was no significant difference between colonization of areas of corals or sand. The first report was further bolstered by Kamalakannan et al. (2010) reporting the unusual appearance of *K. alvarezii* on the brown seaweed*Turbinaria* sp. was observed. The space between the lateral "plates" (or 'cups") of *Turbinaria* sp. provided an ideal settlement surface for *K. alvarezii*, providing protection from wave action and favoring the profuse growth of secondary branches of *K. alvarezii* between the plates, this was directly in contrast to the smothering effect on the top of coral plates.

However, the earliest reports on the invasion of *K. alvarezii* at Kurusadai Is. was disputed by Mandal et al. (2010) which presented findings based on a one-year study (2008–2009) in the same area. Their rigorous sampling revealed that *K. alvarezii* was confined to just two separate patches of 105 m × 55 m and 8 m × 9 m, located at the southeastern part of Kurusadai Island with an actual extent of canopy coverage of 76.7 m^2 – accounting for less than 0.0035% of the total coral reef area. The daily growth rate of the *K. alvarezii* at Kurusadai Is. was very small at 0.7% day^{-1}. *K. alvarezii* was not observed in the coral reef area of the adjoining Pullivasal and Poomarichan Islands or the Palk Bay cultivation sites. The authors claimed that the lack of functional reproductive material, low spore viability and the absence of microscopic phases in the life cycle of this alga coupled with the abundant presence of herbivores may have restricted the spread of this alga; so, on reflection, after much controversy and negative publicity the continued invasion of *K. alvarezii* at Kurusadai Island is considered remote. Today, commercial *Kappaphycus* farming has gained wide acceptance by the coastal fisher-folks as a source of livelihood in India (Krishnan and Narayanakumar 2013; Periyasamy et al. 2014a, b, 2015).

2.8 Future Studies

Although recent molecular studies have emphasized solving fundamental issues pertaining to the identification of commercial *Kappaphycus* and *Eucheuma* (and *Betaphycus*) seaweeds, the authors believed that enhanced applications of molecular tools will provide necessary information for management decisions and options for future development. Taken together, these will be able to bring about further, economically important changes to the seaweed industry. In addition, the advent of efficient molecular techniques and enhanced computing power is expected to expedite molecular research of these seaweeds. Of the potential research that remains to be done, the authors recommend three key areas for improvement:

2.8.1 Wider and Deeper Approach to Sampling

More specimens are undoubtedly required to better understand (any aspects of) *Kappaphycus* and *Eucheuma*. Recent sampling approaches focused on specimen collection from wet markets, seaweed farms or seashores. As many cultivated seaweeds are vegetatively propagated from one particular species (Ask and Azanza 2002; Ask et al. 2003; Hurtado et al. 2001), they share similar, if not identical genotypes. The same applies to specimens collected near seaweed farms or those washed ashore, which often originated from broken thalli of farmed cultivars (and are thus genetically identical). This situation leads to a general under-estimation of genetic diversity of these seaweeds. Therefore, the search for new species or potential cultivars of *Kappaphycus* and *Eucheuma* should involve the collection of specimens from sea beds via scuba-diving, far away from seaweed farms to minimize the chances of collecting broken thalli from farmed seaweeds. Apart from genetic difference, seaweeds collected from the wild are often present in cystocarpic form (rarely seen in cultivars), which is useful in identification or taxonomic description. The various species of *Kappaphycus* and *Eucheuma* originally collected and described from the Coral

Triangle (many of which were not reported again) also imply that the natural diversity has not been adequately described, and there may yet be many species and varieties of carrageenophytes to be collected and evaluated for cultivation. This also applies to the Hawaiian Islands, Africa, etc., where native populations of *Kappaphycus* and *Eucheuma* have been reported in their taxonomic records. In addition, reinvigorated sampling efforts should also be focused at type localities of rare and poorly studied seaweeds, e.g. *K. procrusteanus*, *K. cottonii,E. arnoldii* etc.

2.8.2 Taxonomic Revision and Comprehensive Genetic Studies

The taxonomy of *Kappaphycus* and especially *Eucheuma* require revision, especially when more specimens are collected for analysis. Although the application of molecular methods has greatly aided in the identification and phylogenetic understanding of these seaweeds, further progress has been impeded by the general under-sampling of several important species. This again stresses the importance of more comprehensive sampling. The application of a multi-genetic approach, preferably with genes from different organelles, will not only aid in the identification of (particularly poorly studied) seaweeds, but will better document the evolutionary pathway of *Kappaphycus* and *Eucheuma*seaweeds, which are some of the most important seaweeds in marine agronomy on a global basis. These studies will serve as important supplementary data in taxonomic revisions. The increasing prevalence of genome sequencing methods will also contribute to the field of phylogenomics, genetic selection, breeding and genetic manipulation.

2.8.3 Putting Genetic Information to Use

It is important that the genetic results generated from the molecular studies of *Kappaphycus* and *Eucheuma* be applied on a commercial scale, so as to benefit the carrageenan industry as a whole. Molecular results can be applied under various circumstances: (i) the use of genetic markers in the identification of new strains or varieties of seaweeds, particularly those from the wild (which are often deformed or damaged); (ii) the use of molecular markers in marker-assisted identification and selection of good genotypes for cultivation; (iii) genetic diversity studies to catalog the biodiversity of seaweeds within a region, as well as to monitor the occurrence of invasive seaweed species; (iv) genomic studies to identify genes that govern "quality traits" of seaweeds (e.g. rapid growth, disease resistance, good carrageenan yields and quality, etc.) and to further select them either via breeding and/or genetic engineering. Naturally, the application of scientific knowledge out in the field necessitates a proper platform linking academia to entrepreneurs as well as to the farmers. The improvement of the eucheumatoidseaweed industry will also benefit from research grants, incentives as well as financial aids from local government and authorities. Finally, collaborative efforts will also strengthen research and business ties.

The achievement of these key points will lead to better farm practices and management as well as the cultivation of excellent and robust cultivars. These translate to better consistency and efficiency in the production of quality carrageenan per farm, which in turn generates revenue for the entire carrageenan industry, as well as the economy of the country.

References

Anonymous (2006) Culture of exotic seaweed species hits indigenous ones. The Hindu: (November 1 issue)

Araújo PG, Miranda GEC, Barros-Barreto MB, Fujii MT (2013) Molecular identification of the exotic lineage of *Kappaphycus alvarezii* (Rhodophyta, Solieriaceae) cultivated in the tropical region of Brazil. Phytotaxa 109(1):17

Ask EI, Azanza RV (2002) Advances in cultivation technology of commercial eucheumatoid species: a review with suggestions for future research. Aquaculture 206:257–277

Ask EI, Batibasaga A, Zertuche-González JA, Md S (2003) Three decades of *Kappaphycus alvarezii* (Rhodophyta) introduction to non-endemic locations. Proc Int Seaweed Symp 17:49–57

Bolser RC, Hay ME (1996) Are tropical plants better defended? Palatability and defenses of temperate vs. tropical seaweeds. Ecology 77:2269–2286

Borowitzka MA (1981) Algae and grazing in coral reef ecosystems. Endeavour 5:99–106

Capinpin EC, Corre KG (1996) Growth of the Philippines abalone, *Haliotis asinina* fed an artificial diet and macroalgae. Aquaculture 144:81–89

Chandrasekaran S, Nagendran NA, Pandiaraja D, Krushnankutty N, Kamalakannan B (2008) Bioinvasion of *Kappaphycus alvarezii* on corals in the Gulf of Mannar, India. Curr Sci 94:1167–1172

Chopin T, Kerin BF, Mazerolle R (1999) Phycocolloid chemistry as a taxonomic indicator of phylogeny in the Gigartinales, Rhodophyceae: a review and current developments using Fourier transform infrared diffuse reflectance spectroscopy. Phycol Res 47:167–188

Conklin EJ, Smith JE (2005) Abundance and spread of the invasive red algae, *Kappaphycus* spp., in Kane'ohe Bay, Hawai'i and an experimental assessment of management options. Biol Invasions 7:1029–1039

Conklin KY, Kurihara A, Sherwood AR (2009) A molecular method for identification of the morphologically plastic invasive algal genera *Eucheuma* and *Kappaphycus* (Rhodophyta, Gigartinales) in Hawaii. J Appl Phycol 21(6):691–699

Dang DH, Hoang MH, Ngo HT, Hoang SN, Huynh QN, Tran MD (2008) Analysis of the genetic variation of *Eucheuma* and *Kappaphycus* strains in Vietnam using RAPD markers. In: Phang SM, Lewmanomont K, Lim PE (eds) Taxonomy of southeast Asian seaweeds. Insitute of Ocean and Earth Sciences University of Malaya, Kuala Lumpur

Doty MS (1973) Farming the red seaweed, *Eucheuma*, for carrageenans. Micronesica 9:59–73

Doty MS (1978) *Eucheuma* – current marine agronomy. In: Klauss R (ed) The marine plant biomass of the Pacific. Oregon State University Press, Corvallis, pp 203–214

Doty MS (1985) *Eucheuma alvarezii*, sp. Nov. (Gigartinales, Rhodophyta) from Malaysia. In: Abbott IA, Norris JN (eds) Taxonomy of economic seaweeds: with reference to some Pacific and Caribbean spcies. California Sea Grant College Program, La Jolla, pp 37–45

Doty MS (1988) Prodomus Ad Systematica Eucheumatoideorum: a tribe of commercial seaweeds related to *Eucheuma* (Solieriaceae, Gigartinales). In: Abott IA (ed) Taxonomy of economic seaweeds. California Sea Grant Program, La Jolla, pp 47–61

Doty MS, Alvarez VB (1975) Status, problems, advances and economics of *Eucheuma* farms. Mar Tech Soc J 9(4):30–35

Doty MS, Alvarez VB (1981) *Eucheuma* farm productivity. Int Seaweed Symp 8:688–691

Doty MS, Norris JN (1985) *Eucheuma* species (Solieriaceae, Rhodophyta) that are major sources of carrageenan. In: Abott IA, Norris JN (eds) Taxonomy of economic seaweeds: with reference to some Pacific and Caribbean species. California Sea Grant College Program, La Jolla, pp 47–61

Duffy JE, Hay ME (1990) Seaweed adaptations to herbivory. Bioscience 40:368–375

Dumilag RV, Lluisma AO (2014) Resolving the phylogenetic affinities of *Kappaphycus inermis* within the genus *Kappaphycus* (Gigartinales, Solieriaceae) using mitochondrial and plastid markers. Phytotaxa 162:223–231

Dumilag RV, Liao LM, Lluisma AO (2014) Phylogeny of *Betaphycus* (Gigartinales, Rhodophyta) as inferred from COI sequences and morphological observations on *B. philippinensis*. J Appl Phycol 26:587–595

Ecological success of alien/invasive algae in Hawaii. http://www.botany.hawaii.edu/Gradstud/smith/websites/ALIEN-HOME

Fakhrudin J, Setyaningsih D, Rahayuningsih M (2014) Bioethanol production from seaweed *Eucheuma cottonii* by neutralization and detoxification of acidic catalyzed hydrolysate. Int J Environ Sci Dev 5:455–458

Fredericq S, Freshwater DW, Hommersand MH (1999) Observations on the phylogenetic systematics and biogeography of the Solieriaceae (Gigartinales, Rhodophyta) inferred from *rbc*L sequences and morphological evidence. Hydrobiologia 398/399:25–38

Freshwater DW, Fredericq S, Butler BS, Hommersand MH, Chase MW (1994) A gene phylogeny of the red algae (Rhodophyta) based on plastid *rbc*L. Proc Natl Acad Sci 91:7281–7285

Freshwater DW, Rueness J (1994) Phylogenetic relationships of some European *Gelidium* (Gelidiales, Rhodophyta) species, based on *rbc*L nucleotide sequence analysis. Phycologia 33(3):187–194

Ganzon-Fortes ET, Trono GC, Villanueva RD, Romero JB, Montaño MNE (2012) 'Endong', a rare variety of the farmed carrageenophyte *Eucheuma denticulatum* (Burman) Collins & Hervey from the Philippines. J Appl Phycol 24:1107–1111

Gavio B, Fredericq S (2002) *Grateloupia turuturu* (Halymeniaceae, Rhodophyta) is the correct name of the non-native species in the Atlantic known as *Grateloupia doryphora*. Eur J Phycol 37(3):349–359

Geraldino PJL, Yang EC, Boo SM (2006) Morphology and molecular phylogeny of *Hypnea flexicaulis* (Gigartinales, Rhodophyta) from Korea. Algae 21(4):417–423

Hay ME (1997) Calcified seaweeds on coral reefs: complex defenses, trophic relationships and value as habitats. Proc Int Coral Reef Symp 8:713–718

Hay ME, Kappel QE, Fenical WB (1994) Synergisms in plant defenses against herbivores: interaction of chemistry, calcification and plant quality. Ecology 75:1714–1726

Hayashi L, Yokoya NS NS, Ostin S, Pereira RTL, Braga ES, Oliveira EC (2008) Nutrients removed by *Kappaphycus alvarezii* (Rhodophyta, Solieriaceae) in integrated cultivation with fishes in re-circulating water. Aquaculture 277:185–191

Hayashi L, Bulboa C, Kradolfer P, Soriano G, Robledo D (2014) Cultivation of red seaweeds: a Latin American perspective. J Appl Phycol 26:719–727

Hebert PDN, Cywinska A, Ball SL, de Waard JR (2003a) Biological identifications through DNA barcodes. Proc R Soc B Biol Sci 270(1512):313–321

Hebert PDN, Ratnasingham S, de Waard JR (2003b) Barcoding animal life: cytochrome c oxidase subunit 1 divergences among closely related species. Proc R Soc B Biol Sci 270(Suppl 1):S96–S99

Hebert PDN, Stoeckle MY, Zemlak TS, Francis CM (2004) Identification of birds through DNA barcodes. PLoS Biol 2(10):e312

Holmlund MB, Peterson CH, Hay ME (1990) Does algal morphology affect amphipod susceptibility to fish predation? J Exp Mar Biol Ecol 139:65–83

Hurtado AQ (2013) Different colour morphotypes of *Kappaphycus alvarezii* and *Kappaphycus striatum* used in commercial farming. In: Phang SM, Lim PE (eds) Taxonomy of southeast Asian seaweeds II, Monograph series, vol 15. Institute of Ocean and Earth Sciences, University of Malaya, Kuala Lumpur, pp 83–92

Hurtado AQ, Agbayani RF, Sanares R, de Castro-Mallare MTR (2001) The seasonality and economic feasibility of cultivating *Kappaphycus alvarezii* in Panagatan Cays, Caluya, Antique, Philippines. Aquaculture 199:295–310

Hurtado AQ, Gerung GS, Yasir S, Critchley AT (2014) Cultivation of tropical red seaweeds in the BIMP-EAGA region. J Appl Phycol 26:702–718

Hurtado AQ, Neish IC, Critchley AT (2015) Developments in production technology of *Kappaphycus* in the Philippines: more than four decades of farming. J Appl Phycol 27:1945–1961

Hurtado AQ, Lim PE, Tan J, Phang SM, Neish IC, Critchley AT (2016) Biodiversity and biogeography of commercial tropical carrageenophytes in the southeast Asian region. In: Pereira L (ed) Carrageenans: sources and extraction methods, molecular structure, bioactive properties and health effects. Nova Science Publishers, Hauppauge, New York, pp 67–90, https://www.novapublishers.com

Jinbo U, Kato T, Ito M (2011) Current progress in DNA barcoding and future implications for entomology. Entomol Sci 14(2):107–124

Kamalakannan B, Jeevamani JJJ, Nagendran NA, Pandiaraja D, Krishnan Kutty N, Chandrasekaran S (2010) *Turbinaria* sp. as victims to *Kappaphycus alvarezii* in reefs of Gulf of Mannar, India. Coral Reefs 29:1077

Khambhaty Y, Mody K, Mahesh R, Gandhi MR, Thampy S, Maiti P, Brahmbhatt H, Eswaran K, Ghosh PK (2011) *Kappaphycus alvarezii* as a source of bioethanol. Bioresour Technol:180–185

Krishnan M, Narayanakumar R (2013) Social and economic dimensions of carrageenan seaweed farming in India. In: Valderrama D, Cai J, Hishamunda N, Ridler N (eds) Social and economic dimensions of carrageenan seaweed farming, Fisheries and Aquaculture Technical Paper No. 580. FAO, Rome, pp 163–185. 204 pp

Kronen M (2013) Social and economic dimensions of carrageenan seaweed farming in the Solomon Islands. In: Valderrama D, Cai J, Hishamunda N, Ridler N (eds) Social and economic dimensions of carrageenan seaweed farming, Fisheries and Aquaculture Technical Paper No. 580. FAO, Rome, pp 147–161. 204 pp

Lim PE, Tan J, Phang SM, Nikmatullah A, Dang DH, Sunarpi H, Hurtado AQ (2014a) Genetic diversity of *Kappaphycus* Doty and *Eucheuma* J. Agardh (Solieriaceae, Rhodophyta) in Southeast Asia. J Appl Phycol 26:1253–1272

Lim PE, Tan J, Phang SM, Rahiman AA (2014b) A guide to *Kappaphycus* and *Eucheuma* seaweeds in Malaysia. University of Malaya Press, Kuala Lumpur

Lodge DM, Williams S, Mac Isaac HJ, Hayes KR, Leung B, Reichard S, Mack RN, Moyle PB, Smith M, Andow DA, Carlton JT, McMichael A (2006) Biological invasions: recommendations for US policy and management. Ecol Appl 16:2035–2054

Maili S, Rodrigues KF, Thien VY, Yong WTL, Anton A, Chin GJWL (2016) Development and application of single locus genomic molecular marker for *Kappaphycus* and *Eucheuma* (Solieriaceae, Rhodophyta) seaweeds. Aquat Bot 128:26–32

Mandal SK, Mantri VA, Haldar S, Eswaran K, Ganesan M (2010) Invasion potential of *Kappaphycus alvarezii* on corals at Kurusadai Island, Gulf of Mannar, India. Algae 25:205–216

Marine invasives in Hawaii. http://www.botany.hawaii.edu/invasive/default.htm

Meier R, Zhang G, Ali F (2008) The use of mean instead of smallest interspecific distances exaggerates the size of the "barcoding gap" and leads to misidentification. Syst Biol 57(5):809–813

Meinita MDN, Kang JY, Jeong GT, Koo HM, Park SM, Hong YK (2012) Bioethanol production from the acid hydrolysate of the carrageenophyte *Kappaphycus alvarezii* (cottonii). J Appl Phycol 24:857–862

Meyer CP, Paulay G (2005) DNA barcoding: error rates based on comprehensive sampling. PLoS Biol 3(12):e422

Msuya FE, Buriyo A, Omar I, Pascal B, Narrain K, Ravina JJM, Mrabu E, Wakibia JG (2014) Cultivation and utilisation of red seaweeds in the Western Indian Ocean (WIO) region. J Appl Phycol 26:699–705

Namukose M, Msuya FE, Ferse SCA, Matthew J, Slater MJ, Kunzmann A (2016) Growth performance of the sea cucumber *Holothuria scabra* and the seaweed *Eucheuma denticulatum*: integrated mariculture and effects on sediment organic characteristics. Aquac Environ Interact 8:179–189

Neish IC (2013) Social and economic dimensions of carrageenan seaweed farming in Indonesia. In: Valderrama D, Cai J, Hishamunda N, Ridler N (eds) Social and economic dimensions of carrageenan seaweed farming, Fisheries and Aquaculture Technical Paper No. 580. FAO, Rome, pp 61–89. 204 pp

Paul VJ (1997) Secondary metabolites and calcium carbonate as defenses of calcareous algae on reefs. Int Coral Reef Symp 8:707–712

Pereira N, Verlecar XN (2005) Is Gulf of Mannar heading for marine bioinvasion? Curr Sci 89:1309–1310

Periyasamy C, Anantharaman P, Balasubramanian T (2014a) Social upliftment of coastal fisher women through seaweed (*Kappaphycus* alvarezii (Doty) Doty) farming in Tamil Nadu, India. J Appl Phycol 26:775–781

Periyasamy C, Anantharaman P, Balasubramanian T, Subba Rao PV (2014b) Seasonal variation in growth and carrageenan yield in cultivated *Kappaphycus alvarezii* (Doty) Doty on the coastal waters of Ramanathapuram, Tamil Nadu, India. J Appl Phycol 26:803–810

Periyasamy C, Anantharaman P, Subba Rao PV (2015) Experimental farming of *Kappaphycus alvarezii* (Doty) Doty with income estimates at different sites in the Mandapam region, Palk Bay, southeast coast of India. J Appl Phycol 27:935–944

Phang SM, Yeong HY, Lim PE, Nor ARM, Gan KT (2010) Commercial varieties of *Kappaphycus* and *Eucheuma* in Malaysia. Malays J Sci 29(3):214–223

Porse H, Rudolph B (2017) The seaweed hydrocolloid industry: 2016 updates, requirements, and outlook. J Appl Phycol . doi:10.1007/s10811-017-1144-0

Rodgers SK, Cox EF (1999) The distributions of the introduced rhodophytes *Kappaphycus alvarezii, Kappaphycus striatum* and *Gracilaria salicornia* in relation to various physical and biological factors in Kane'ohe Bay, O'ahu, Hawai'i. Pac Sci 53:232–241

Rodrigueza MRC, Montaño MNE (2007) Bioremediation potential of three carrageenophytes cultivated in tanks with seawater from fish farms. J Appl Phycol 19:755–762

Russell DJ (1983) Ecology of the red imported seaweed *Kappaphycus striatum* on Coconut Island, Oahu, Hawai'i. Pac Sci 37:87–107

Sen S (1991) Seaweed collection and culture in Tanzania. Food and Agriculture Organization of the United Nations, Harare, Zimbabwe. Retrieved from http://www.fao.org/documents/show_cdr.asp?url_file=/docrep/005/AC986E/AC986E06.htm

Smith JE, Cynthia L, Hunter CL, Smith CM (2002) Distribution and reproductive characteristics of nonindigenous and invasive marine algae in the Hawaiian Islands. Pac Sci 56:299–315

Tan J, Lim PE, Phang SM, Hong DD, Sunarpi H, Hurtado AQ (2012) Assessment of four molecular markers as potential DNA barcodes for red algae *Kappaphycus* Doty and *Eucheuma* J. Agardh (Solieriaceae, Rhodophyta). PLoS One 7 (12):e52905

Tan J, Lim PE, Phang SM (2013) Phylogenetic relationship of *Kappaphycus* Doty and *Eucheuma* J. Agardh (Solieraceae, Rhodophyta) in Malaysia. J Appl Phycol 25(1):13–29

Tan J, Lim PE, Phang SM, Rahiman AA, Nikmatullah A, Sunarpi H, Hurtado AQ (2014) *Kappaphycus malesianus* sp. nov.: a new species of *Kappaphycus* (Gigartinales, Rhodophyta) from Southeast Asia. J Appl Phycol 26:1273–1285

Tewari A, Eswaran K, Subba Rao PV, Jha B (2006) Is *Kappaphycus alvarezii* heading towards marine bioinvasion? Curr Sci 90:619–620

Vijayalakshmi E (2007) PepsiCo endangers biodiversity hot spot in Gulf of Mannar. Down to Earth, 7 October

Williams SL, Smith JE (2007) A global review of the distribution, taxonomy, and impacts of introduced seaweeds. Annu Rev Ecol Evol Syst 38:327–359

Woo M (1999) Ecological impacts and interactions of the introduced red alga *Kappaphycus striatum* in Kaneohe Bay, Oahu. Master thesis, Department of Botany, University of Hawaii, Honolulu

Yang EC, Kim MS, Geraldino PJL, Sahoo D, Shin JA, Boo SM (2007) Mitochondrial cox1 and plastid rbcL genes of *Gracilaria vermiculophylla* (Gracilariaceae, Rhodophyta). J Appl Phycol 20(2):161–168

Zemke-White WL, Smith JE. 2006. Environmental impacts of seaweed farming in the tropics. See Critchley et al. 2006

Zhao S, He P (2011) Molecular identification based on ITS sequences for *Kappaphycus* and *Eucheuma* cultivated in China. Chin J Oceanol Limnol 29(6):1287–1296

Zuccarello GC, Burger G, West JA, King RJ (1999) A mitochondrial marker for red algal intraspecific relationships. Mol Ecol 8:1443–1447

Zuccarello GC, Critchley AT, Smith J, Sieber V, Lhonneur GB, West JA (2006) Systematics and genetic variation in commercial *Kappaphycus* and *Eucheuma* (Solieriaceae, Rhodophyta). J Appl Phycol 18(3–5):643–651

Reproductive Biology and Ecophysiology of Farmed *Kappaphycus* and *Eucheuma*

3

Rhodora V. Azanza and Erick Ask

Abstract

The impetus to enhance the quality and quantity of industrial-scale supplies of raw material and extract production drives the needs for research and monitoring of the reproductive biology and physiology of commercial species/strains/variants of *Kappaphycus* and *Eucheuma*. Publications and advancements in these fields are reviewed in this chapter. These seaweeds have now become true marine, agronomic crops which have been farmed for several decades principally through cuttings or vegetative reproduction. The scale and importance of this industry requires that further studies are needed imminently in relation to the changing environments within which varied methods of cultivation are practiced. Research on the potential acclimation of various seedstocks when exposed to a varied range of culture practices and changing ecological conditions, including Integrated Multi-Trophic Aquaculture (IMTA), are required to be implemented more vigorously in order to assess sustainability. Descriptive and predictive modeling using available data from biological and physio-ecological experiments together with additional studies required are highlighted here, to serve as inputs to the collective goal of successful and sustained cultivation of these carrageenophytes.

3.1 Introduction

For decades carrageenan, an economically important natural product/biopolymer, has been derived from members of the genera *Kappaphycus* and *Eucheuma*. These have been farmed successfully by virtue of their distinct biological and cultivation characteristics. Basic and applied studies on their biology and ecology were driven by the necessity to increase productivity and enhance carrageenan quality and yield. Preliminary studies on the reproductive biology of *Kappaphycus* and *Eucheuma* were essential in order to improve their commercial-scale cultivation (as had been done for *Porphyra* (*Pyropia*) Nori), have been surprisingly few and sporadic. It is well known that the work of Dr. Kathleen Drew Baker on the "Conchocelis" filamentous stage, of the life history of *Porphyra* was a necessary step becoming the basis of the now multi-billion-dollar Nori industries of Japan, Korea and China.

This chapter reviews the work so far on reproduction of the carrageenophyte species which are still grown vegetatively in many areas in the tropical world. Thallus production from tetraspores and carpospores has been initiated but not practiced on a commercial scale. Research published on the physiology/ecophysiology of these organisms are likewise reviewed and assessed in relation to their relevance to commercial cultivation operations. Much of the work, to date, has focused on light, temperature, salinity and nutrients (growth regulators) as single and/or interacting factors which can affect growth (productivity) and also the more important aspects of carrageenan yields and quality. These factors underpin the generation of "greater and better yields" in the field (cultivation sites), in the most economically valuable,

R.V. Azanza (✉)
The Marine Science Institute, University of the Philippines, Diliman, Quezon City 1101, Philippines
e-mail: rvazanza@up.edu.ph

E. Ask
FMC Bio Polymer, 1735 Market St., Philadelphia, PA 19103, USA

A.Q. Hurtado et al. (eds.), *Tropical Seaweed Farming Trends, Problems and Opportunities*, Developments in Applied Phycology 9, DOI 10.1007/978-3-319-63498-2_3

efficient and environmentally-friendly manner. *K. alvarezii* has been cited as the most commonly cultured red seaweed (marine plant) for decades now and is the most studied carrageenophyte species.

3.2 Reproduction and Life Cycles of *Kappaphycus* and *Eucheuma* in Natural and Farmed Areas

3.2.1 Populations

Evidence derived from natural and farmed populations have shown that *Kappaphycus* and *Eucheuma* undergo a *Polysiphonia*-type of life history, i.e., involving male and female gametophytes (N), a carposporophyte (2 N) and a tetrasporophyte (2N; Fig. 3.1) (Doty 1987; Azanza-Corrales 1990; Ask and Azanza 2002). The tetrasporophyte generation has been found to dominate natural beds and in the farms. The tetrasporophytic thalli are morphologically (i.e. isomorphic) the same as the gametophytes but appear to be "sturdier". The tetrasporangia are zonate (Fig. 3.2a, b) appearing as translucent, whitish spots on the thalli. Male gametophytes are reported (Dawes 1979; Doty 1985) and have been found in the farms (Azanza-Corrales 1990; Azanza-Corrales and Sa-a 1990; Azanza-Corrales et al. 1992, 1996). The male thalli are relatively smaller and more fragile than the female gametophyte generations. It is easier to distinguish the reproductive, female gametophytes which are morphologically the similar again more robust, as compared to the other N and 2N phases.

Cystocarps are visible to the naked eye and found numerously scattered on the surface of the female gametophyte. These haploid females, carry the (parasitic) carposporophyte generation (2N) which is the result of the *in situ* fertilization of sedentary, oogonial cells. Mature cystocarps (Fig. 3.2c) produce numerous carpospores (2N), thereby capitalizing by internal, mitotic multiplication on the event of syngamy; the diploid spores, when released each has the potential to produce (2N) tetrasporophyte thalli (Azanza-Corrales 1990).

For several decades, *Kappaphycus* and *Eucheuma* farming has been primarily based on vegetative regeneration or self-propagation, (Ask and Azanza 2002; Hurtado et al. 2015). However, interesting developments in innovations to the industry have been reported such as: micropropagation, carpospore release and regeneration and clone cultivation (Doty 1987; Mairh et al. 1986; Azanza-Corrales et al. 1996; Azanza-Corrales and Aliaza 1999; Azanza and Ask 2003; Bulboa et al. 2007; Hayashi et al. 2008; Luhan and Sollesta 2010).

The majority of applied reproductive biology studies on *Kappaphycus* and *Eucheuma* utilized *in vitro* carpospore release and subsequent germination. In *Kappaphycus alvarezii* tetrasporophytes have been suggested to be more robust than gametophytes and hence useful for optimal strain selection – where consistent and reliable methods in maintenance of tissue ploidy levels and reproductive stages are very important (Ask and Azanza 2002). Luhan and Sollesta (2010) observed that carpospore progenies produced mature tetrasporophytes which then develop high-temperature-resistant haploid or gametophytic progenies. The preliminary study undertaken in Indonesia revealed that several

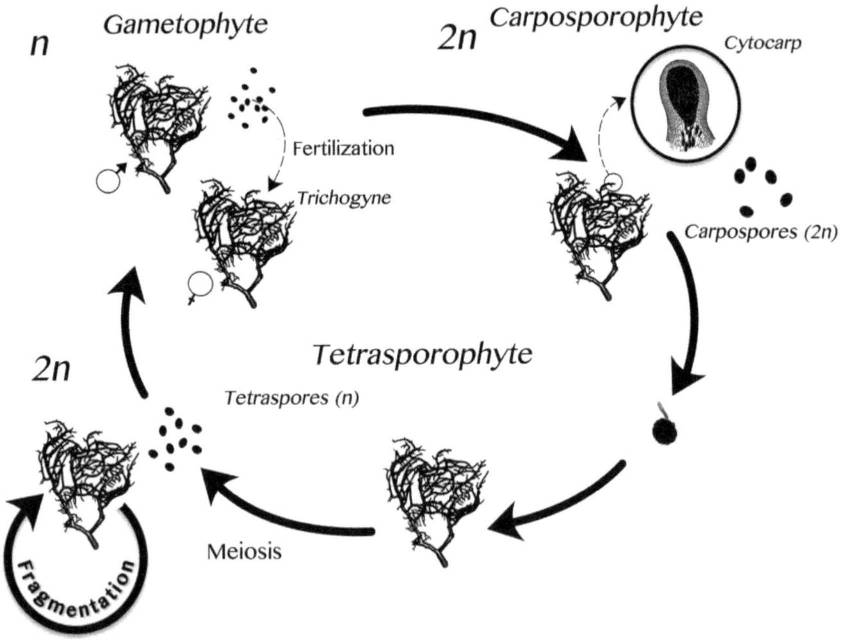

Fig. 3.1 Life cycle of *Kappaphycus* and *Eucheuma* (Illustration by: Nero M. Austero from materials of RV Azanza)

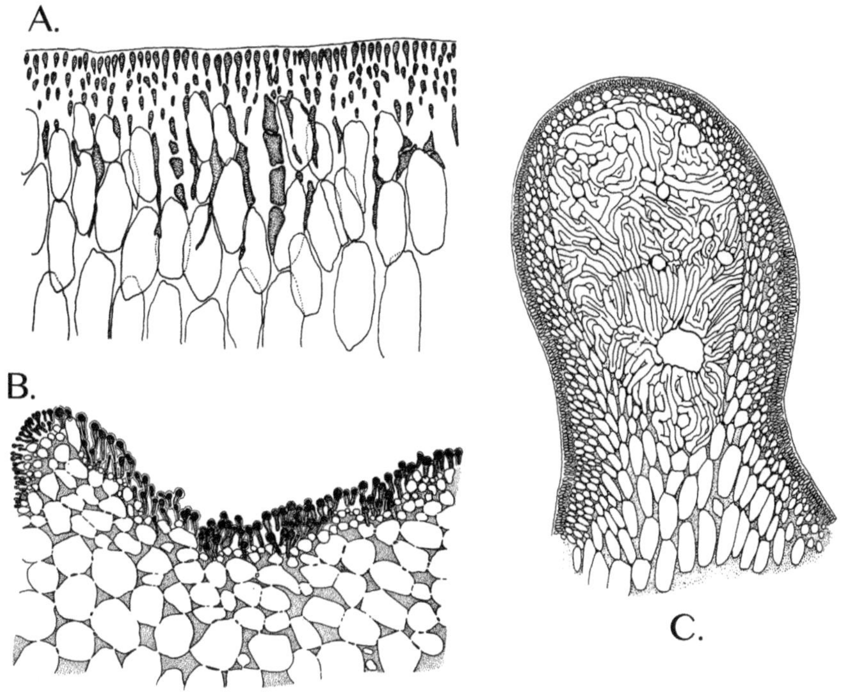

Fig. 3.2 Reproductive structures of *K. alvarezii*. (**a**) L.S. of spermatangial sorus, (**b**) L.S. of cystocarp, and (**c**) L.S. of tetrasporangia (Original drawn plates from prepared slides of RV Azanza, first published in J Appl Phycol 1990 2:57–62)

types of light quality (i.e. white, yellow, blue, green) could induce the release of carpospores from mature thalli, e.g. 5000–10,000 lux irradiance with a 14:10 L:D photo-period (Syamsuddin 2013); these observations should be followed up with more applied investigations due to the economic value of the outcome.

K. alvarezii tetraspores which were germinated in the natural environment of the south-east coast of Brazil resulted in low survival rates and absence of recruitment (Bulboa et al. 2008). Earlier, Bulboa et al. (2007) reported high viability and 79% germination of *K. striatum* tetraspores, resulting in robust plantlets, with large morphological differences and satisfactory growth rates. Zitta et al. (2012) recommended the use of confocal fluorescence microscopy in addition to karyotyping. The method stimulated nuclear fluorescence of chloroplasts that allowed for the analysis of ploidy providing a first for the rapid, accurate identification and segregation of thalli of haploid gametophyte and diploid tetrasporophyte generations.

It should be mentioned that the knowledge and understanding of the genetics of farmed eucheumatoids and/or their close relatives is highly relevant to crop improvement, management and strain selection. There are two main goals in the farming of carrageenophytes in relation to their genetics: (1). Strain selection from wild and farmed varieties and (2). Breeding programs. The best strains for farming are those with higher: (a). yields of high quality carrageenan (e.g. percent yield, viscosity and gel strength) and (b). sustained seasonal growth rates (Ask and Azanza 2002).

The varieties of *Kappaphycus which* have been farmed recently in the Philippines and Malaysia (Tan et al. 2012; Villanueva et al. 2011; Rodrigueza and Montaño 2007) represent morphologically distinct and genetically related representatives of *K. alvarezii*, *K. striatus* and *K. inermis* (Dumilag and Lluisma 2014; Tablizo and Lluisma 2014). These farming varieties have been grown in number of regions and called a variety of colloquial names by farmers in order to make differentiations and selections for cultivars in the field. The majority of these varieties likely represent the range of morphological expression of the genetic material, in particular as related to the varied farming localities, local conditions and methods applied. Whether the corresponding variations affect physiological (functional) expressions, particularly of the carrageenan quality, is however, not clear and genetic as well as studies of reproductive biology are required.

The complete mitochondrial (mt) DNA sequence of *Kappaphycus striatus* (Green Sacol) was determined and its overall genome structure and organization reported to be highly similar with other fully sequenced mt genomes of members of the Florideophyceae (Dumilag and Lluisma 2014). A trend was observed, however, that *K. striatus* mt DNAs tended to encode a lower number of protein-coding genes, albeit not smaller in size, as the thalli increased in morphological complexity (Tablizo and Lluisma 2014).

Use of genetic engineering for land plants has been practiced and products have been placed in selected markets. The general view that genetically modified crops could be less acceptable due to possible health or environmental risks might preclude this development being widely applied to seaweeds. Hence, natural strain selection and breeding programs, used for other seaweeds of commercial value, e.g. *Porphyra* and *Chondrus* (van der Meer 1990) could provide guidance and inspiration for strain selection and improvement of eucheumatoid crops in a marine agronomic setting.

Branch culture, micropropagation and callus production techniques for *K. alvarezii* and *E. denticulatum* were already well developed some time ago (Dawes et al. 1993). Cheney et al. (1998) claimed to have created new strains of *E. denticulatum* and *K. alvarezii* with superior growth rates of up to 14% per day in the laboratory, however, practical applications of these studies are not readily available. Laboratory experiments were recently initiated on *Kappaphycus* carpospore coalescence, in order to reveal if they could be useful for further strain selection; more research is required to verify the promising results (Azanza et al. 2014; Azanza unpub. Data).

There is a need to further harness reproductive materials and its inherent variability in natural and farmed *Kappaphycus* and *Eucheuma* populations for the enhancement and sustainability of these marine crops, which remain major sources of carrageenan and whose production technology remains almost unchanged (Ask and Azanza 2002; Hurtado et al. 2014, 2015).

3.3 The Physiology of Farmed *Kappaphycus* and *Eucheuma*

Primary production/photosynthesis and how carrageenan yields (quality and quantity) are impacted in *Kappaphycus and Eucheuma* crops are primary concerns in their commercial production. Light, temperature, salinity, nutrients and water motion are major abiotic factors affecting the physiology of these crops.

3.3.1 Light

Photosynthetic responses of *K. alvarezii* and *E. denticulatum* are frequently studied to determine the optimal conditions for photosynthesis, growth and more recently commercial, quality characteristics such as carrageenan production. The levels of photosynthetically active radiation (PAR) vary depending on: amongst others, distance from the shore and depth and turbidity of seawater (Fig. 3.3). Research has shown that the photosynthetic rates of *K. alvarezii* and *K. striatus* are better adapted to the intertidal environment (Ask and Azanza 2002; Glenn and Doty 1992; Dawes 1992) (where they are commonly grown) with a greater photosystem II reaction center tolerance and slower dehydration and faster rehydration rates. Synergism with other parameters, such as surface seawater temperature (SST) and relative electron transport rate, affect the optimum and saturated levels of PAR. These levels have been documented for *E. denticulatum* and *K. alvarezii* by Lideman et al. (2013), in crops from Indonesia.

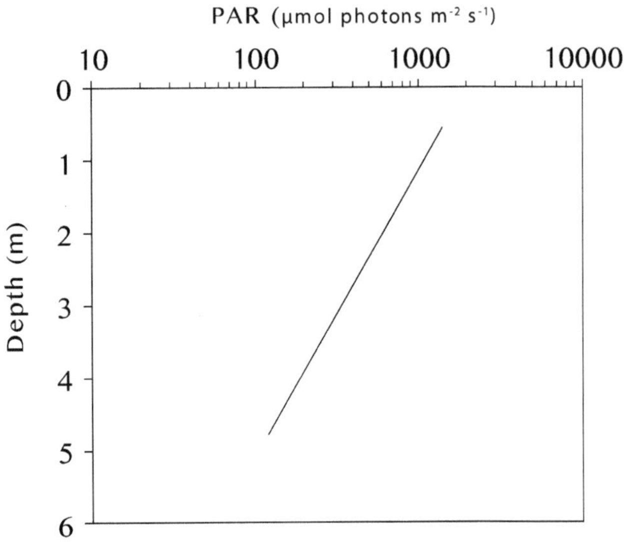

Fig. 3.3 Photosynthetically Active Radiation (PAR) documented for eucheumatoids as a function of water depth (Redrawn from Lideman et al. 2013)

Earlier studies on the photosynthesis of *Kappaphycus striatum* showed it follows a diurnal pattern with a peak at noon (Glenn and Doty 1981) with maximum growth rates at 12:12 L/D cycle at 6000 lx (Mairh et al. 1986). *Kappaphycus* sp. exhibited a diurnal and circadian rhythm in continuous, white light at an irradiance of 100–1000 μmol photons m^{-2} s^{-1} (Granbom et al. 2001), although Schubert et al. (2004) earlier observed that mitotic activity (which gives rise to more cells), indicated as growth/growth rate in a practical sense, was maximal during the dark or night phase. *K. alvarezii* exhibited a lower photosynthetic activity at high irradiance levels, usually at noon, then recovering during the night, implying photo-inhibition had taken place (Terada et al. 2015). It was found that a photosynthetic rhythm held true to both growing and non-growing parts of the thallus indicating that photosynthesis was not dependent on the cell cycle (Schubert et al. 2004); this could also mean that farming of these plants could even make use of continuous lighting (as in the photo-bioreactor concept). However, what was not reported and requires further attention, was the impact of continuous light on the important characteristics of carrageenan production and its properties which are more the products of mature cells.

Excessive light can have damaging effects on *K. alvarezii,* such as bleaching by increasing the plastoquinone pool availability (at an irradiance of: 504 ± 13 μmol photon m^{-2} s^{-1}), thereby affecting the over-production of reactive oxygen species, and also chilling injuries when tissue was exposed to low temperatures (Barros et al. 2006). However, excessive irradiance at 200 μmol m^{-2} s^{-1} can help stabilize the efficiency of Photosystem II. Under low light conditions, Photosystem II yield efficiency declined during the first

4 days then increased the following days – an indication of a low-light-adapted species with regulatory mechanisms in response to extreme levels of light (Guan et al. 2013). The impact of ultra-violet (UV) radiation on both micro- and macroalgae have been studied (Wood 1987; Worrest 1983; Karentz et al. 1991; Premkumar et al. 1993). Exposure to UV radiation in *K. alvarezii* reportedly changed cortical and sub-cortical cellular structure including increased cell wall thickness and plastoglobuli, reduced intracellular spaces, changes in cell contour, destruction of chloroplast internal organization and 50% decreased relative growth rate (Schmidt et al. 2009, 2010a, b). Effects on the carrageenan content/quality were not included in these very similar studies.

Pigment composition also plays a major role in light responses as absorption varies depending on the nitrogen storage compounds in the pigments. As such, different color morphotypes of *E. denticulatum* showed ecotypic differentiation and variation in photosynthetic efficiency have been reported (Dawes 1992; Ask and Azanza 2002). The over-expression of pigments in *K. alvarezii* morphotypes do not necessarily increase photosynthetic efficiency, nor growth rates (von-Wobeser et al. 2001). Seaweed diseases, i.e. *ice-ice*, pitting, tip darkening and discoloration, can affect photosynthetic responses by decreasing the concentration of host/plant pigments – i.e. chlorophyll a, r-phycoerythrin and r-phycocyanin (Ganzon-Fortes et al. 1993). Pang et al. (2011) reported that the presence of epiphytes (i.e. *Neosiphonia savatieri*) could reduce the productivity of *Kappaphycus* because of stress effects, but the host could compensate by acclimating to the lowered light conditions.

Eucheumatoid growth rates or their photosynthetic rates, as affected by light as a single factor, interacting with other factors (such as nutrient availability) have been well studied (Glenn and Doty 1981; Lideman et al. 2013) Practical applications of this knowledge for optimal carrageenan yield and growth rates during periods of extreme field conditions should be considered, for example during prolonged El Niño years in the Pacific.

3.3.2 Temperature

Surface seawater temperature is one of the most important factors in controlling the growth rates of eucheumatoids (Bulboa and de Paula 2006), since it can directly and indirectly affect a range of physiological processes. These responses have been well studied both in laboratory and field (Ask and Azanza 2002). Growth rates, photosynthetic efficiency, disease vulnerability and Reactive Oxygen Species (ROS) over-production have been some of the parameters used to determine the optimum range for eucheumatoids, its morphotypes and varieties (see Table 3.1).

Growth rates peaked for *E. denticulatum* (brown strain) and *K. striatum* (green strain) at 25 °C (Gerung and Ohno 1997). Tolerance to water temperature variability, rapid growth and high biomass production for both *Kappaphycus* and *Eucheuma* have been reported during months characterized by generally warm SST (25–30 °C) (Trono and Ohno 1989; Araujo et al. 2014). In terms of optimum photosynthetic rates, *E. denticulatum* and *K. alvarezii* require 23–32 °C (Lideman et al. 2013) and 28–32 °C respectively (Terada et al. 2015); whereas, optimum PAR decreased at 32–34 °C (Fig. 3.3).

Table 3.1 Optimum temperature conditions of eucheumatoids

Eucheumatoid	Optimum temperature (°C)	References
K. alvarezii	25–28	Ohno et al. (1994)
	25–30	Trono and Ohno (1989)
	20–24	Araujo et al. (2014)
	27–30	Preisig and Hans (2005)
Sumba strain	22–33	Lideman et al. (2013)
E. amakusaensis	18–25	Ohmi and Shinmura (1976)
E. denticulatum	23–32	Lideman et al. (2013)
	25–30	Trono and Ohno (1989)
	27–30	FAO (2005)
E. isiforme	18–25	Mathieson and Dawes (1974)
K. striatum	23–30	Mairh et al. (1995)
	22.8–29.2	Ohno and Orosco (1987)

Stressful conditions of SST may result in either or both lethal and sub-lethal effects to eucheumatoids. At higher SST of 33–35 °C, wide-scale whitening or *ice-ice* leads to heavy damage to the thalli (Largo et al. 1995). *K. alvarezii* fragments tend to have a high tolerance to temperature changes (20–28 °C) during the "dry season", as compared to fragments collected during the rainy season (Araujo et al. 2014). Alternatively, cultivation during winter (16–18 °C) may lead to lethal effects on *K. alvarezii*. In order to maintain seaweed cultures during this period, the use of AMPEP (Acadian Marine Plant Extract Powder) as a "preventive action" (prophylactic) for the cultivation of seedlings in tanks and open sea has been recommended (Hurtado et al. 2009; Loureiro et al. 2012, 2014).

3.3.3 Salinity

Salinity is a major environmental factor affecting osmoregulation since it interferes with cell turgor pressure that can cause damage to membranes, organelles and enzymatic processes (Lobban and Harrison 1994).

Kappaphycus spp. and *K. alvarezii* grow optimally at 30–33 and 25–35 psu respectively, with a high chlorophyll a content and 3.4–3.5% day^{-1} growth rate (Preisig and Hans 2005; Araujo et al. 2014). *K. alvarezii* infected with the epiphyte, *Neosiphonia* sp. could still grow optimally at 35 psu (Borlongan et al. 2011). *Eucheuma isiforme* has a broad respiratory maximum at 30–40‰ while *E. uncinatum* and *E. denticulatum* both had a maximum at 30‰ (Ask and Azanza 2002). In *K. alvarezii,*carrageenan yield was reported to be highest at 25 psu but growth rates were lower, as compared to salinities above 25 psu, (Hayashi et al. 2011; Reis et al. 2011).

Hypersaline conditions (i.e. 45–55 psu) can increase the levels of phycobiliproteins in *K. alvarezii* (Araujo et al. 2014), while mortality has been observed in *E. striatus* when exposed to high salinity conditions beyond 7–14 days (Mairh et al. 1986). Low salinity levels of 15 psu increased cellular turgidity with increasing cell wall thickness (Hayashi et al. 2011). Largo et al. (1995) however reported that in *K. alvarezii* 'ice-ice' whitening was induced at 20% salinity and low irradiance (50 μmol photon m^{-2} s^{-1}). Mairh et al. (1986) recorded low survival of *K. striatum* at a salinity below 24 psu. *K. alvarezii* fragments tend to have a higher tolerance to salinity changes during the dry season, as compared to those fragments collected in the rainy season (Araujo et al. 2014), this is primarily because *K. alvarezii* can cope with hypo-osmotic stress conditions through the up-regulation of an ATP-binding cassette transporter and sodium phosphate symporter genes (Liu et al. 2016).

More research on single and synergistic effects of salinity with other factors in *Eucheuma* spp. and *Kappaphycus* spp. are required to address knowledge gaps relating to growth and carrageenan production. As pointed out as early as 2000 (Ask and Azanza 2002), this important research aspect is required since the crops are exposed to rapid changes in salinity due to heavy downpours and intermittent exposure to extremely high SST and other stressors/changes at the farming site.

3.3.4 Nutrients

According to Leibeg's law of the limiting factor, when all other factors are optimal: "the nutrient available to the plant in the smallest quantity, in relation to what it needs, will limit its growth". de Boer and Rhyter (1977), as cited by Lobban and Harrison (1994) considered availability of N as the most limiting nutrient affecting seaweed growth (considering the N cycle in the marine environment). This was an observation that has been proven by many studies in the field and in the laboratory for example, Li et al. (1990) noted a doubling of *K. alvarezii* growth rates upon using NH_4 nutrient dips. *K. alvarezii* tends to exhibit surge ammonium uptake and nitrogen limitation, a mechanism used to survive environments with low or erratic nutrient availability (Dy and Yap 2001). The observation that *K. alvarezii* has a lower N requirement than *K. striatus* had been reported earlier by Trono and Lluisma (1992).

AMPEP, a commercial extract from the brown alga, *Ascophyllum nodosum*, contains the major macronutrients: total nitrogen N (0.8–1.5%), available phosphoric acid P_2O_5 (1–2%) and soluble potash K_2O (17–22%), for the growth and development of eucheumatoids (Hurtado et al. 2009). *In vitro* and field studies using AMPEP extract showed promising results on epiphyte- and pathogen-resistance, growth rate and carrageenan yield (Borlongan et al. 2011; Loureiro et al. 2012; Zuldin and Shapawi 2015). Further studies with extracts of other algal origins might be useful.

In a field experiment by Glenn and Doty (1992), water motion was found to influence the growth responses of *K. alvarezii,K. striatus* and *E. denticulatum*. Of the three species, *K. alvarezii* had the greatest response to water motion, with higher growth in summer than winter. Growth rates were observed to increase up to the highest water motion velocities, i.e. 15 cm s^{-1}. Reports on pigment loss were also documented by Neish (2008) in areas where there was low water-flow.

No significant differences in growth rates were observed between nutrient-rich and nutrient-poor *K. alvarezii* and *E. denticulatum* however, this factor was reported to influence the quantity of carrageenan in *E. isiforme* (Freile-Pelegrin and Robledo 2006); the growth rate of *K. striatus* was increased by nutrient enrichment as reported by Mairh et al. (1995). Studies by Rodrigueza and Montaño (2007); Hayashi et al. (2008) and Msuya et al. (2011) investigated the use of *K. alvarezii* to treat or remove effluents from nutrient-rich seawater. The maximum removal of effluents was reported as the following rates: nitrate – 18.2%; nitrite – 50.8%; ammonium – 70.5% and phosphate – 26.8% thereby indicating the feasibility of using cultivated biomass as a biofilter or sorbent for some (i.e. dissolved) fish cultivation effluents and if balanced correctly, reducing eutrophication. Rathod et al. (2014) noted the biosorption potential of *K. alvarezii*biomass for phosphates providing positive environmental services at selected, culture sites. The uptake of phosphates was reported as being up to 59.77 mg/g at 25 °C, with an increase in sorption at increasing temperature, at pH 6.0.

Further studies on the multi-factorial and integration (over different physical scales) of the abiotic influences of water motion, temperature, salinity and nutrients from the marine environment and fertilizers have been recommended (Ask and Azanza 2002) and surprisingly remain insufficiently addressed – even in 2017. The effects of nutrients are only significant upon inter-actions with seasonal variations such as temperature, irradiance and water motion; these are relevant issues in the future when seasonality of typhoon and storm surges could be considerable. It has also been shown that eucheumatoids have the tendency to absorb and/or concentrate heavy metals (FAO 2005); hence, site selection is of utmost importance.

3.3.5 pH

Environmental perturbation and oceanic acidification, resulting from high atmospheric carbon, affects biochemical processes of marine organisms, including seaweeds (Gutow et al. 2014; Hurd et al. 2009). Unsurprisingly, these abiotic stressors influence the development of macroalgae, and may lead to reduction in growth rates (see Chap. 7 in this book on the effects of climate change and use of biomass as a CO_2 sink).

A study by Zitta et al. (2012) subjected *K. alvarezii* to pH 6–9 reported higher daily growth rates at pH 8.4, followed by pH 9 and 6; taken together meaning that growth inhibition occurred under acidic seawater conditions. Further findings in this study indicated the feasibility of Rubisco as a potential biomarker for pH-induced abiotic stress. Decreasing the pH in both healthy and epiphyte-infected *K. alvarezii* increased its photosynthetic rates (Borlongan et al. 2016). However, the underlying reasons and mechanisms to such a response need further study.

3.4 Summary and Recommendations

In *Eucheuma* and *Kappaphycus* culture (indeed as with all other seaweeds/marine plants) irradiance and SST ranges (which are themselves determined by water depth) are major factors for consideration. Salinity and nutrient quality/availability are affected by water movement and will also determine, to a great extent, epiphytic growth and siltation. The crop's biological organization and potential for acclimation in new and changing culture environments are species (or even strain) specific and, hence, should be the clarified for culturists.

The eucheumatoids are now a major marine crop and many of the principles of terrestrial agriculture may be applied to marine agronomy. Plant scientists and agriculturists took millennia to get where we are with land plant production. However, the science of seaweeds and marine agronomy are at most very much in their infancy. There is a need to develop quickly as the needs for feeding the nine billion people in world by 2050 is approaching fast and the marine environment/marine plants could very well address these needs (FAO 2016).

Knowledge/science-based cultivation of seaweeds (and of any other organism) ensures economic and environmental sustainability of the value-chain since an understanding of the biology/physiology of the crop is fundamental and can help prevent and/or mitigate farming problems and unnecessary losses. Sustainability of marine agronomy can also be achieved by having sufficient data to produce reliable and predictive models and risk analyses and information from the laboratory and the field. Likewise, the political will of the community and the government especially, would help in the achievement of sustainable, marine agronomy goals. Clarifications/inputs into public policies related to environmental concerns can also be attained.

Kappaphycus alvarezii remains the major carrageenophyte of the times and is apparently the most well-studied. However, there are still, surprising gaps in knowledge of its physiology (and those of other related species) that can be elucidated to achieve better or high quality yield per unit area, per unit time. The use of sporulation techniques in commercial-scale farming should be pursued for production economy and efficiencies.

References

Araujo PG, Ribeiro ALNL, Yokoya NS, Fujii MT (2014) Temperature and salinity responses of drifting specimens of *Kappaphycus alvarezii* (Gigartinales, Rhodophyta) farmed on the Brazilian tropical coast. J Appl Phycol 26:1979–1988

Ask EI, Azanza RV (2002) Advances in cultivation technology of commercial eucheumatoid species: a review with suggestions for future research. Aquaculture 206:257–277

Azanza-Corrales RV (1990) The *Eucheuma* species in Danajon Reef, Philippines: vegetative and reproductive structures. J Appl Phycol 2:57–62

Azanza-Corrales RV, Aliaza TT (1999) *In vitro* carpospores release and germination in *Kappaphycus alvarezii* (Doty) Doty from Tawi-Tawi, Philippines. Bot Mar 42:281–284

Azanza-Corrales RV, Sa-a P (1990) The farmed *Eucheuma* species (Gigartinales, Rhodophyta) in Danajon Reef, Philippines: carrageenan properties. Hydrobiologia 204(205):521–525

Azanza-Corrales RV, Mamauag SS, Alfiler E, Orolfo MJ (1992) Reproduction in *Eucheuma denticulatum* (Burman) Collins and Hervey and *Kappaphycus alvarezii* (Doty) Doty farmed in Danajon Reef, Philippines. Aquaculture 103:29–34

Azanza-Corrales R, Aliaza TT, Montano NE (1996) Recruitment of *Eucheuma* and *Kappaphycus* on a farm in Tawi-Tawi, Philippines. Hydrobiologia 326(327):235–244

Azanza RV, Ask E (2003) *Kappaphycus alvarezii* (Doty) Doty carposporeling growth and development in the laboratory. In: Anthony RO, Chapman et al (eds) Proceedings of the international seaweed symposium, vol 17, pp 95–99

Azanza RV, Bobadilla J, Luciano D (2014) Strain development/hybridization in *Kappaphycus* utilizing spores. Submitted to Philippine Council for Agriculture, Aquatic and Natural Resources Research and Development, Terminal Report

Barros MP, Necchi OJ, Colepicolo P, Padersén M (2006) Kinetic study of the plastoquinone pool availability correlated with H_2O_2 release in seawater and antioxidant responses in the red alga *Kappaphycus alvarezii* exposed to single or combined high light, chilling and chemical stresses. Biochim Biophys Acta 1757:1520–1528

Borlongan IAG, Tibubos KR, Yunque DAT, Hurtado AQ, Critchley AT (2011) Impact of AMPEP on the growth and occurrence of epiphytic *Neosiphonia* infestation on two varieties of commercially

cultivated *Kappaphycus alvarezii* grown at different depths in the Philippines. J Appl Phycol 23(3):615–621
Borlongan IAG, Luhan MRJ, Padilla PIP, Hurtado AQ (2016) Photosynthetic responses of '*Neosiphonia* sp. epiphyte-infected' and healthy *Kappaphycus alvarezii* (Rhodophyta) to irradiance, salinity and pH variations. J Appl Phycol 28:2891–2902
Bulboa CR, de Paula EJ (2006) Introduction of non-native species of *Kappaphycus* (Rhodophyta, Gigartinales) in subtropical waters: comparative analysis of growth rates of *Kappaphycus alvarezii* and *Kappaphycus striatum in vitro* and in the sea in south-eastern Brazil. Phycol Res 53(3):183–188
Bulboa CR, de Paula EJ, Chow F (2007) Laboratory germination and sea-out planting of tetraspore progeny from *Kappaphycus striatum* (Rhodophyta) in subtropical water of Brazil. J Appl Phycol 19:357–363
Bulboa CR, de Paula EJ, Chow F (2008) Germination and survival of tetraspores of *Kappaphycus alvarezii* var. *alvarezii* (Solieriaceae, Rhodophyta) introduced in subtropical waters of Brazil. Phycol Res 56:39–45
Cheney DP, Metz B, Levine I, Rudolph B (1998) Genetic manipulation and strain improvement of seaweeds for aquaculture. Book of Abstracts, *Aquaculture* '98 Las Vegas, Nevada, p 105
Dawes CJ (1979) Physiological and biochemical comparisons of species of *Eucheuma* yielding iota carrageenan from Florida and the Gulf of California with *E. denticulatum* from the Pacific (Rhodophyta). In: Jenson AJ Stein (eds) Proceedings of the 9th international seaweed symposium. Science Press, Princeton, pp 199–208
Dawes CJ (1992) Irradiance acclimation of the cultured Philippines seaweeds, *Kappaphycus alvarezii* and *Eucheuma denticulatum*. Bot Mar 35:189–195
Dawes CJ, Trono GC, Lluisma AO (1993) Clonal propagation of *Eucheuma denticulatum* and *Kappaphycus alvarezii* in the Philippines. J Appl Phycol 6:21–24
DeBoer JA, Ryther JH (1977) Potential yields from a waste-recycling algal mariculture system. In: Krauss RW (ed) The marine plant biomass of the Pacific northwest coast. Oregon State University Press, Corvallis, pp 231–249
Doty MS (1985) *Eucheuma alvarezii*, sp. Nov. (Gigartinales, Rhodophyta) from Malaysia. In: Abbott LA, JN Norris (eds) Taxonomy of economic seaweeds: with reference to some Pacific and Carribean Species. California Sea Grant College Program. Rep. T-CSGCP-011, La Jolla, California, pp 37–45
Doty MS (1987) The production and use of *Eucheuma*. In Doty MA, Caddy JF, Santalices B (eds) Case studies of seven commercial seaweed resources. FAO fisheries technical paper, 281 Rome, pp 123–161
Dumilag RV, Lluisma AO (2014) Resolving the phylogenetic affinities of *Kappaphycus inermis* within the genus *Kappaphycus* (Gigartinales, Solieriaceae) using mitochondrial and plastid markers. Phytotaxa 162(4):223–231
Dy DT, Yap HT (2001) Surge ammonium uptake of the cultured seaweed, *Kappaphycus alvarezii* (Doty) Doty (Rhodophyta: Gigartinales). J Exp Mar Bio Ecol 265(1):89–100
Food and Agriculture Organization (FAO) (2005) Cultured aquatic species information programme. *Eucheuma* spp. In: FAO Fisheries and Agriculture Department [online]. Rome, Italy
Food and Agriculture Organization (FAO) (2016) The state of world fisheries and aquaculture. 24 pp (Online) Rome, Italy
Freile-Pelegrin Y, Robledo D (2006) Carrageenan of *Eucheuma isiforme* (Solieiaceae Rhodophyta) from Yucatan, Mexico. II. Seasonal variations in carrageenan and biochemical characteristics. Bot Mar 49(1):72–78
Ganzon-Fortes ET, Azanza-Corrales R, Aliaza T (1993) Comparison of photosynthetic responses of healthy and "diseased" *Kappaphycus alvarezii* (Doty) Doty using P vs I curve. Bot Mar 36:503–506
Gerung GS, Ohno M (1997) Growth rates of *Eucheuma denticulatum* (Burman) Collins et Harvey and *Kappaphycus striatum* (Schmitz) Doty under different conditions in warm waters of Southern Japan. J Appl Phycol 9:413–415
Glenn EP, Doty MS (1981) Photosynthesis and respiration of the tropical red seaweeds, *Eucheuma striatum* (tambalang and elkhorn varieties) and *E. denticulatum*. Aquat Bot 10:353–364
Glenn EP, Doty MS (1992) Water motion affects the growth rates of *Kappaphycus alvarezii* and related red seaweeds. Aquaculture 108:233–246
Granbom M, Pedersén M, Kadel P, Luning K (2001) Circadian rhythm of photosynthetic oxygen evolution in *Kappaphycus alvarezii* (Rhodophyta): dependence on light quantity and quality. J Phycol 37:1020–1025
Guan X, Wang J, Zhu J, Yao C, Liu J, Qin S, Jiang P (2013) Photosystem II photochemistry and phycobiliprotein of the red algae *Kappaphycus alvarezii* and their implications for light adaptation. Biomed Res Int, Article ID 256549, 9 p. http:dx.doi.org/10.1155/2013/256549
Gutow L, Rahman MM, Bartl K, Saborowski R, Bartsch I, Wiencke C (2014) Ocean acidification affects growth but not nutritional quality of the seaweed *Fucus vesiculosus* (Phaeophyceae, Fucales). J Exp Mar Biol Ecol 453:84–90
Hayashi L, Yokoya NS, Ostini S, Pereira RTL, Braga ES, Oliveira EC (2008) Nutrients removed by *Kappaphycus alvarezii* (Rhodophyta, Solieriaceae) an integrated cultivation with fishes in recirculating water. Aquaculture 277(3–4):185–191
Hayashi L, Faria GSM, Nunes BG, Zitta CS, Scariot LA, Rover T, Felix MRL, Bouzon ZL (2011) Effects of salinity on the growth rate, carrageenan yield, and cellular structure of *Kappaphycus alvarezii* (Rhodophyta, Gigartinales) cultured in vitro. J Appl Phycol 23(3):439–447
Hurd CL, Hepburn CD, Currie KI, Raven JA, Hunter KA (2009) Testing the effects of ocean acidification on algal metabolism: considerations for experimental designs. J Phycol 45:1236–1251
Hurtado AQ, Yunque DA, Tibubos K, Critchley AT (2009) Use of Acadian marine plan extract powder from *Ascophyllum nodosum* in tissue culture of *Kappaphycus* varieties. J Appl Phycol 21, 633
Hurtado AQ, Gerung GS, Yasir S, Critchley AT (2014) Cultivation of tropical red seaweeds in the BIMP-EAGA region. J Appl Phycol 26:707–718
Hurtado AQ, Neish IC, Critchley AT (2015) Developments in production technology of *Kappaphycus* in the Philippines: more than four decades of farming. J Appl Phycol 27:1945–1961
Karentz D, Cleaver JE, Mitchell DL (1991) DNA damage in the Antarctic. Nature 350:28
Largo DB, Fukami K, Nashijima T, Ohno M (1995) Laboratory-induced development of the *ice-ice* disease of the farmed red algae *Kappaphycus alvarezii* and *Eucheuma denticulatum* (Solieriaceae, Gigartinales, Rhodophyta). J Appl Phycol 7(6):539–543
Li R, Li J, Wu CY (1990) Effect of ammonium on growth and carrageenan content in *Kappaphycus alvarezii* (Gigartinales, Rhodophyta). Hydrobiologia 204–205:499–503
Lideman GN, Nishihara TN, Terada R (2013) Effect of temperature and light on the photosynthesis as measured by chlorophyll fluorescence of cultured *Eucheuma denticulatum* and *Kappaphycus* sp. (*Sumba* strain) from Indonesia. J Appl Phycol 25:399–406
Liu H, Li D, Li Y, Hou T (2016) Atomistic molecular dynamics simulation of ATP-binding cassette transporters. Comp Mol Sci 6(3):255–265
Lobban CS, Harrison PJ (1994) Seaweed ecology and physiology. Cambridge University Press, Cambridge, p 36
Loureiro RR, Reis RP, Berrongain FD, Critchley AT (2012) Extract powder form the brown alga *Ascophyllum nodosum* (Linnaeus) Le Jolis (AMPEP): a "vaccine-like" effect on *Kappaphycus alvarezii* Doty (Doty) ex P.C. Silva. J Appl Phycol 24(3):427–432

Loureiro RR, Reis RP, Marroig RG (2014) Effect of the commercial extract of the brown alga *Ascophyllum nodosum* Mont. on *Kappaphycus alvarezii* (Doty) Doty ex P.C. Silva *in situ* submitted to lethal temperatures. J Appl Phycol 26(1):629–634

Luhan MR, Sollesta H (2010) Growing the reproductive cells (carpospores) of the seaweed, *Kappaphycus striatum,* in the laboratory until outplanting in the field and maturation to tetrasporophyte. J Appl Phycol 22(5):579–585

Mairh OP, Soe-Htun U, Ohno M (1986) Culture of *Eucheuma striatum* (Rhodophyta, Solieriaceae) in subtropical waters of Shikoku, Japan. Bot Mar 29:185–191

Mairh OP, Zodape ST, Tewari A, Rajyaguru MR (1995) Culture of marine red alga *Kappaphycus striatum* (Schmitz) Doty on the Saurashtra region, west coast of India. Indian J Mar Sci 24:24–31

Mathieson AC, Dawes CJ (1974) Ecological studies of Floridian *Eucheuma* (Rhodophyta, Gigartinales): II. Photosynthesis and respiration. Bull Mar Sci 34:274–285

Msuya FE, Kyewalyanga MS, Bleicher-Lhonneur G, Lampin T, Lhonneur JP, Mazoyer J, Critchley AT (2011) Seasonal variation in growth rates and carrageenan properties of *Kappaphycus alvarezii* and *Eucheuma denticulaum* cultivated with and without additional nutrients, in Uroa, Zanzibar, Tanzania. TaJONAS 2(2):409–423

Neish IC (2008) SEAPLant.net Monograph no. HB2F 1008 V3 GAP

Ohno M, Largo DB, Ikumoto T (1994) Growth rate, carrageenan yield and gel properties of culture kappa-carrageenan producing red alga *Kappaphycus alvarezii* (Doty) Doty in the subtropical waters of Shikoku, Japan. J Appl Phycol 6:1–5

Ohno M, Orosco CA (1987) Growth rate of three species of *Eucheuma*, commercial red algae from the Philippines. In: Umezaki I (ed) Scientific survey of marine algae and their resources in the Philippine Islands. Laboratory of Fishery Resources, Graduate School of Agriculture, Kyoto University, pp 77–81

Ohmi H, Shinmura I (1976) Growth of *Eucheuma amakusaensis* in field culture. Bull Jap Soc Phycol 24:98–102

Pang T, Liu J, Liu Q, Li W (2011) Changes of photosynthetic behaviors in *Kappaphycus alvarezii* infected by epiphyte. Evidence-Based Complementary and Alternative Medicine http://dx.doi.org/10.1155/2011/658906

Preisig R, Hans R (2005) Historical review of algal culturing techniques. Academic, Zurich, pp 1–12

Premkumar A, Eswaran K, Kulandaivelu G (1993) Growth, biochemical and photosynthetic responses of greengram seedlings under water stress to ultraviolet-B radiation. Plant Physiol Biochem 20:31–35

Rathod M, Mody K, Basha S, 2014. Efficient removal of phosphate from aqueous solution by red seaweed, *Kappaphycus alvarezii.* J Clean Prod 84, 484–493

Reis R, Loureriro RR, Mesquita FS (2011) Does salinity affect growth and carrageenan yield of *Kappaphycus alvarezii* (Gigartinales/Rhodophyta)? Short communication. Aquac Res 42:1231–1234

Rodrigueza MR, Montaño NE (2007) Bioremediation potential of three carrageenophytes cultivated in tanks with seawater from fish farms. J Appl Phycol 19(6):775–762

Schmidt EC, Scariot LA, Rover T, Bouzon ZL (2009) Changes in ultrastructure and histochemistry of two red macroalgae strains of *Kappaphycus alvarezii* (Rhodophyta, Gigartinales), as a consequence of ultraviolet B radiation exposure. Micron 40:860–869

Schmidt EC, Maraschin M, Bouzon ZL (2010a) Effects of UVB radiation on the carragenophyte *Kappaphycus alvarezii* (Rhodophyta, Gigartinales): changes in ultrastructure, growth, and photosynthetic pigments. Hydrobologia 649(1):171–182

Schmidt EC, Nunes BG, Maraschin M, Bouzon ZL (2010b) Effect of ultraviolet-B radiation on growth, photosynthetic pigments and cell biology of *Kappaphycus alvarezii* (Rhodophyta, Gigartinales) macroalgae brown strain. Photosynthetica 48(2):161–172

Schubert H, Gerbensdorf S, Titlyanov E, Titlyanova T, Granbom M, Pape C, Luning K (2004) Circadian rhythm of photosynthesis in *Kappaphycus alvarezii* (Rhodophyta): independence of the cell cycle and possible photosynthetic clock targets. Eur J Phycol 39:423–430

Syamsuddin R (2013) Spore discharge and development and carrageenan content of seaweed *Kappaphycus alvarezii* illuminated with different light colours. http://repository.unhas.ac.id/handle/123456789/7090

Tablizo FA, Lluisma AO (2014) The mitochondrial genome of the red alga *Kappaphycus striatus* ("Green Sacol" variety): complete nucleotide sequence, genome structure and organization, and comparative analysis. Mar Genomics 18:155–161

Tan J, Lim PE, Phang SM, Hong DD, Sunarpi H, Hurtado AQ (2012) Assessment of four molecular markers as potential DNA barcodes for red algae *Kappaphycus* Doty and *Eucheuma* J. Agardh (Solieriaceae, Rhodophyta). PLoS One 7(12):e52905

Terada R, Vo TD, Nishihara GN, Shioya K, Shimada S, Kawaguchi S (2015) The effect of irradiance and temperature on the photosynthesis and growth of a cultivated red alga *Kappaphycus alvarezii* (Solieraceae) from Vietnam, based on *in situ* and *in vitro* measurements. J Appl Phycol 28(1):457–467

Trono GC, Ohno M (1989) Seasonality in the biomass production of the *Eucheuma* strains in Northern Bohol, Philippines. In: Umekazi I (ed) Scientific survey of marine algae and their resources in the Philippine Islands. Monbusho International Scientific Research Program

Trono GC, Lluisma AO (1992) Differences in biomass production and carrageenan yields among from strains of farmed carrageenophytes in Northern Bohol, Philippines. Hydrobiologia 247:223–227

van Der Meer JP (1990) Genetics. In: Cole KM, Sheath RG (eds) Biology of the red algae. Cambridge University Press, Cambridge, pp 103–122

Villanueva RD, Romero JB, Montaño MNE, de la Peña PO (2011) Harvest optimization of four *Kappaphycus* species from the Philippines. Biomass Bioenergy 35:1311–1316

von Wobeser EA, Figueroa FL, Cabello-Pasini A (2001) Photosynthesis and growth of red and green morphotypes of *Kappaphycus alvarezii* (Rhodophyta) from the Philippines. Mar Biol 138:679–686

Wood WF (1987) Effect of solar ultraviolet radiation on the kelp *Ecklonia radiata.* Mar Biol 96:143–150

Worrest RC (1983) Impact of solar ultraviolet-B radiation (290–320 nm) upon marine microalgae. Physiol Plant 58:428–434

Zitta CS, Oliviera EM, Bouzon ZL, Hayashi L (2012) Ploidy determination of three *Kappaphycus alvarezii* strains (Rhodophyta, Gigartinales) by confocal fluorescence microscopy. J Appl Phycol 24:495–499

Zuldin WH, Shapawi R (2015) Performance of red seaweed (*Kappaphycus* sp.) cultivated using tank culture system. J Fish Aquat Sci 10(1):1–12

The Cultivation of *Kappaphycus* and *Eucheuma* in Tropical and Sub-Tropical Waters

4

Leila Hayashi, Renata P. Reis, Alex Alves dos Santos, Beatriz Castelar, Daniel Robledo, Gloria Batista de Vega, Flower E. Msuya, K. Eswaran, Suhaimi Md. Yasir, Majid Khan Majahar Ali, and Anicia Q. Hurtado

Abstract

Kappaphycus and *Eucheuma* species have been successfully cultivated in Southeast Asia since the early 1970s. The increasing global demand for carrageenan in processed foods and thereby the need for industrial-scales of biomass to be provided to feed an extraction industry, exceeded wild stock availability and productivity and commercial demands could only be achieved through extensive cultivation of selected carrageenophytes. This unprecedented situation led to the introduction of carrageenophyte species and cultivars around the world; combined production of *Eucheuma* and *Kappaphycus* is one of the largest for seaweed biomass in the world.

The activity of, and economic benefits accrued from, seaweed farming are indeed responsible for significant changes in rural, coastal communities in a number of important countries. Such activities generating new activity and income are often the only source of cash

L. Hayashi (✉)
Aquaculture Department, Universidade Federal de Santa Catarina (UFSC), Rodovia Admar Gonzaga, 1346, Itacorubi, 88034-001 Florianópolis, Santa Catarina, Brazil

Integrated Services for the Development of Aquaculture and Fisheries (ISDA) Inc., McArthur Highway, Tabuc Suba, Jaro, Iloilo City 5000, Philippines
e-mail: leila.hayashi@ufsc.br

R.P. Reis
Instituto de Pesquisa Jardim Botânico do Rio de Janeiro (JBRJ), Rua Pacheco Leão, 915, Jardim Botânico, 22460-030 Rio de Janeiro, Brazil

A.A. dos Santos
Empresa de Pesquisa Agropecuária e Extensão Rural de Santa Catarina (EPAGRI), Centro de Desenvolvimento de Aquicultura e Pesca, Rodovia Admar Gonzaga, 1181, Itacorubi, 88034-901 Florianópolis, Santa Catarina, Brazil

B. Castelar
Fundação Instituto de Pesca do Estado do Rio de Janeiro (FIPERJ), Av. das Américas, 31501, Guaratiba, 23032-050 Rio de Janeiro, Brazil

D. Robledo
Cinvestav Unidad Mérida, Departamento de Recursos del Mar, A.P.73, Cordemex, 97310 Mérida, Yucatán, Mexico

G.B. de Vega
Director de Investigación y Desarrollo (I+D) de Gracilarias de Panamá S.A., and Facultad de Ciencias Naturales y Tecnología, Universidad de Panamá, Panamá

F.E. Msuya
Institute of Marine Sciences, University of Dar es Salaam, Mizingani Road, P.O. Box 668, Zanzibar, Tanzania

K. Eswaran
Division of Marine Biotechnology & Ecology, CSIR-Central Salt & Marine Chemicals Research Institute, Bhavnagar 364021, India

S.M. Yasir
Seaweed Research Unit (UPRL), Faculty of Science and Natural Resources, Universiti Malaysia Sabah, 88400 Kota Kinabalu, Sabah, Malaysia

M.K.M. Ali
School of Mathematical Sciences, Universiti Sains Malaysia, 11800 Gelugor, Penang, Malaysia

Seaweed Research Unit (UPRL), Faculty of Science and Natural Resources, Universiti Malaysia Sabah, 11800 Kota Kinabalu, Sabah, USA

A.Q. Hurtado
Integrated Services for the Development of Aquaculture and Fisheries (ISDA) Inc., McArthur Highway, Tabuc Suba, Jaro, 5000 Iloilo City, Philippines

A.Q. Hurtado et al. (eds.), *Tropical Seaweed Farming Trends, Problems and Opportunities*, Developments in Applied Phycology 9, DOI 10.1007/978-3-319-63498-2_4

to some of the farmers. However, in spite of the enormous size and value of the industry the techniques applied to cultivation of carrageenophytes has remained almost unchanged in the commercial farms. Seedlings (or vegetatively propagated fragments of older, larger, mature thalli) are still most commonly planted along ropes held in various configurations, in most of the major production centers, e.g. mostly classified as simple stakes in the ground (off-bottom planting to floating rafts of various design). Some technological developments have been made in Brazil and India with the implementation of tubular nets for planting, which allows a degree of mechanization on the farms; both countries are developing mechanical harvesters. Further to co-production of biomass and harvesting, the drying process is still a limiting problem for production, and drying costs are high, in countries with high rainfall, this segment of the process is discussed in detail by Ali et al. Chap. 8). Diseases such as "ice-ice" and blooms of epiphytes and endophytes (see Chap. 6 by Loureiro et al.), as well as present and impending climate change effects (see Chap. 7 by Largo et al.) are also very big challenges.

In general, the production of carrageenophytes in tropical and sub-tropical regions is very dependent on the weather: as in prevailing conditions during the dry and wet seasons or surface seawater temperatures (SSTs) during "cold" winters and/or "hot" summers. For the future, efforts need to be made to increase the productivity and resistance of the selected strains of carrageenophyte seedlings against environmental instabilities.

4.1 Introduction

Seaweed farming has been increasing in economic significance over recent years, and at the time of writing, is considered the second largest production in aquaculture. FAO usually includes seaweeds in the group called "aquatic plants", of which in 2014, 27.3 million MT were produced, with a value of US$5.6 billion. Seaweeds were responsible for almost 98% of this production, the remaining 2% being represented by microalgal biomass and other aquatic plants (FAO 2016). According to Porse and Rudolph (2017), the major sources of carrageenan are *Kappaphycus* (170,000 MT, dwt), followed by *Eucheuma* (4,500 m MT, dwt), *Gigartina radula* and *G. skottsbergii* (14,000 MT, dwt) and finally, *Chondrus crispus* and others (3,200 MT, dwt). *Kappaphycus and Eucheuma* are mainly cultivated in the Southeast Asian region and Tanzania and only small volumes are produced in the Latin American countries, while the rest of the carrageenophytes are produced in temperate countries.

Eucheumatoids (which include the *Kappaphycus* and *Eucheuma* genera, with multiple species and varieties), have been given names of convenience by industry, viz. the commercial names: "cottonii" and "spinosum", respectively are the main cultivated stocks. In only 4 years, the production of these changed from 5,629 million MT (2010) to 10,992 million MT (2014). The main producers, the Philippines, were subsequently overtaken by Indonesia. However, due to the success of cultivation and value of the biomass they are now being farmed in more than 50 countries (FAO 2016). Figure 4.1 presents the main countries involved in commercial cultivation.

Both genera are predominantly cultivated in tropical and sub-tropical waters. They are the major raw material sources of *kappa* and *iota* carrageenans traditionally used for their rheological properties as emulsifiers, binders and for suspension or stabilization in a wide range of processed food products, pharmaceutical carriers and the cosmetic industries (Bixler and Porse 2011).

The cultivation of *Kappaphycus* and *Eucheuma* displaced wild-gathered *Chondrus crispus* as a major source of *kappa*carrageenan, since it was commercially grown in the Philippines and other tropical countries, mainly because of the lower cost of labor. From the beginning of commercial cultivation in early 1970s until 2008, the Philippines was the major supplier of *Kappaphycus* species. In 2008, Indonesia expanded their seaweed farms and, because they have more geographic area, became the largest producer at the time of writing.

Seaweed farming became a complementary income for fisheries where commercial cultivation was established, changing the social profile mainly in small villages in Africa. Tanzania, especially the Zanzibar Islands, are good examples of successful commercial production from seedlings introduced from the Philippines and now commercial farms are spread along the Eastern African coast and India.

The recent increase in demand for carrageenan (particularly kappa for its rheological properties) promoted the development of these cultivation activities, mainly in the tropics. Southeast Asian Countries and Pacific Islands are trying to achieve the same success modelled after the Philippines and Indonesia. Similarity in environmental conditions makes all such regions very promising for the expansion of seaweed farms.

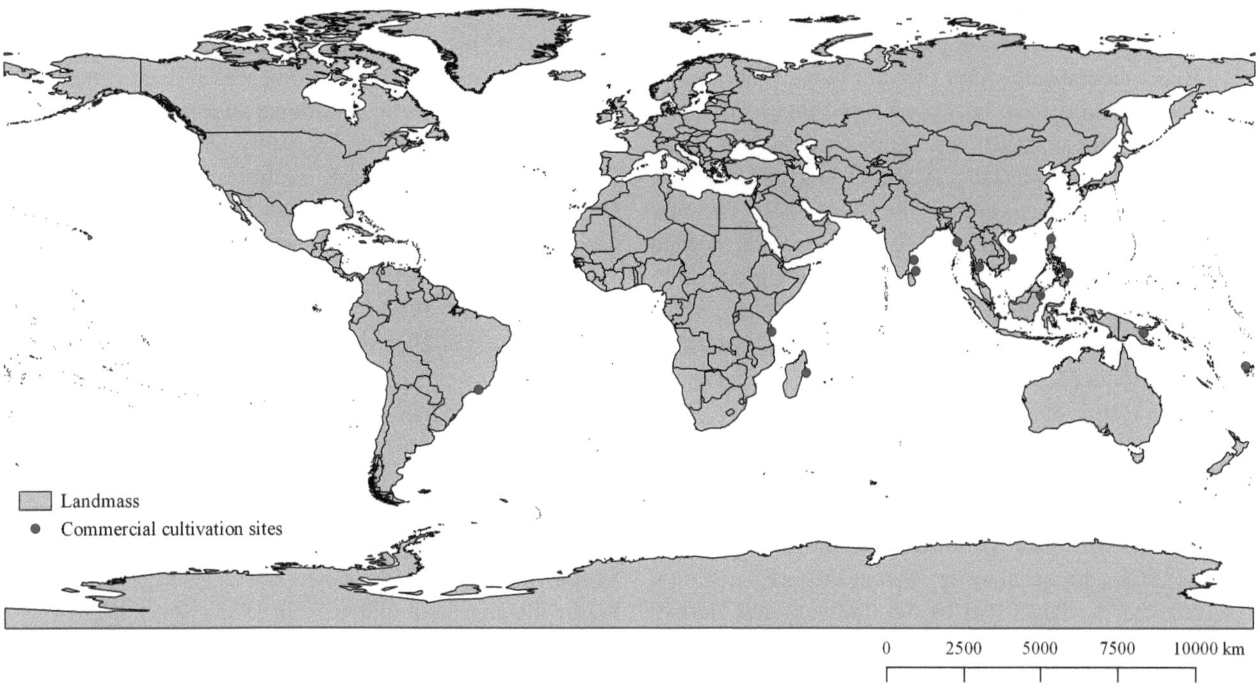

Fig. 4.1 Countries involved in commercial cultivation

On the Caribbean coast of México and Central America, including the Caribbean Islands, wild stocks of *Eucheuma* (viz. *Eucheuma isiforme*) had been sporadically exploited, particularly for local consumption – as a vegetable. Depletion of natural stocks and insufficient volumes for commercial extraction from local species led to the introduction of non-native *Eucheuma/Kappaphycus* varieties to be tested in research, pilot-farms from approximately 1981 in this region. In South America, the lack of success into the research for valorization of local (indigenous) species which could be commercialized, stimulated the introduction of varieties of both genera in 1994. However, until recently there was no officially permitted, commercial cultivation. We present some of the past and recent experiences on the use and cultivation of *Eucheuma* and *Kappaphycus* around the world, including an assessment of socio-economic benefits for coastal communities and future, potential development.

4.2 Cultivation in Tropical and Sub-tropical Waters

4.2.1 Southeast Asia and China

The geographical locations of the Philippines, Indonesia and Malaysia, within the Coral Triangle, where natural populations of *Kappaphycus* and *Eucheuma* are found, are also well suited to open-water, extensive, commercial cultivation of these seaweeds. These areas provide multitudes of shallow reef areas and also access to deeper, off-shore areas with cooler, surface seawater temperatures; sandy to coral debris bottom substrata; clear and clean water and also moderate to strong water movement. Such conditions favor robust and luxuriant growth of eucheumatoid seaweeds. The absence of tropical cyclones in Indonesia and Malaysia are a bonus to the stability of farming operations there and furthermore contributed to high production rates. The geographic location of the Philippines is unfortunately often beset with many weather disturbances including super typhoons which can cause havoc for farming rigs (anchors, floats, lines and nets).

4.2.1.1 The Philippines

Marine Colloids, Inc., a USA seaweed-processing firm, established the first *Kappaphycus* ("cottonii" type; originally called *Eucheuma*) cultivation research program in the Philippines in 1969. This initiative was later joined by the Bureau of Fisheries and Aquatic Resources (BFAR, formerly called the Philippine Fisheries Commission), and the University of the Philippines. The late Dr. Maxwell S. Doty, Vicente Alvarez, and Dr. Gavino C. Trono Jr. were pioneer academic practitioners in advancing *Kappaphycus* cultivation in the Philippines and later around the world (Doty 1973; Trono 1977; Ricohermoso and Deveau 1979; Doty and Alvarez 1981).

Intensive field surveys and collection of data were conducted during 1967–1970 in order to assess the best cultivation sites. Though the first cultivation trials were made in Panagatan Is., Caluya, Antique, and Ilin Is. Occidental Mindoro, Philippines, in 1969, it was a satellite cultivation trial in Tapaan Is., Siasi, Sulu in 1971 which provided the most promising indications for suitability of farming, with an average growth rate of 1.5–5.5% day^{-1} (Doty 1973; Parker 1974). Experimental cultivation sites in Caluya, Antique and Occidental Mindoro were later abandoned due to frequent typhoons and poor management. Other areas such as Sacol Is., Zamboanga City and Garza Is. and Occidental Mindoro, started to cultivate *Kappaphycus* in 1971–1972. The first commercial quantities of *Kappaphycus* (then called *Eucheuma* and 'cottonii' – a source of kappa carrageenan) were harvested from the cultivation areas in 1972. Since then, commercial cultivation of *Kappaphycus* has been successfully introduced, not only in Mindanao, but also the Visayas and Luzon, as seen in Fig. 4.2 (BAS 2001).

4.2.1.2 Indonesia

The first attempt to farm '*spinosum*' (now known as *Eucheuma denticulatum* (Burman) Collins et Harvey, a source of *iota*carrageenan) in Indonesia was made at Thousand Islands, Pulau Pari by Soerjodinoto and Hariadi Adnan in 1967; however, it did not progress further. Prompted by insufficient and unsustainable supplies of wild stocks, the True Blue Expedition was initiated in order to further focus on seeking additional sites suitable for cultivation. This was organized by Auby (Sanofi), Copenhagen Pectin (CP Kelco), Marine Colloids Inc. (MC) and the Indonesian Fisheries, LPPL. The CP Kelco and MC farming projects at Pulau Samaringa, Sulawesi and Riau Islands in 1975–1977 were inconclusive, hence the project moved to Bali. From 1978–1984, CP Kelco continued farming in Bali and success was eventually achieved at Nusa Lembongan (Neish 2013).

Free seedlings and farm materials were provided to fishermen and the resulting production of "*spinosum*" biomass (for iota carrageenan) steadily increased to commercial levels. From 1985–1986, CP Kelco initiated farms in Indonesia to supplement those initiated by Marine Colloids Inc. and Sanofi (now Cargill). Commercial 'cottonii' seaweed farming finally arrived in Indonesia in 1985 (Neish 2013).

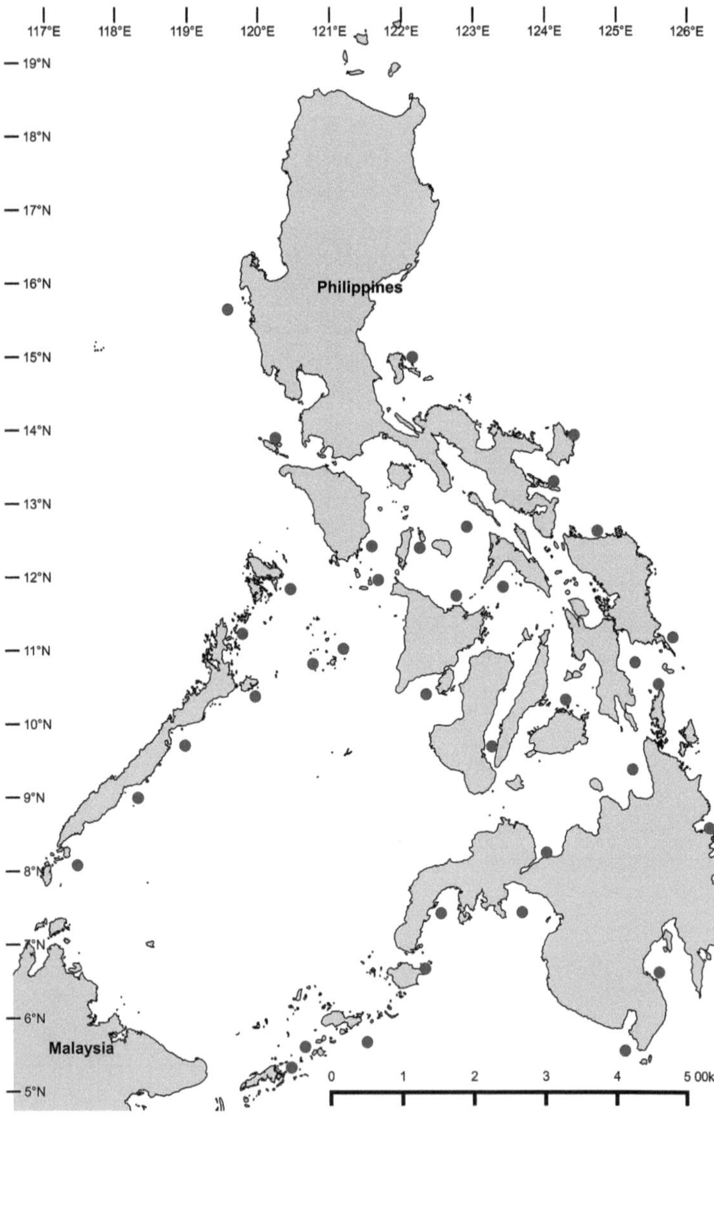

Fig. 4.2 Cultivation sites in the Philippines

From 1985–2009, Indonesian seaweed farm development was driven by the farmers themselves and local trader/collectors in a reflective, "bottom-up" manner. A market need for raw material for processing was made known to the prospective seaweed farmers by value-chain stakeholders on the demand side; farmers were exposed to the initially very simple technology that was required for growing these eucheumatoid seaweeds and, with facilitation from a variety of development organizations, i.e. the Australian Agency for International Development (AusAID), Canadian International Development Agency (CIDA) and the International Finance Corporation (IFC), to name a few, local seaweed farmers were able to build their businesses within the context of village norms, mores and structures (Neish 2013). Seaweed farming became integrated into the social fabric of the farmer/villagers to the point where it now appears to be a traditional, economic activity – even though it did not begin until the "relatively recent" mid-1980s, as further asserted by Neish (2013). Seaweed farming in Indonesia was also seen as being complementary to, and compatible with, other economic activities in the coastal villages, such as fishing and farming land crops. Ready cash from seaweed farming also had a noticeable multiplier effect in the local economies. Shops, support services for seaweed farming and village infrastructure all benefited notably from the cash derived from seaweed activities flowing through the local, village economies.

In 2007, Indonesia became the world's number one producer of *Kappaphycus* production by volume (i.e. 1.7 M MT fwt), when it overtook the Philippines (1.5 M MT fwt, FAO 2010). Almost all of the Indonesian islands (Fig. 4.3) are currently planted with different strains or varieties of either *Kappaphycus* and *Eucheuma* using fixed-off bottom, hanging long-line and various raft methods, involving tens of thousands of coastal families (Neish 2013; Hurtado et al. 2014).

4.2.1.3 Malaysia

A carrageenophyte farming industry was introduced to Malaysia in 1978; Sabah is the state producing the commercial volumes (Sade et al. 2006; Fig. 4.4). Farming was introduced in Semporna, east coast of Sabah by the Bajau and the Suluk peoples from Tawi-Tawi, Philippines (Eranza et al. 2015) and operated at the commercial scale by Aquatic Resources Limited (an American company). Since 1980, the management of the seaweed farming project on Semporna was solely under the responsibility of the Department of Fisheries, Sabah (DOFS). In the 1990s, support was extended to further seaweed cultivation as the new economic engine for local development with several initiatives receiving government support. The private sector was invited to participate in this development. In early 2000, a Seaweed Industrial Zone (SIZ), listing seaweed as a high-value product from aquaculture was established by the National Aquaculture Program. In 2010, the Seaweed Commercial Project Approach (SCPA) was established and further accompanied by a New Innovation Technology and Environment (NITE) strategy under the auspices of the Government Transformation Program (GTP) to strengthen Research and Development and Commercialization (RDC; Azhnar 2016).

For decades, Malaysia worked on several continuous economic plans each of which was intended to help the local economy and eradicate poverty. Seaweed farming was one of the National Key Economic Activities (NKEA) of the 10th Malaysia Plan. Unfortunately annual production of eucheumatoids was low, as compared to neighboring countries such as the Philippines and Indonesia. This was in spite of the existence of extensive, favorable farming areas for *Kappaphycus* (approximately 38,500 ha). The recent report of Eranza et al. (2015) claimed that there were three major issues which beset the Malaysian seaweed industry which had to be addressed, namely: (1) seaweed farming activities

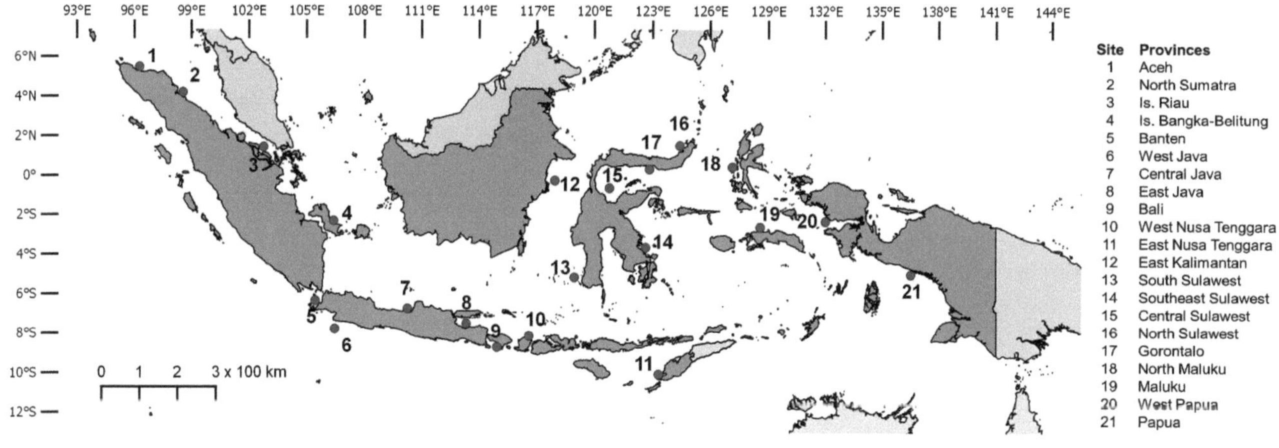

Fig. 4.3 Provinces of Indonesia involved in cultivation

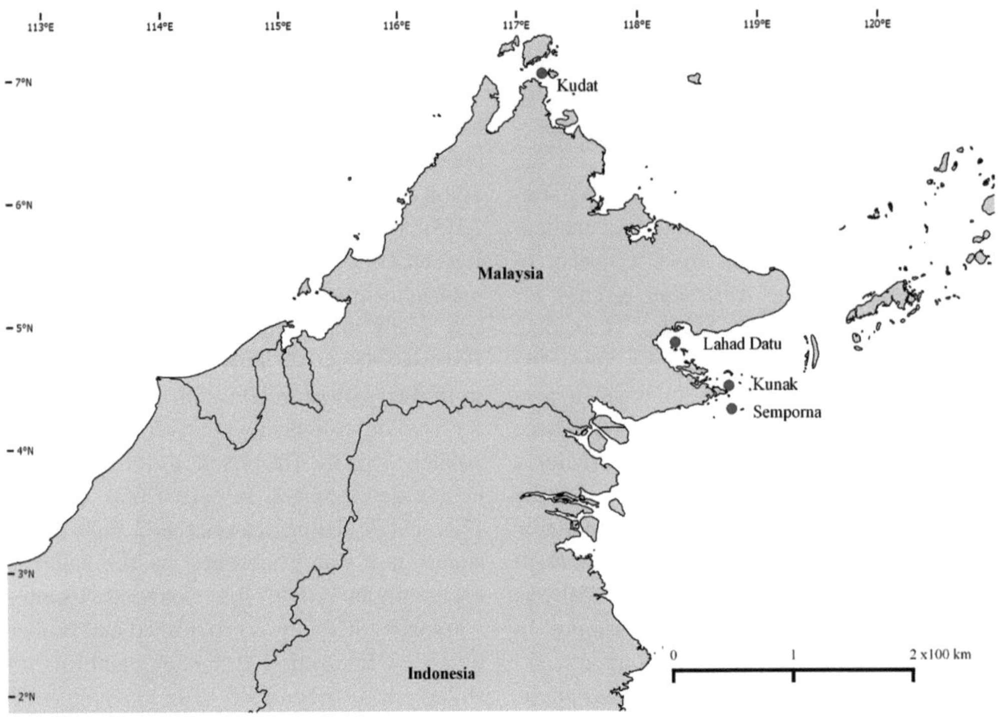

Fig. 4.4 Cultivation sites in Malaysia

between Malaysian islanders and immigrant seaweed farmers (Hussin and Khoso 2017); (2) seaweed farmers and the middlemen; and (3) seaweed quality, pricing and semi-refinery companies. The dilemma is now that less progressive seaweed farming methods in Semporna have resulted in Malaysia no longer being one of the main global producers (Eranza et al. 2015). The same author concluded that although the economic returns from the industry through farming, harvesting and post-harvest handling to the refining stages can be profitable, Malaysia has some major issues to resolve before this can become a reality.

4.2.1.4 Vietnam

Kappaphycus alvarezii has been reported as having been introduced from the Philippines to the south of Ninh Thuan province in Vietnam by the Institute of Applied Material Technology in Nha Trang and the Fisheries Extension Department in Ninh Thuan, via Kochi University, Japan in 1993 (Ohno et al. 1995, 1996). The first experimental evaluation was undertaken by Ohno et al. (1996) in ponds, lagoons, inlets and off-shore. The authors reported that the best daily growth rate (DGR) of plants grown in the lagoon areas was 9–11% day^{-1} in May–June (i.e. the cold season); those grown in inlets ranged from 7–9% day^{-1} in June; and those grown in ponds ranged from 5–6% day^{-1} in January–July, but decreased to less than 4% day^{-1} in August (i.e. the hot season). Grazing by fish was observed to occur in the areas thereby reducing productivity. These values were similar to those obtained from *K. alvarezii* cultivated in the Philippines and Indonesia.

Three years after its introduction, commercial farming of *Kappaphycus* started in the central and southern regions (Nang and Dinh 1998; Tsutsui et al. 2005) and it was regarded as an important resource for the coastal fishery (Le 2011; Fig. 4.5). However, after 2003, productivity declined, allegedly because of the deteriorating quality of the seedling material (i.e. after the old stock had been used continuously for vegetative production for over 10 years).

Dung and Nhan (2001) reported on the successful experimental cultivation of *K. alvarezii* in Haiquan Lagoon and Cat Ba Bay, Vietnam utilizing open-water, floating rafts and tie-tie with plastic cultivation lines. The growth rates observed from the open sea were higher than those achieved in the sheltered bays and ranged from 5.2–9.8% day^{-1} and 5.3–7.9% day^{-1}, respectively. In 2003, *Kappaphycus* was also introduced to Khanh Hoa Province, in Bich Dam on Hon Tre Island (facing the Hon Mun Marine Protected Area). This attempt was however abandoned because of pressure from fish predation, a problem that can largely be mitigated only by technical options such as cages, which were not known in Vietnam at that time. In Cam Ranh and Van Phong bays, located in the same province, *Kappaphycus* and *Eucheuma* were found to grow well and the area occupied had been expanding there (Dung and Nhan 2001).

K. alvarezii is considered a highly efficient cultivar, the cultivation of which can generate jobs and supplemental

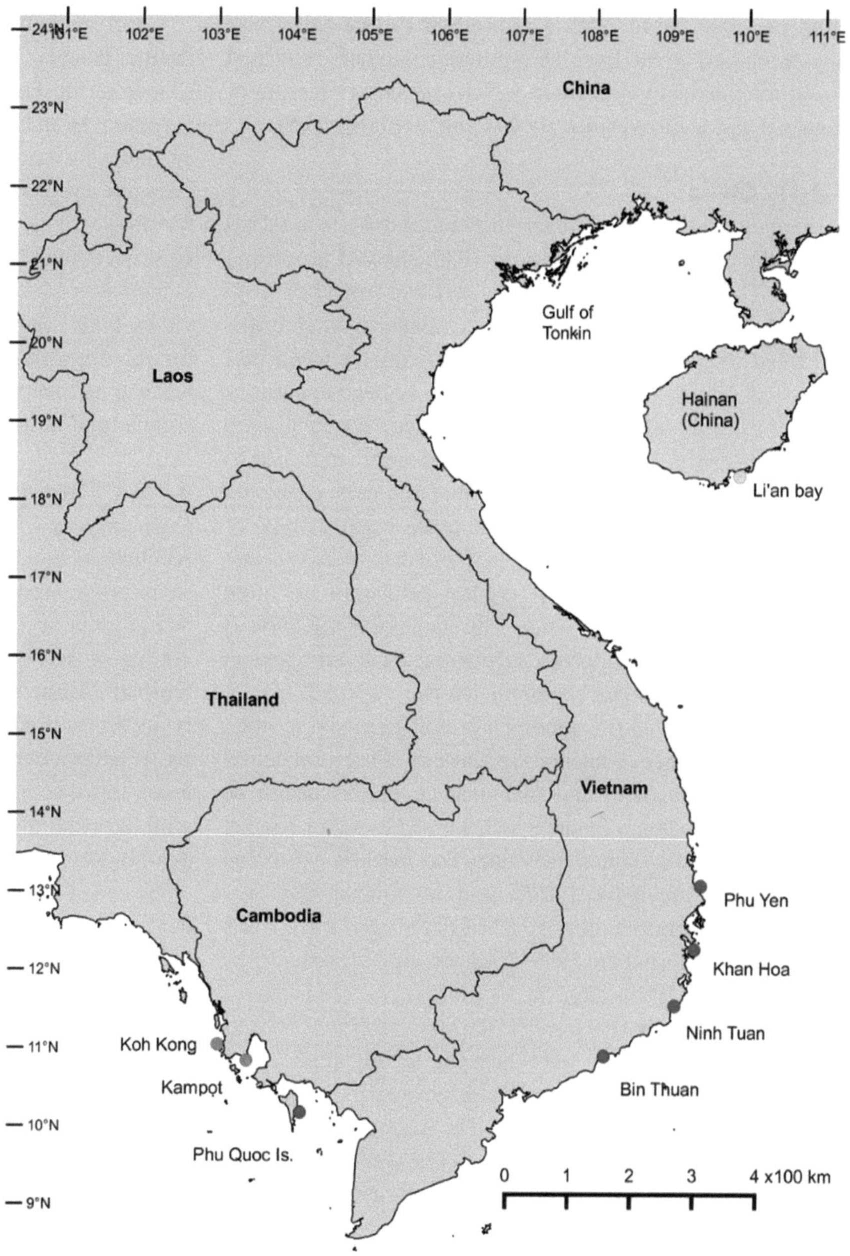

Fig. 4.5 Cultivation sites in Vietnam (red dots), Cambodia (orange dots) and China (yellow dot)

income for poor Vietnamese coastal areas (Nang 2005). Additionally, the nutrient extraction capacity of the cultivated biomass can also provide significant contributions to the ecological balance and bioremediation of eutrophic coastal areas. In this region, cultivation frames were set on the bottom of shallow areas during the cool season, with rapid water movement, high winds and waves; during the warm season there were sunny conditions but with limited water movement, minimal winds and wave energy. The cultivation of *Kappaphycus* led to reductions in the concentrations of ammonia, nitrite, nitrate, phosphate and phosphorus which widely ranged from 10–80% (Nang 2005). However, the quality of Vietnamese cultivated *Kappaphycus* remained high, meeting the required standards for sustainable commercial cultivation, particularly in southern Vietnamese waters.

4.2.1.5 Cambodia

The first company to plant *Kappaphycus* in Cambodia was the Korean-owned, Star Private, which was established in 1999 (The Cambodia Today 2005). Seedlings from Malaysia were first planted in Kampot. Several years later, production was high in 2003 and 2004 at 6,000 MT and 16,000 MT, fwt, respectively. However, by 2006, only 6,800 MT fwt were harvested. In 2007, the farmers were demotivated as to why their seaweeds were not growing as well as the previous years. As a result, production declined even further to almost nothing. The farmers allegedly blamed the construction of a nearby seaport for pollution which led to the poor growth of the seaweed (The Phnom Penh Post 2008). Three companies were engaged in farming, namely, Kampot-Star Private Enterprise, Cambodia Seaweed and Kampot Seaweed operating on 9,000 ha. In 2007, Kampot-Star Private Enterprise was purchased by the Enviro Corporation and later relocated its farming areas to Koh Khong. No reported production of seaweed has occurred since 2006 (Lang 2015; Fig. 4.5).

4.2.1.6 China

A two-year study of *K. alvarezii* introduced to Hainan Island (Fig. 4.5), from the Philippines in 1985, showed promising results. The average daily growth in weight reached 10% day^{-1} at a favorable SST of about 26 °C. Experimental results indicated that timely cutting of the growing seaweed biomass was an effective way to obtain new vegetative cuttings (seedlings) for cultivation which also accelerated growth (Wu et al. 1989). New cuttings (0.05 kg) were suggested to be suitable as "seeds" and that the remaining parts of the old plant were then to be harvested. Cultivation with a type of underwater raft designed for places where there was a strong, prevailing wind was used to expand cultivation to more regions within Hainan Province. The status of *Kappaphycus* farming in Hainan supports a substantial local carrageenan industry (Pang, personal communications)

Sullivan et al. (2001) reported a prototype model to illustrate the importance of integrating seaweed cultivation activities with management and maintenance of the health of coastal embayments, a model which was explored for Xincun Bay, southeastern Hainan Province. The nutrient extraction capacity of *Kappaphycus* cultivated in Xincun Bay was reported to be able to remove 53.8 MT of nitrogen and 3.7 MT of phosphorus during the 1999–2000 growing season.

4.2.1.7 Myanmar

Kappaphycus seedlings were brought to Myanmar from Cambodia in 2008 and reportedly originated from Vietnam (Ohno, personal communications) The Mergui Archipelago was the major site of seaweed farming (Fig. 4.6). MSC Korea, a private company engaged in agar and carrageenan processing, established seaweed farms in the locality using single bamboo rafts and later adopting the hanging long-line method (Kim, personal communications). Monsoon rains from May–October limited production, this was in addition to the annual occurrence of 'ice-ice' and also endo-epiphytism. It was only during the period November–April when farming was found to be favorable, however, production has remained at a very limited level (Ohno, personal communications).

4.2.2 The Pacific Islands

A history of introductions of *Kappaphycus* to the Pacific has been described by several authors such as Solly and Booth 1977; Why 1985; Luxton et al. 1987; Nelson 1988; Adams and Foscarini 1990; Smith 1991; Ram 1991; McLachlan 1992; South 1993; Eldredge 1994; Luxton and Luxton 1999; Ask et al. 2003a, b.

Carrageenophyte farming areas, as introduced in the Pacific Islands, are shown in the Fig. 4.7. Amongst the islands in the Pacific, Kiribati, Papua New Guinea and the Solomon Islands were described as having suitable sites available for eucheumatoid farms. However, there were no local nor national markets for the raw materials grown, there was however an increasing demand from international markets (i.e. demand in US and Europe; Table 4.1). The Solomon Islands was a candidate for good, commercial viability, providing high export earnings, together with a good livelihood for the farmers and thereby alleviation of their poverty. Kiribati had the same potential to adopt and be successful at commercial farming (Table 4.2).

4.2.2.1 The Cook Islands

Kappaphycus alvarezii was introduced to the Islands of Aitutaki in the late 1980s from Fiji (Eldredge 1994), but never progressed to a successful mariculture activity. A second attempt at re-kindling interest was made in 2001 when 700 kg of seaweed was brought from Tabuaeran Lagoon in Kiribati destined for Tongareva, Rakahanga and Pukapuka. However, in Tongareva Islands, this activity was refused by the Island Council at the last moment for fear of inadvertently introducing unwanted foreign marine organisms along with the seedlings, a fear which was used to protect the local pearl oyster industry.

Seaweed farming was tested in Pukapuka and Penrhyn but unfortunately it did not take off. It is likely that labor availability was a limiting factor and/or too expensive and that the costs of freight and farming equipment were also too expensive for the modest volumes that could be produced in the Northern Islands. Palmerston was considered the better location because of existing, established transportation/trade links due to parrot fish (Luxton 2001). However, despite a second attempt in 2001, the introduction of *Kappaphycus* farming to the Cook Islands was unsuccessful.

4.2.2.2 The Fiji Islands

Kappaphycus farming was introduced to Fiji in 1976 (Ram 1991), when *Kappaphycus* seedstocks were brought from the Philippines (Solly and Booth 1977). Field cultivation trials were carried out and maintained for more than two years at Telau Island, near Suva (Prakash 1990). Sadly, these farms failed due to their destruction by a cyclone in 1980 (Ram 1991).

Kappaphycus was reintroduced in 1984 by the Fiji Fisheries Division with the cooperation of Coast Biologicals (NZ) Ltd., with funding support from the Commonwealth Fund for Technical Cooperation (CFTC). *K. alvarezii* seedstock was imported from Tonga and trials were conducted successfully at Tavua, Rakiraki and Verata in Tailevu (Ram 1991; South 1993). The first commercial production began in 1986 at Tavua, Rakiraki, Kaba, Kiuva and Rewa. Considerable expansion took place in 1987, with farms established in new

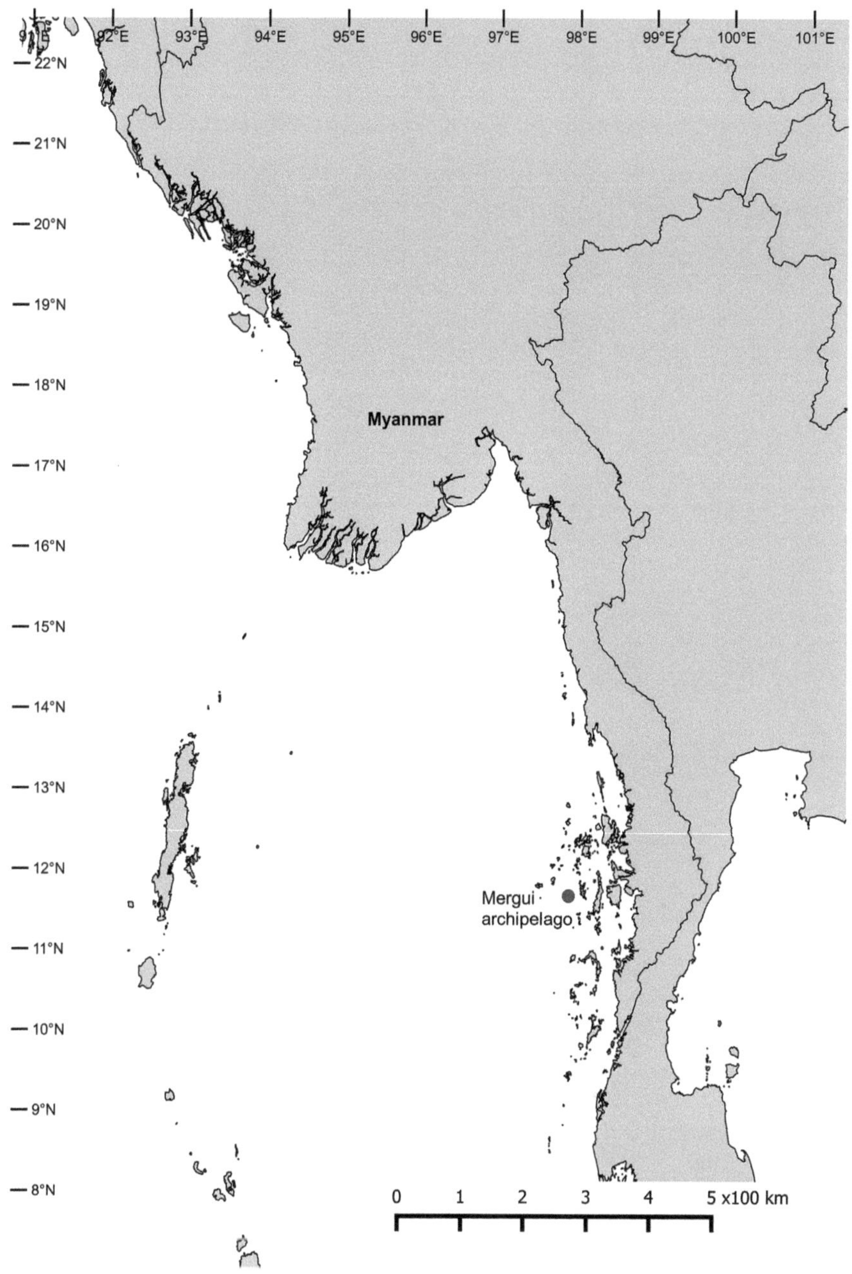

Fig. 4.6 Cultivation sites in Myanmar

areas such as: Moturiki, Ovalau, Bua, Batiki, Vanuabalavu, Fulaga and Ogea (Ram 1991; South 1993). However, there were several set-backs for *Kappaphycus* farming in these areas due to the effects of bad weather on cultivation and crop losses due to damage by cyclone-induced, high tides and heavy swells (Pickering 2005).

In 2005, there were four sites under cultivation: two were in the Lau Group, on the islands of Ono-i-Lau and Namuka-i-Lau; one was in Kadavu, near Kabariki Village and one was at Vanua Levu, near Namuka Village. Whilst commercial production is still carried out in these areas, the volumes are very small (i.e. less than 20 MT). It was projected that production would be around 100 MT (Lal and Vuki 2010).

4.2.2.3 The Kiribati Islands

Kiribati is regarded as having enormous marine fishery resources and extensive biological diversity (Adams and Foscarini 1990; Ram 1991). The economic and social development of the islands have been directly dependent on the availability and sustainability of marine resources. Developments in marine aquaculture such as seaweed farming, have been encouraged in order to minimize fishing pressure especially on the outer islands and villages with approximately 300 villagers.

K. alvarezii and *Eucheuma denticulatum* from Hawaii, which originally came from the Philippines, were first introduced to Kiritimati (Christmas) Island in 1977 by Dr.

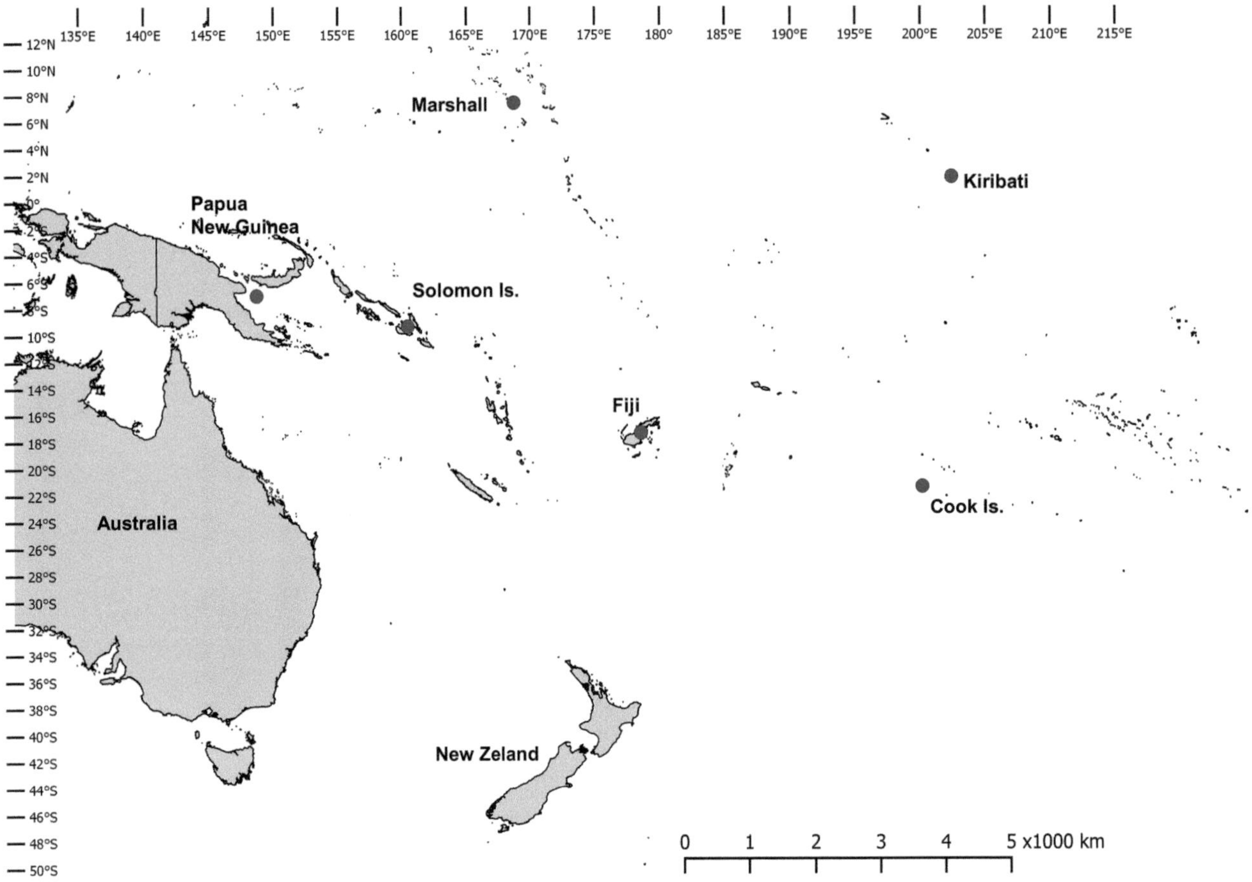

Fig. 4.7 The Pacific Islands involved in cultivation

Table 4.1 Opportunities for eucheumatoid seaweed cultivation in the Pacific Islands

	Commercial viability	Import substitution	Export earnings	Livelihood and poverty alleviation	Food security
Cook Is.	Low-medium	Zero	High	Good	Low
Fiji Is.	Fair	Zero	High	Good	Low
Kiribati	Good	Zero	High	Good	Low
Marshall Is.	Good?	None	Fair	Fair	Low
Papua New Guinea	Good	Low	Fair	Fair	Low
Solomon Is.	Good	Low	High	Good	Low

After www.hambreyconsulting.co.uk

Table 4.2 Potential contribution of eucheumatoid seaweed cultivation to development objectives

Islands	Site availability	Local markets	National markets	International markets	Comparative advantage	Production and market risks
Cook Is.	Fair	None	None	Good	Low	Medium
Fiji Is.	Fair	None	None	Good	Fair	Low; medium
Kiribati	Good	None	None	Good	Fair	Medium
Marshall Is.	Fair	Poor	None	Good	Low	Low; medium
Papua New Guinea	Good	None	None	Good	Fair	Low; medium
Solomon Is.	Good	None	None	Good	High	Low; medium

After www.hambreyconsulting.co.uk

Maxwell Doty (Why 1985; Uan 1990; South 1993; Luxton and Luxton 1999) and to Tabuaeran (Fanning Is., also from the Hawaiian seedstock) by Russell in 1977 (Luxton and Luxton 1999). Its commercial production began in 1986 on six of the Gilbert Group of islands namely: Tarawa, Butaritari, Aranuka, Abemama, Abaiang and Beru. Production peaked in 1991 at 1,020 MT, fwt, however, this declined to 400 MT in 1993. The reasons for the decline were cited mostly to be as a result of adverse weather conditions associated with the 'El Niño' effect which induced strong westerly winds and destroyed many farms. Other important reasons contributing to the decline in activities were late payments to farmers, lack of production inputs and weak production and marketing management (Bergschmidt 1997).

The Atoll Seaweed Company (ASC), owned by the Government of Kiribati, purchased seaweed from the farmers and shipped it to the capital of Tarawa, where it was pressed into bales for onward shipping to New Zealand and Europe. A number of institutions such as the Atoll Seaweed Company (ASC) Ltd., the European Development Fund (EDF), Central Pacific Producers Ltd. (CPPL) and some private companies, e.g. CP Kelco, Dragon World, the Asil Group Ltd., Honjin Intl. Ltd. and Betio Mini Mart Ltd., were involved in purchasing and marketing dried seaweed produced by the farmers in Kiribati (Komarudin and Hussein 2015). In 2008–2009, the Ministry of Fisheries Kiribati distributed seedlings to farmers in Abaiang; these originated from the same island and Onoto Island. In 2010, six villages, or about 300 households, were involved in farming seaweeds and contributed to the national production which increased to more than 400 MT from 2010 to early 2011. Since 2011, seaweed production decreased further; it was not because of the price, but rather due to technical, biological and environmental issues. The seedstock no longer continued to prosper in the local conditions in Abaiang, possibly due to continued vegetative propagation and loss of strain vigour.

Family-based farming was the norm in Kiribati; whole families were involved together in all farming-related activities. Men usually went to the seashore for planting, maintenance and harvesting, whilst women and children had the responsibility of drying, binding seedlings and sorting the dried seaweed.

4.2.2.4 The Marshall Islands

E. denticulatum and *K. alvarezii* were introduced into Mil and Likiep in Majuro Lagoon in 1990 (Eldredge 1994). There is no other information available relating to those early trials. In 2002, the Marshall Islands began a new seaweed farming project with seedlings imported from Kiribati under the expertise of Mr. James Uan from Kiribati. The Food and Agriculture Organization (FAO), Marshall Islands Marine Resources Authority (MIMRA), Marshall Islands Development Bank (MIDB) and the College of Marshall Islands (CMI) supported this initiative with the training of staff and community members. This resulted in the establishment of trial farms on several of the atolls. However, farming activities did not progress that much; the latest country report made by Hambrey Consulting group in 2011a (www.hambreyconsulting.co.uk) reported the following opportunities and constraints:

Opportunities

1. Trials suggested a number of locations to provide adequate conditions for farming.
2. Dried produce not so susceptible to unreliable shipping
3. Simple, low-cost technology, relatively suited to rural farmers
4. There was an established international market for dried seaweed for processing into a range of natural gels

Constraints

1. A number of technical issues remained to be resolved in terms of site selection and husbandry including the control of grazing and negative impacts on sharks affected by grazing control measures.
2. Biosecurity concerns relating to the introduction of non-native seedlings
3. Comparative disadvantage related to large-scale and well-organized production in other production centres such as Indonesia.
4. RMI production would be minute on an international scale and it would be difficult to ensure that a reasonable market value was paid for relatively small volumes of local product.

4.2.2.5 Papua New Guinea

In 2010 a new venture was established by a Malaysian Company at Milne Bay Province. There were initial problems with rapid SST fluctuations at one site, but with a change of locations to Suau Island and Trobriand Islands, the imported seedlings were reportedly growing well (www.hambreyconsulting.co.uk). The company invested significantly in supplying materials and providing technical assistance to the farmers through the provision of Malaysian technicians living on site.

There appeared to be fair opportunities for cash generation which could then be used for poverty alleviation. Foreign exchange earnings were expected to flow through seaweed cultivation as an activity in favorable locations around Papua New Guinea. Reasons cited for the optimism included: a reasonable infrastructure for transportation; suitable environmental conditions; limited alternative income generating opportunities; low labor costs and positive economic ambitions of the coastal fisherfolks. Investments by a Malaysian company in Milne Bay Province suggested that there was both a strong market demand and a significant role for the private sector to play in facilitating the start-up of family/community-based production (Hambrey 2011b, see www.hambreyconsulting.co.uk). The economics of carrageenophyte cultivation appeared to be favorable, although the cost of local freight (i.e. from the growing site to local collection/shipping center) were suggested to be relatively high. The outcome of this enthusiasm remains unknown.

4.2.2.6 The Solomon Islands

Carrageenophyte farming was tested by the UK ODA (Overseas Development Agency) in the late 1980s in Munda, Vona Vona and Rarumana, but suffered from rabbit fish grazing and low prices/low volumes produced and was ultimately abandoned in the early 1990s (Hambrey 2011c, see www.hambreyconsulting.co.uk). At that time, product was shipped to Fiji before being forwarded to the Philippines. Under the EU Rural Fisheries Development (RDF) project in the 1990s and early 2000s, fisheries centers were upgraded in several locations and opportunities for diversification and increased use were explored. As a result, carrageenan-bearing seaweed production was re-started at Rarumana using residual stocks from the previous initiative; 19 MT, fwt was produced in 2001–2002. This aroused the interest of a local trading company (traditionally engaged mainly in collecting copra) which began to buy and export seaweed, with an exclusive license. Production was then extended to Wagina – supported/facilitated by the Fisheries Department, the Nature Conservancy and private sector traders.

In 2010, four additional trading companies obtained licenses to export 'cottonii' seaweed, and this, along with more favorable international markets, saw beach prices rise significantly from US\$ 3–4 kg^{-1} to US\$ 5–6 kg^{-1}, generating significantly increased interest and activity in several locations. Seaweed cultivation became a relatively well-established enterprise at Wagina, with smaller amounts also produced in Rarumana, Buena Vista (by the Sandfly Group). Production in 2010 amounted to "several hundred (metric) tons". This represented a significant income to the communities involved and in some cases, smaller groups or individuals involved.

Prices for *Kappaphycus* or *Eucheuma* are relatively good and artisanal seaweed farming was shown to be clearly economically viable, with a potential to make significant positive contributions to poverty alleviation, gross domestic product (GDP) and foreign exchange. Good sites were known to be available in the island grouping (e.g. good access, shelter, water exchange, with a lack of grazers), however, testing new sites was a time consuming and risky activity. Switching activities and efforts between seaweed cultivation, bêche-de-mer, other fishery products and copra, each according to their price and availability was a characteristic of the innovative and adaptable coastal village economies. Mechanisms were badly needed in order to maintain eucheumatoid seedstock during periods of relatively low prices and reduced activity (Hambrey 2011c, see www.hambreyconsulting.co.uk)

Kronen (2013) reported a brief, but comprehensive, history of *Kappaphycus* farming on the Solomon Islands with the following milestones:

1988 – First trials undertaken by the Overseas Development Agency (ODA) of the United Kingdom at Vona Vona Lagoon and Rarumana village in the Western Province, in cooperation with the Ministry of Fisheries and Marine Resources (MFMR). The one-year project demonstrated good growth for *Kappaphycus alvarezii* seedlings imported from Fiji (Tiroba and McHugh 2006) however, most trials were negatively affected by fish grazing.

2000 – The Aquaculture Division of the MFMR was established and started collecting seedstock remaining from the 1988 growth trials in Vona Vona Lagoon.

2001 – The Aquaculture Division carried out growth trials in Rarumana and after 2 years, 600 kg of dried seaweed was produced in this area.

2002 – The Rural Fishing Enterprise Project (RFEP) funded by the European Union (Member Organization) became involved in seaweed farming. In cooperation with the Secretariat of the Pacific Community (SPC) and MFMR, the RFEP conducted a seaweed training workshop in November, targeting 30 fisheries officers. Successful growth trials under the RFEP at Rarumana were completed in 2003.

2003 – The remaining funds (SBD1.5 million, EU STABEX funds) from the RFEP project were allocated to provide further support in the framework of a one-year seaweed farm development project. The project provided farm materials, outboard motors and a warehouse in Rarumana; and the first PFnet (broadband, e-mail) system was set up to ensure communication between producers and buyers.

2004 – In July, a warehouse was built in Wagina where the second PFnet was set up. A feasibility study for further support from the European Union (Member Organization) was carried out.

2005 – By early 2005, there were about 130 farmers in Rarumana and the Shortland Islands (Western Province) plus 300 farmers in Wagina, Choiseul Province; seaweed farming had also expanded to Malaita and Makira-Ulawa. About seven export licenses were approved; however, only one exporter (Solomon Seaweed) renewed its license in 2006. Agents were paid a commission based on production. Export of seaweeds was tax-free.

2005 – In July 2005, a 3-year SBD15 million seaweed commercialization project (CoSPSI) funded by the European Union (Member Organization; STABEX funds) began, which was extended until January 2009, with a focus on sites in Ontong Java Atoll, Reef Islands, Malaita. The project also supported continued seaweed farming development at Wagina and Rarumana.

2005 – With assistance from CoSPSI, the International Waters Project (IWP; Global Environmental facility, SPREP) established a seaweed farming operation in the eastern Marovo Lagoon, within the framework of its community development approach, which remained one of the major producers.

2006 – May, the farm-gate price for seaweed dropped from SBD 2.00 kg^{-1} to SBD 1.50 kg^{-1}, due to increased fuel prices, which in turn increased national and international freight costs.

2007 – A severe earthquake and associated tsunami on April 2, resulted in the loss of some of the best seaweed farming areas in the Western Province (Rarumana), and an estimated loss of 20–30% of production within the CoSPSI project.

2008 – In July, the farm-gate price for carrageenophyte biomass increased to SBD 3.10 kg^{-1}.

2009 – The remaining CoSPSI funds were used to support a seaweed farming adviser based at the MFMR for 1 year (April 2009–March 2010).

At the time of writing, the Solomon Islands was the biggest, single producer of non-edible seaweed in the Pacific region (Algae World News 2015), mainly because of the strong support provided by the local government (MFMR, Ministry of Fisheries and Marine Resources), as seen from the following factors: (1) Training of 609 seaweed farmers and 25 seaweed buying agents; (2) Writing and distribution of a training manual for all seaweed farmers; and (3) Assessment of 15 new seaweed farming sites and development of another 15 new seaweed farms; and provision of seaweed farming material including canoes, outboard motors and water tanks to seaweed farmers. The MFMR added that the total production of seaweed in 2014 was 1,520 MT, worth about SBD$5.6 Million to the seaweed farmers (Algae World News 2015).

4.2.3 Western Indian Ocean

In the Western Indian Ocean Region, although seaweed farming was documented in 1970s and further experimented in mid-1980s, it was not until 1989 that commercial farming started (Fig. 4.8). Commercial farming of *Eucheuma* and *Kappaphycus* in Tanzania was initiated by private entrepreneurs from Denmark and the Philippines in 1989 using an imported strain of the then *Eucheuma* genus (which was a collective name given to *Eucheuma* and *Kappaphycus* species and variants) from the Philippines. It was necessary to import seedstock because the local, indigenous strain could not provide the yield required by the need for cultivated, commercial volumes (Msuya 2013). Private entrepreneurs situated the first farms on the East Coast of the Zanzibar Islands, in the villages of Paje and Jambiani.

To ensure the largest biomass production (it is interesting to note that the quality of the carrageenan produced was not considered at that time), varieties of *Eucheuma* and *Kappaphycus* were imported from the Philippines, under strict quarantine regulations. The first exports of dried carrageenophyte seaweeds from the Western Indian Ocean was in 1990 (one year after the first commercial farms were established) when 808 MT dry were exported from Zanzibar (Msuya 2013).

4.2.3.1 Tanzania

The success of seaweed farming in the two pilot villages on Zanzibar Island triggered the rapid expansion of farming practices to other areas including the sister island of Pemba (Lirasan and Twide 1993), as well as nearby areas of the mainland, such as Tanga and Bagamoyo (between 1992–1994). In 1995, it was found necessary to begin seaweed farming in the southern part of mainland Tanzania. This was executed by Dr. Flower Msuya of the University of Dar es Salaam, Institute of Marine Sciences who was contracted by a project called: "Rural Integrated Project Support (RIPS)", funded by the Finnish International Development Agency (FINNIDA) to farm the first seaweed seedlings at Mtwara, Lindi and Kilwa Districts. A feasibility study was undertaken and nursery farms were established in 1995 (Msuya 1995); in 1996 farming expanded to more than ten villages (Msuya 1996). The chronology from the beginnings to expansion of seaweed farming from Zanzibar to other parts of Tanzania has been documented by multiple authors, viz. Eklof et al. (2012), Msuya (2010), Msuya (1995, 1996) and URT (2005). By 1997, carrageenophytes were farmed in all coastal areas of Tanzania. From the first export of 808 MT recorded in 1990, seaweed production continued to rise maximizing at 15,000 MT in 2012. In the ensuing years, production fluctuated between 11,000–13,000 MT annually.

4.2.3.2 Madagascar

From Tanzania, eucheumatoid farming expanded to other countries in the Indian Ocean. The first country to pick up commercial farming was Madagascar starting in 1997. Support for commercial production in Madagascar came from a project aimed to improve the income of coastal communities (Améliorationdes Revenus des Populations Littorales), this was funded by the European Development Fund (EDF) and implemented by the Food and Manufacturing Company (FMC) of USA, working with local institutions (Msuya et al. 2014). In 1998, strains of two cultivars were imported from Zanzibar to Madagascar – for exactly the

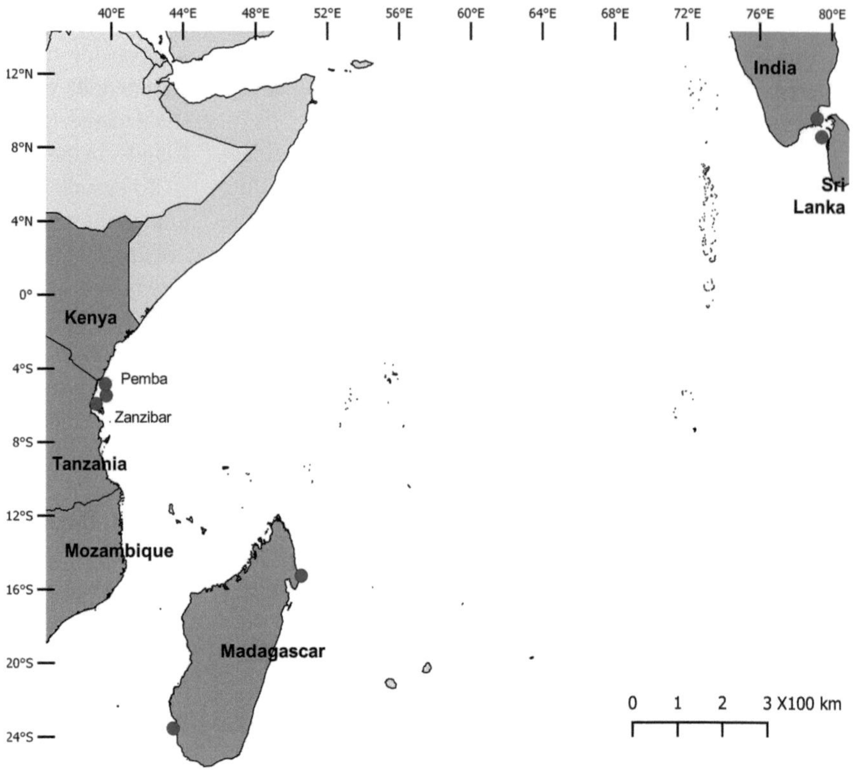

Fig. 4.8 Cultivation sites in the Western Indian Ocean region

same reasons that Zanzibar had imported seedstocks – the indigenous species could not produce enough volume to sustain a commercial business (Randriambola and Rafalimanana 2005).

4.2.3.3 Mozambique

After Madagascar, it took some time for other Indian Ocean countries to begin commercial farming of carrageenan-bearing seaweeds. In 2002, farming operations commenced in Mozambique. Cultivation began in Pemba, in 2002 and expanded to Nampula, both sites located in northern Mozambique.

4.2.3.4 Kenya

Kenya also joined the industry of producing carrageenophyte biomass for processing, starting commercial farming in 2010 on the southern coast. Efforts to support commercial production were made by a number of organizations such as ACT Kenya, SUDRA, and East African Wildlife Society (EAWS). In some instances the self-help groups system was used. Since 2014, commercial farming has been supported by the Kenya Coastal Development Project (KCDP), funded by the World Bank. At the time of writing (Jan 2017), there was commercial production in three villages; FAO supported commercial production in Kenya under a Technical Cooperation Project with the Kenyan Government. Kenya also had a local buyer, a company called East African Seaweeds. This solved the problem of dependency on buyers from Tanzania; there is potentially a bright future for commercial seaweed farming in Kenya.

4.2.3.5 India

Cultivation of commercial carrageenophytes began in India in 2000 when Pepsi Foods Ltd. (a subsidiary of Pepsico India Holdings Ltd.) became interested in the cultivation of *Kappaphycus alvarezii*. The economic viability of open-sea cultivation was proven and the technology of commercial cultivation was transferred to PepsiCo, India Holdings Ltd., in 2001 who then initiated pilot-scale farming along the Palk Bay side of the Mandapam coast (Eswaran et al. 2002). The cultivation business rights for *K. alvarezii* were transferred by Pepsi Foods Ltd. to Aquagri Processing Pvt. Ltd., a company formed by a group of entrepreneurs led by a former PepsiCo executive, Mr. Abhiram Seth – as a tripartite agreement in August 2008. This company was successful in expanding cultivation activities to Tamil Nadu.

The farming systems in India was seen to evolve in concert with innovations in biomass processing. The bio-economic analysis for individual and artisanal culture of *K. alvarezii* in India reported a positive, net-present value only 6-months after inception of the business, with an internal rate of return at 210% (Jara and Riffo 2016). This analysis further endorsed the added benefits that *K. alvarezii* farming could lead the diversification of coastal livelihoods where opportunities were known to be sorely limited. The socio-economic impact of *K. alvarezii* farming in India was investigated in depth as a part of the study sponsored by FAO (Krishnan and Narayanakumar 2010).

4.2.3.6 Sri Lanka

In 2008, the Sewalanka Foundation, a leading Sri Lankan NGO, in collaboration with Abeysinghapura-Panama Fisheries Cooperative Society and the United Deep Sea Fisheries Cooperative Society in Pottuvil, started a pilot assessment of *Kappaphycus* (*alvarezii*) using a cage method at Panama and Ullei (in Pottuvil) on Sri Lanka's east coast. There was initial success with good growth, however, grazing, drying and marketing were perceived as problems (Samarakoon 2009). Though the Foundation envisioned this aquaculture activity to bring economic benefits to the coastal communities, they also believed that an ecological impact assessment was required in order to prevent potential, future degradation of the marine ecosystem.

The germplasm of *K. alvarezii* was imported to Sri Lanka in 2011 from Malaysia, following procedures for introduction and quarantine. Material was acclimated to Sri Lankan waters by maintaining trial plots in various locations of the Kilinochi and Jaffana regions. Commercial farming of *K. alvarezii* began in 2012 after carrying out thorough feasibility studies using floating bamboo raft and off-bottom, monoline methods (Shanmugam et al. 2016). The average daily growth rate ranged between 4.0–5.5% day^{-1}, a good indication that commercial farming was indeed viable. As of 2017, only 50–60 seaweed farmers were still engaged in this activity, with a total production of 120 MT for 2014–2016. The possibility for the expansion of *K. alvarezii* cultivation to other coastal parts of Sri Lanka remained promising; Shanmugam et al. (2016) claimed that engagement in the activity could create an alternative livelihood for a large number of coastal people, which could also earn foreign exchange through either the export of dried seaweed biomass or by adding value through products such as agricultural biostimulants and production of the processed food additive carrageenan.

Since 2014, Sri Lanka's Hayleys Aquagri has been active in promoting seaweed cultivation in the western part of the country, i.e. Vallaipadu and Eranamatha Nagar in the Mannar Sea Basin by supplying propagules, rafts, drying frames, training to the families and assistance in the marketing of the dried seaweed. In June 2015, the same company and the United Nations Development Programme (UNDP) agreed to collaborate in expanding seaweed farming to the islands of Analaitivu, Eluvaitivu and Nainaitivu in order to provide livelihood opportunities for 40 families, eight of which were households with women at their head. This project encouraged communities to diversify livelihood opportunities in an environmentally sustainable manner, based on a: 'Community-Based Aquaculture for Livelihood Diversification' strategy. As of 2017, Sri Lanka had vigorously adopted *Kappaphycus* farming in suitable areas.

4.2.4 Central America

For this chapter, we use the term "Central America" not in its purest geographical sense, but to delimit those countries sharing a common coast with the Caribbean Sea, from México to Panama and also including the Caribbean Islands. At present, there is no facility for the industrial extraction of phycocolloids in the Caribbean. In this region, the history of utilization of seaweeds, particularly members of the genus *Eucheuma*, is relatively recent in comparison to other geographical areas. *Eucheuma isiforme* received much attention as being the main genus providing biomass for the commercial production of *iota*carrageenan. This alga can be found naturally occurring throughout the Caribbean where it is locally abundant (Cheney 1988) and commonly found seasonally, as beach-cast material during the north winds. Seaweeds have been cultivated for food in the Caribbean, this is a tradition dating back to the first half of the nineteenth century. These activities began in St. Lucia using techniques and methods of cultivation developed by the Fisheries Management Unit/International Development Research Centre: 'Seamoss Project' (Smith 1992). Various species of *Gracilaria* were grown successfully, but their susceptibility to epiphytism and poor agar qualities made their further utilization impossible. In 1991, preliminary results with *E. isiforme*, which originated from Belize, showed higher growth rates (i.e. doubling time of 10–12 days, Allan Smith unpublished data) and provided more resistant seaweeds. Therefore, commercial farmers in St. Lucia began to use it on their farms. By 2000, *E. isiforme* became the principal farmed crop. With the exception of *E. isiforme*, no other domestication of a seaweed species was undertaken in the region. The native species, had been harvested only for local consumption on some of the Caribbean Islands. However, after 1991, select carrageenophytes were introduced by people aware of their potential benefits and relative ease of cultivation in several countries including Mexico (Muñoz et al. 2004), Panama (Batista de Vega 2009) and several Caribbean Islands (Fig. 4.9).

Here we review the published information available for countries where native and introduced species of *Kappaphycus* and *Eucheuma* have been studied so far.

4.2.4.1 The Caribbean Islands

Until the 1980s, "sea moss" was marketed almost exclusively in a dried, bleached form. In English-speaking islands and a few Central American locations, including Honduras and Panama, various red algae were harvested for the preparation of traditional drinks and puddings (Smith 1997).

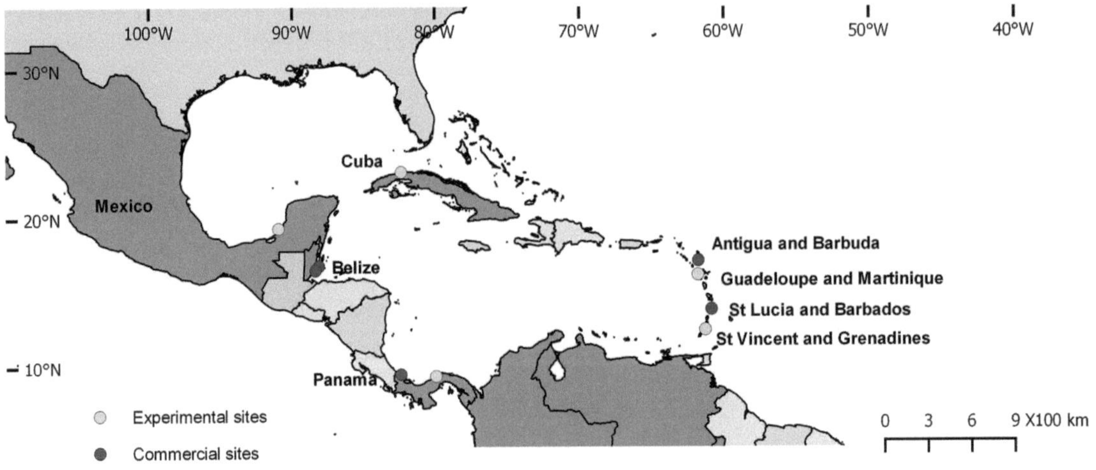

Fig. 4.9 Experimental and commercial sites of cultivation in Central America

Seaweeds and their drinks were already known in most Caribbean Islands as "sea moss", as well as "Irish Moss" (but not *Chondrus*) in Jamaica, or simply as "seaweed" in Belize. It was sold locally, or exported to the more populous islands of the region where wild stocks had been depleted due to un-restricted, over-harvesting. After 2001, processing facilities were established or expanded on a number of islands, including Jamaica, Barbados, Grenada, Antigua and Barbuda, Trinidad and Tobago, St. Vincent and the Grenadines, St. Lucia, and Dominica. As a result, a wide variety of packaged and bottled "sea moss" products became available. However, the supply of commercial species from wild stocks also declined over the same period and this limited the output from the smaller processing facilities which existed on many islands. Jamaica rose to be the region's largest processor of sea moss products and imported an average of 13 MT of *K. alvarezii per* year from the Philippines to meet the demand (Smith and Rincones 2006). Given the shortage of raw material for processed products in the region there was an evident potential for further expansion of seaweed mariculture. The development of cultivation to provide raw materials to the extract industry was limited to those areas with a depressed economy, with very low incomes; due to the value afforded raw material, it is unlikely that production for the phycocolloid industries would ever be viable in the countries where seaweeds are already used for food.

4.2.4.2 Guadeloupe and Martinique

The first introduction of carrageenophytes from the Indo-Pacific into the Neo-tropical Realm, in a natural environment, occurred in Guadeloupe, French West Indies, in 1981 (Areces et al. 2014). Olivier Barbaroux reported the introduction of two *Eucheuma* species (e.g. *E. denticulatum* and *E.cottonii*) in Guadeloupe (Barbaroux and Perez 1983). Barbaroux et al. (1984) further described cultivation trials in the West Indies where a raft system was developed, providing an effective surface area of 2,500 m^2 consisting of a flexible rope framework, which operated at depths greater than 10 m. Nylon culture ropes would be seeded with 19,000–22,000 plants. The net suspension technique was also used, taking an idea from Amat (1981); a small mesh net (3 cm) was maintained 1 m below the surface by polystyrene floats. The yield was comparable to the raft method, except there were effectively no losses, whereas the floating unit had approximately 16–20% loss of plants over the cultivation period. It should be noted that this method has also been utilized with great success in Brazil (Goes and Reis 2011). Growth rates recorded from the Guadeloupe pilot-project were however low, when compared to other sites of commercial cultivation (viz. 1.5–2% day^{-1} for *E. denticulatum* and 4% day^{-1} for *E.cottonii*). Perez et al. (1992) concluded that the economic return on effort for the cultivation of the carrageenophytes would be too low for the biomass production and hence not competitive with Asian sources of raw materials.

4.2.4.3 Santa Lucia and Barbados

In the mid-1990s, a program was conducted in St. Lucia to identify sea moss species suitable for the market that would be less susceptible to the seasonal appearance of epiphytes, which reduced the quality of the common *Gracilaria* spp. produced in many areas. Preliminary results with *Eucheuma isiforme* from Belize showed that under cultivation conditions, plants remained completely free of silt and epiphytes and furthermore maintained high and stable growth rates. Accordingly, *E. isiforme* had already replaced *Gracilaria* spp. as the principal farmed crop for sale in a dried, bleached form. Coastal communities on the east coast of St. Lucia formed the Praslin Seamoss Farmers Association and also established a processing and bottling facility aimed at both local and export markets (Brown 1999). Although processing produced traditional concentrates and beverages, business developed informally and there were no accurate records of the quantities involved. However, the dried, bleached seaweeds commanded high prices and the harvest and local processing was viewed as being economically important to many coastal communities (Smith 1998).

4.2.4.4 Antigua and Barbuda

E. isiforme was also harvested commercially in Antigua and Barbuda until the 1980s, primarily from the Codrington Lagoon on Barbuda. It was used locally and exported for traditional food applications in the region, however this enterprise was discontinued because of severe over-harvesting. The most widespread use of seaweeds in this Caribbean region was for food, as in other Caribbean Islands (Smith 1998).

4.2.4.5 St Vincent and Grenadines

Both *Eucheuma isiforme* and *Kappaphycus alvarezii* were the focus of sea moss producers on Union Island, St. Vincent and the Grenadines. The project: "Developing the Economic Viability, Human Capacity, and Environmental Sustainability of a Seamoss Cooperative" on Union Island, St. Vincent and the Grenadines, funded by GEF-PNUD Small Grant Program (2010–2012), was intended to improve the viability of the "sea moss" industry through cultivation of both of these genera. The potential was thoroughly investigated by research and experimentation, capacity development, improved monitoring of human impacts on the marine environment and improvement of community awareness of the connections between environmental pollution and livelihoods. The project had the expected outcome of creating a sustainable sea moss industry in the area.

4.2.4.6 Belize

Eucheuma isiforme was harvested commercially in Belize; approximately 800 kg (air-dry weight) were exported annually to the USA for use in health-food applications during the 1990s (Smith 1998). Unfortunately, this exportation ceased, but harvesting continues to supply local consumption. Natural *E. isiforme* populations grew close to the mainland; they were however, severely depleted making the scarce biomass even more "valuable". Seaweed farms off the shore of the Twin Cayes in the South Water Caye Marine Reserve were supported by a non-profit group called the: "Dangriga Development Initiative" (DDI). The project was also supported by the Belize Fisheries Department with the intention to provide a supplementary source of income for the residents and fishermen surrounding the South Water Caye Marine Reserve. Potentially, the large-scale production of this product was a feasible option since dried seaweed was already sold at the central market in Dangriga Town. Clean, dry and bleached sea moss fetched relatively high prices locally, commonly retailing for around US\$8 kg^{-1} (Smith 1998). The establishment of marine protected areas which included existing harvesting sites on the Belize Barrier Reef led to renewed interest in cultivation in order to provide alternative sources of economic activity and income. The project entitled: "Sustainable and Environmentally Responsible Production of Seaweed (*Eucheuma isiforme* and *Gracilaria* spp.)" in the marine areas adjacent to Placencia Village, was supported through GEF UNDP Small Grant Program to revitalize the Placencia Cooperative. By establishing 10 sustainable seaweed farms near Placencia they attempted to resolve two problems: the decline of commercial fishing stocks in and around the Southern Belize Barrier Reef System World Heritage Site (Gladden Spit & Silk Cayes Marine Reserve and Laughing Bird Caye National Park) and the reduced productivity of the Placencia Producers Co-operative. The outcome of these initiatives remain unknown at the time of writing.

4.2.4.7 Cuba

After trials with native species (i.e. *Gracilaria domingensis* and *Bryothamnion triquetrum*), the introduction of *Kappaphycus striatum, K. alvarezii* and *E. denticulatum* was made to Cuba, from the Philippines, in 1991. Cultivation trials on the coast north west of Havana were conducted in order to determine if the appropriate environmental conditions for cultivation were present (Areces and Céspedes 1992). With the exception of *E. denticulatum,* the other two introductions were found to adapt successfully to Caribbean conditions. In field tests, two long-line approaches were used for farming with ropes placed at either, or near, to the surface, or along the bottom. When the experimental farming was extended at pilot-scale, only the raft techniques were used over a two-year continuous cycle of *Kappaphycus* production. Harvesting and re-seeding of the lines occurred every 45 or 60 days; daily growth rates for both *Kappaphycus* species exceeded 6% day^{-1} (Serpa-Madrigal et al. 1997; Serpa-Madrigal and Areces 1998). Cuba also conducted tests to evaluate the potential ecological risks of these introduced species Many of the studies by Cuban researchers into quarantine procedures and the ecological impacts of introductions remain unpublished. A summary of research results on the development of guidelines for quarantine procedures and site testing was published by Areces et al. (2014). As a result of the recommendations, *Kappaphycus* remains banned from cultivation in Cuba, the reasons are unknown at time of writing, but likely due to political/environmental concerns, reasons rather than the potential to cultivate.

4.2.4.8 Panama

In Panama, around the Province of Colon, as in other parts of the Caribbean, the harvesting and use of sea moss seaweeds has been part of the daily tradition and diet of the local Afro-Antillan communities. The "pearl moss" (*Chondrus crispus*) from Europe was exported to the Caribbean by soldiers who went to the Panama Canal Zone in order to support the efforts of the First and Second World Wars. *Chondrus* was also called "Isinglass", from the Dutch word meaning fish gelatin. At the time of writing (March, 2017), at least two ethnic Panamanian groups continue to consume seaweed: the Gunas and the West Indians (Batista de Vega and Connor 1990). A total of 37 species of seaweed were recorded as being commonly used by the Gunas people for medicinal and cultural purposes (Batista de Vega 2006).

A seaweed program of the Institute of Tropical Research, Smithsonian Institution (SI) operated at the Galeta Point Marine Laboratory from 1979–1982 and reported a list of nearly 1,500 species, some of commercial importance (Hay 1981). Amongst the species selected for culture, *Gracilaria* sp. was used to establish the first experimental farm in Central America within the community of La Playita, in the city of Colon. In 1995, an experimental seaweed farm around the District of Colon, at the northern entrance of the Panama Channel was established. In 1996, after the introduction of *Kappaphycus alvarezii*, Gracilarias de Panama S.A. involved Guna Indians and Afro-descendant communities of the surrounding areas to expand the cultivation practices. The seaweed cultivation project was organized into 14 farms called "polygons" by the Maritime Authority of Panama (MAP) and monitoring was provided by the Ministry of Environment (MiAmbiete), the autonomous entity of Panama with a responsibility for natural resources in conjunction with the Aquatic Resources Authority of Panama (ARAP). The area chosen for these activities remained one of the most monitored areas in the tropics, with more than 80 years of data collected by the Panama Channel Authority and long-term monitoring projects by SI (http://www.stri.si.edu/sites/esp/data_access_policy.htm). Fourteen environmental parameters were recorded directly at the farming sites, e.g. seawater temperature (both at the surface and depth), salinity, pH, velocity and direction of the currents, were part of the evaluation since they could be directly correlated to growth and the quality of the carrageenan obtained from the cultivated biomass. These data were analyzed by Contreras and Rosenthal (2003). Monitoring from 1999–2009, by the Heltmholtz-Zentrum Geesthacht Center for Material and Costal Research in Germany was undertaken, they also installed an oceanographic buoy in Bahia Las Minas, near the seaweed farms which further contributed to the collection of data.

Experiments with *K. alvarezii* were initiated with indigenous material donated by fishermen from around La Playita and these were conducted within a lagoon open to the sea, surrounded by mangrove ecosystems, with a sandy bottom. A rectangular, semi-fixed structure constructed from wooden stakes, secured into the substratum and ropes alternating with buoys to form a grid and fixed structures constructed using PVC pipes, were all evaluated for farming purposes. *In situ* growth rates were measured for a series of experiments in order to establish the effects of various environmental parameters and seasonality on the daily growth rate. These studies were also conducted in relation to planting structures (fixed vs. semi-fixed); the presence or absence of epiphytes and initial seeding density (e.g. 50 and 100 g). Daily growth rates were compared in five weather conditions during the annual seasonal cycles of the area, viz. "dry season" (December 15–March 15); "dry to wet season" (March15–July 28); "very wet season" (July 28–August 15); "wet season" (August 15–October 15) and "wet to dry season" (October 15–December 15). Significant differences in the growth rates were observed between seasons and by site. Moreover, significant differences between daily growth in relation to current direction were also found, as well as in relation to the fixed structures used; the study period was extended from 2005–2008 and was also conducted at multiple sites. The percentage yield of carrageenan derived from the cultivated biomass was found to vary in relation to the duration of the cultivation (i.e. number of days since planting). The molecular weight of the extracted carrageenan was found to increase in relation to culture duration. These results allowed the farmers to extend the cultivation model to other places, with the intention of planting at lower cost with a greater production (Batista, own obs).

Kappaphycus farming in Panama evolved into another successful commercial endeavor called: Panama Sea Farms (PSF), which was a company working in Bocas del Toro with Indian communities. Gracilarias de Panama S.A. and PSF joined forces to make carrageenophyte seaweeds an export product which could generate a cash income for the inhabitants of coastal areas; encouragingly, with the potential great social impact and also revenue for the country. Both Gracilarias de Panama and PSF worked locally to improve the drying process, transforming the wet alga from 92 to 32% moisture content, which was stable point for storage and handling. At the time of writing (April, 2017), these two companies purchased all production available from local producers and export the dried seaweed to Malaysia (Teejoy Marine Company, in partnership with PSF) in order to produce semi-refined carrageenan (SRC). Despite the high demand for carrageenans, Panamanian seaweed scientists, farmers, buyers and processors still lack access the tools required to accurately and inexpensively assess the carrageenan content of their own seaweed crops throughout all stages of growth and refining. A recent technique capable of measuring and characterizing the carrageenan content *in vitro*, using confocal laser scanning microscopy was published to identify and broaden the knowledge of the carrageenan content located in the cell wall of *K. alvarezii* (Batista de Vega et al. 2016).

4.2.4.9 Mexico

In Mexico, the most studied carrageenophyte has been *E. isiforme*, both in terms of the carrageenan content, seasonality and chemical characteristics (Robledo 2006); further research is also underway. A biomass assessment at the Campeche Banks, west coast of the Yucatan Peninsula, dates to the late 1970s and reported a density of *E. isiforme* at approximately 6.7 kg m^{-2}, with an estimated total biomass of 35,794 MT fresh weight. Thereafter, an incipient exploitation of *Agardhiella* sp. and *E. isiforme* occurred from 1977–1980 in some locations of the Campeche coast and the material was exported to Denmark to produce ι-carrageenan.

Two forms belonging to the genus *Eucheuma*, i.e. *E. isiforme* and *E. isiforme* var. *denudatum*, both native to the Gulf of Mexico and the Caribbean, were studied and cultivated on the southern coast of Florida (Dawes et al. 1974, 1977) and in the Yucatan Peninsula (Perez-Enriquez 1996a). The distribution and abundance of *E. isiforme* was characterized by its seasonality. Biomass availability of *E. isiforme,* over an annual cycle, ranged from 0.98–3.20 kg m^{-2}. These data suggested that the populations were amenable to commercial exploitation in some areas. The largest *E. isiforme* standing stock occurred during the cold season (November–December) when high concentrations of nutrients and lower levels of light were to be found off the coast of Yucatan. This period also coincided with the largest yield of carrageenan (>40%; Freile-Pelegrin and Robledo 2006). The maximum growth of *E. isiforme* in the area occurred during October–April, with no growth during summer, when low densities were reported (approx. 20 g m^{-2}) which were too low to be harvested for commercial purposes, suggesting that recruitment took place at the end of summer (Perez-Enriquez 1996b). These results should be taken into consideration for the management of natural populations, as well as for mariculture of this species. More recent results (Robledo and Freile-Pelegrín 2011) on their cultivation reported higher growths in April (1.2–3.8% day^{-1}) being significantly affected by surface seawater temperatures (SST), whilst the carrageenan yields ranged from 34.8–37.1%. The seasonal variation in the biochemical compositions were similar to variations in carrageenan yield, suggesting that biochemical composition could be related to SST and hence to growth, while incident light and nutrients, influenced carrageenan quantity. In general, the carrageenan yield and viscosities found in *E. isiforme* were within the levels required for *Eucheuma* species for industrial applications, suggesting commercial/industrial potential. However, particularities in relation to the geographical location of raw material and extraction conditions are frequent. In this regard, the yield and physico-chemical properties of *E. isiforme*carrageenan was found to vary from 44.6% carrageenan content for "native" extraction, to between 31.8–35.3%, depending on the alkali concentration and time applied. Similar effects were seen for carrageenan viscosity values, with the lowest being for native extracts (e.g. 39–57.0 cPs), whilst alkali extractions with 1% KOH, for 3–4 h increased viscosity (e.g. 160.0–161.3 cPs). The yield and physico-chemical properties of native and alkali-treated carrageenans from *Eucheuma isiforme* was also found to depend upon the geographical origin of the harvested material, e.g. *Eucheuma isiforme* from Nicaragua produced higher yields, both with native (57.2%) and alkali treatments (43.5%), viscosities were also significantly different between the native (144.6 ± 3.3 cPs) and alkali-treated carrageenan (113.9 ± 2.6 cPs). In general, chemical and FTIR analysis of *Eucheuma isiforme* revealed a preponderantly iota-carrageenan extract. Commercial exploitation of *E. isiforme* off the Yucatan Coast, cannot be fully achieved without a management plan, or proven cultivation practices (Robledo and Freile-Pelegrín 2011). Unlike other Latin American countries, such as Brazil, to date, there have been no official initiatives to promote seaweed farming in Mexico. Techniques have been successfully tested in the region on different carrageenophyte species and varieties suited to various types of environment, however, practical experience on an adequate scale required to establish the costs and detailed activities of the operation are also of paramount importance (Dominguez-May et al. 2015). Recently, the domestication of *E. isiforme* was proven to be possible and cultivars were being produced in experimental IMTA systems (Robledo, unpublished data).

Although *Kappaphycus alvarezii* was not a native of the Gulf of Mexico, one species of the genus *Eucheuma,* previously described as *E. inerme,* from Campeche, was later transferred to the genus *Kappaphycus* (Ortega et al. 2001). Early laboratory experiments to observe the acclimation temperatures of *Kappaphycus* (*Eucheuma*) *isiforme* and *Kappaphycus alvarezii,* found that the former presented a greater capacity to acclimate to changes in parallel to the local, seasonal SST changes, which was attributed to the habitat characteristics (Dawes 1989). A commercial cultivar of *K. alvarezii* was first introduced to Mexico for evaluation in cultivation in 1999; the cultivar was provided by Dr. Edison J. de Paula, which had originated from the Philippines and was later cultured in the laboratory in Brazil. Whilst the technical viability to produce *K. alvarezii* was proven (Robledo et al. 2013), using a commercial, non-indigenous species, there were questions on its sustainability. Nevertheless, the carrageenan yield of the *K. alvarezii*biomass produced, i.e. 32–41% and harvesting time-management as a parameter that conditions both the carrageenan quality and its economic feasibility have been evaluated (Valderrama et al. 2015, Dominguez-May et al. 2015), suggesting that further efforts to support this activity in the area could bring social and economic benefits, if planned in a sustainable manner.

4.2.5 South America

Eucheuma and *Kappaphycus* were introduced in some South American countries to supply the national demand for carrageenan (Paula et al. 1999; Smith and Rincones 2006; Hayashi et al. 2007, 2011, 2014; Góes and Reis 2011; Leon and Tirado 2011; Reis et al. 2016).

Even with the long history of seaweed introductions, seaweed farming in South America (SA) is still incipient, despite the potential, with 29,199 km of coastline and a collective exclusive economic zone of 10.43 million km^2, from ten countries. South America has different climatic zones, from tropical to polar, but for the most-part being tropical and warm-temperate (SEDAC 2016). Four main, warm ocean currents warm the South-American coastal zone: (a) South Equatorial Current, (b) Brazil Current, (c) Guyana Current, which warms almost all of the South America Atlantic coast of Colombia, Venezuela, Guyana, Suriname, French Guyana and most parts of the Brazilian coast, and d) Pacific Equatorial Counter-Current which warms the northwestern parts of the Pacific Coast – including Colombia and Ecuador. The similarities of the environmental and social characteristics in these warm water countries under-pins a similar history of eucheumatoid production.

As of 2017, more than 50% of the population of SA was reported to live in the coastal zone, there is a forecast of a 32% increase within 100 km of the shore by 2025 (Duedall and Maul 2005). A large number of these people live in traditional, coastal communities, comprising low-income, artisanal fisher-folk with low levels of formal education, high rates of illiteracy and other social barriers which make them most vulnerable to limitations of natural, resource-dependent activities, such as fishing (Salas et al. 2011). These similarities were, or are, shared with other countries where cultivation of eucheumatoid species has been successfully implemented, e.g. Southeast Asia and Africa, however, this success has not yet been achieved in South America (Reis et al. 2016). Regardless of many introductions being made from the 1990s onwards, commercial and experimental farming is only reported in Brazil, Venezuela, Colombia and Equador (Leon and Tirado 2011; Hurtado et al. 2014; FAO 2016; Fig. 4.10).

4.2.5.1 Brazil

In 1995, *K. alvarezii* and *K. striatus* were introduced to a sea farm in São Paulo State, Brazil, after 10 months of laboratory, quarantine conditions (Paula et al. 2002; Bulboa and Paula 2005; Bulboa et al. 2007; Oliveira 2006; Silva et al. 2010). The original seedlings were from the Philippines, but had also been cultivated in Japan (Silva et al. 2010). During the experimental period only *K. striatus* produced viable spores, *K. alvarezii* did not produce spores. As a result, the material was removed from the sea and its production forbidden in Brazil (Bulboa and Paula 2005; Silva et al. 2010). *K. alvarezii* produced spores but further studies concluded that they were not viable under normal, prevailing sea conditions

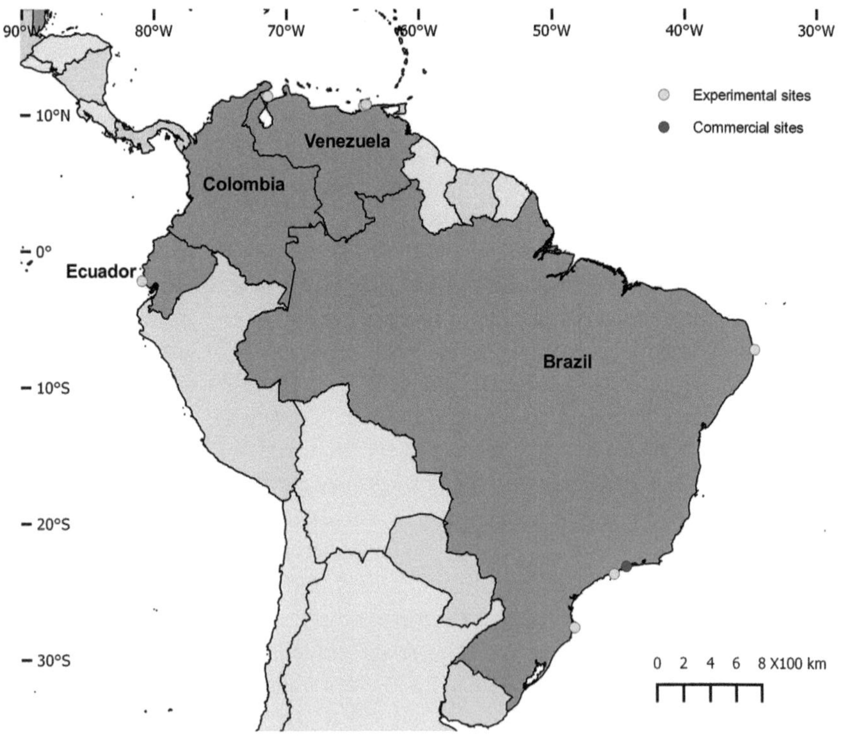

Fig. 4.10 Cultivation sites in South Amercia

(Paula et al. 1999). Neither spores nor sporangia were observed in other studies in the Brazilian South such as Santa Catarina State (Hayashi et al. 2011), Southeast Rio de Janeiro State (Castelar et al. 2009a, b) and the Northeast Region – Paraíba State (Araújo 2013) where the species were also introduced.

In 1998, a new introduction was reported to Brazil. After 4 months of quarantine, *K. alvarezii* from Venezuela was introduced to Ilha Grande Bay, Rio de Janeiro State, thereby starting the first commercial cultivation in Brazil (Castelar et al. 2009a, b). In 2004, the species was further introduced in Sepetiba Bay which became the largest Brazilian farm (see also Chap. 1). The farm facilities employed more than 100 workers and had approximately 120 floating rafts; each raft was 150 m long by 3 m. An average productivity of 39.3 g dw day^{-1} and an average daily growth rate of 3.8% day^{-1} were reported (Goes and Reis 2012; Reis et al. 2016). In 2008, seaweed rafts installed at Marambaia Bay were transferred to areas nearby at Itacuruçá City (Rio de Janeiro State; Alexandre Feder and Henrique Geromel de Góes, personal communications). However, in 2012, activities related to carrageenan processing from these farms were closed, it is presumed due to lack of profitability, but no reasons were apparent at the time of writing. Recently, another factory was opened, producing carrageenan from ten seaweed farms located in Sepetiba Bay and Ilha Grande Bay (Alexandre Feder, personal communication).

In 2008, *K. alvarezii* from São Paulo State was legally introduced in to Santa Catarina State (Silva et al. 2010) for experimental purposes. Santa Catarina is located in southern Brazilian, with sub-tropical conditions with the coolest SST occurring in winter. This was in contrast to São Paulo and Rio de Janeiro, where commercial cultivation was already allowed. Farming was successful in spring, summer and autumn, with growth rates close to 3.5% day^{-1}. Also, since Santa Catarina was the main mollusc producer (responsible for 95% of Brazilian production), technologies which amalgamated mollusc and seaweed production were co-developed to minimize production costs and offer potential mechanisation and productivity benefits to the local farmers (Santos 2014). At the time of writing (April 2017), Santa Catarina State region was still undertaking experimental studies before scaling-up to commercial cultivation.

K. alvarezii was also introduced in to north-eastern Brazil, in Ceará, Rio Grande do Norte, Paraíba and Pernambuco States and possibly in Bahia State (Oliveira et al. 2009; Silva et al. 2010). These areas have been indicated as having high environmental risks by Castelar et al. (2015). Araújo (2013) reported on small rafts in Paraiba State. However, at the time of writing, things are reported to be changing rapidly, see elsewhere in this book (Neish et al. on Reflections on the commercial development of *Kappaphycus* chapter) for a further update on the cultivation and processing of carrageenophytes in Brazil.

4.2.5.2 Venezuela

In Venezuela, *K. alvarezii* and *E. denticulatum* were both introduced in 1996 in two areas: Araya Peninsula and Coche Island. The seedlings came from the Bohol Sea, the Philippines and were cultivated on floating rafts, or an off-bottom system. However, in spite of the high growth rates reported (i.e. between 4.5–7.0% day^{-1}, depending on the season) and consequent high productivity, concerns arose related to bio-dispersion of introduced, non-native species and coral reef-bleaching which stopped any further activity (Smith and Rincones 2006; Barrios et al. 2007). According to Perez et al. (2003), the introductions were made without the required environmental impact studies and the Autonomous Service of Fisheries and Aquatic Resources (SARPA) approved cultivation after the species were already introduced into Venezuelan waters.

4.2.5.3 Colombia

In Colombia, *K. alvarezii* was introduced in late 1990s from Cuban seedlings. Cultivation experiments were made using floating cages in Centro de Investigación, Educación y Recreación – CEINER, located in Islas del Rosario (Cartagena) and Acuario Mundo Marino from Universidad Jorge Tadeo Lozano in Santa Marta (Magdalena) as part of a study for an undergraduate project (Rincones and Tirado 2011). In 2010, native *Eucheuma isiforme* was cultivated experimentally on the Guajira Peninsula to evaluate its economic and environmental suitability for the region. At that time, seedlings were cultivated on an off-bottom system, or floating rafts (Rincones and Tirado 2011). Cultivation was stopped in Colombia due to the lack of necessary, environmental impact assessments required to evaluate the introduction (Sepulveda, personal communications).

4.2.5.4 Ecuador

In 2011, seedlings of *K. alvarezii* which originated from Brazil, were introduced into shrimp (*Litopenaeus vannamei*) farms in Ecuador and cultivated by tie-tie on mono-lines, at an experimental scale (Sepulveda 2016). In 2015, seedlings of *K. alvarezii* var. *tambalang* from Panama were introduced to the Santa Elena Peninsula, Salinas, Ecuador. A pilot-scale cultivation was operated for 9 months using floating rafts. The first results reported growth rates of 4.5–8% day^{-1}, over production cycles of 45–60 days. At the time of writing (April 2017), an environmental impact survey and a socio-economic feasibility study are being conducted as requirement for commercial cultivation (Rincones et al. 2016, Rincones, personal communications).

4.3 *Kappaphycus/Eucheuma* Species Commercially, or Experimentally Farmed

The main species cultivated around the world are presented here (Figs. 4.11a–d, 4.12, 4.13, 4.14, 4.15 and 4.16) However, for more color morphotypes of seaweeds cultivated in the Philippines, please see Chap. 2 of this book).

4.4 Farming Systems

Different farming techniques are employed by each country from shallow (<5 m during the lowest tide) to deep water (>5 m during the lowest tide). These vary from simple, fixed-off-bottom (FOB) and single floating rafts in shallow waters, to free-swing (FS), hanging long-line (HLL) and complex structures such as the multiple-raft, long-line (MRLL), spider web (SW), triangular hanging long line (THLL), tubular net bag (TNB), basket net bag (BNB) and octagonal rafts (OR; Fig. 4.17a, 4.17b, 4.17c, 4.17d, 4.17e, 4.17f, 4.17g, 4.17h, 4.17i, 4.17j and Table 4.3).

The FOB technique of culturing *Kappaphycus* remains the simplest and most traditional and has been practiced since seaweed farming was first introduced to the Philippines in the early 1970s. Farmers use flat binders or polyethylene ropes as cultivation lines, with soft plastic strips (called "tie-ties") to tie the vegetatively propagated branches or fragments, called propagules (also known as 'seedlings') on to the lines; wooden stakes are used to peg both ends of the cultivation line into the sandy/muddy substrata. However, whilst this method is still widely used, over the years, innovations have been made such that the structures were better able to withstand bad weather and have longer economic life-span. As such, for floating structures in particular, PVC piping has been used instead of bamboo and kuralon twine instead of the unsustainable plastic soft tie-tie.

Earlier farming techniques were described and reported in detail by Hurtado et al. (2008); Hayashi et al. (2010); Hurtado et al. (2014); Reis et al. (2015). Surprisingly, it has taken quite some time for innovations to occur in eucheumatoid cultivation and only those newer farming techniques will be detailed further in this chapter.

The basket method of cultivating *K. alvarezii* was introduced in Semporna, Sabah, Malaysia in mid-2000 with the objective of minimizing grazing by herbivorous fishes such as siganids and parrotfish and also the highly voracious turtles. This type of seaweed cultivation has not been widely adopted by the farmers due to the high cost of materials and also the labor required for cleaning and removing attached animals, in particular barnacles (K. Lewanomont, personal communications).

The "tie-tie" method is slowly being replaced by tubular netting cultivation (Góes and Reis 2011; Reis et al. 2015). This technique alone allowed farms to increase their productivity and the planting and harvesting process became easier and also allowed for opportunities for future potential mechanization.

In Rio de Janeiro State, Brazil, tubular nets are detached from the main rafts and transported to land by small kayaks, and then to the drier facilities by trucks, where the seaweeds are stripped out of the tubes. Some producers use big round sacks made of nylon netting with stainless steel rings of 1 m diameter. Once the sacks are filled, they are attached to buoys and kept in the water until they can be collected in one trip for efficiency, and transported to a wharf. The sacks are transferred to a truck with the help of a winch and transported to the drying facility, where the seaweed biomass is stripped out. The truck winch has a scale in order to measure the fresh seaweed biomass to be sold.

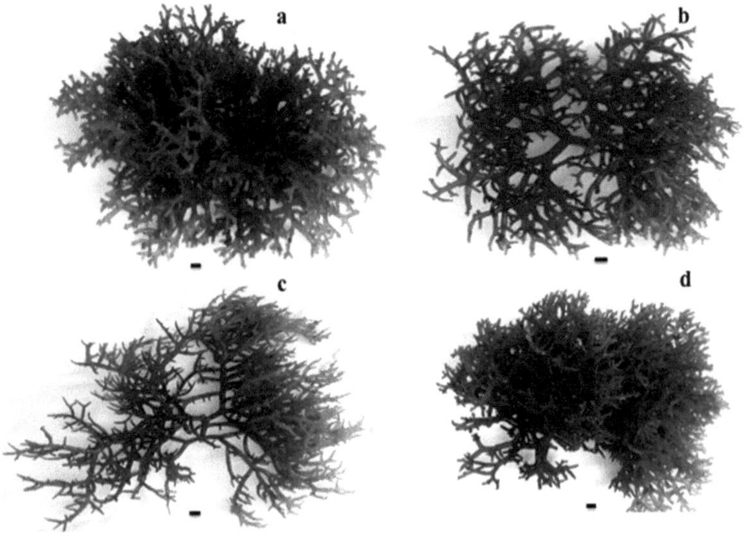

Fig. 4.11 (a–d) Some cultivars of *K. striatus* in the Philippines (Photos courtesy of AQ Hurtado)

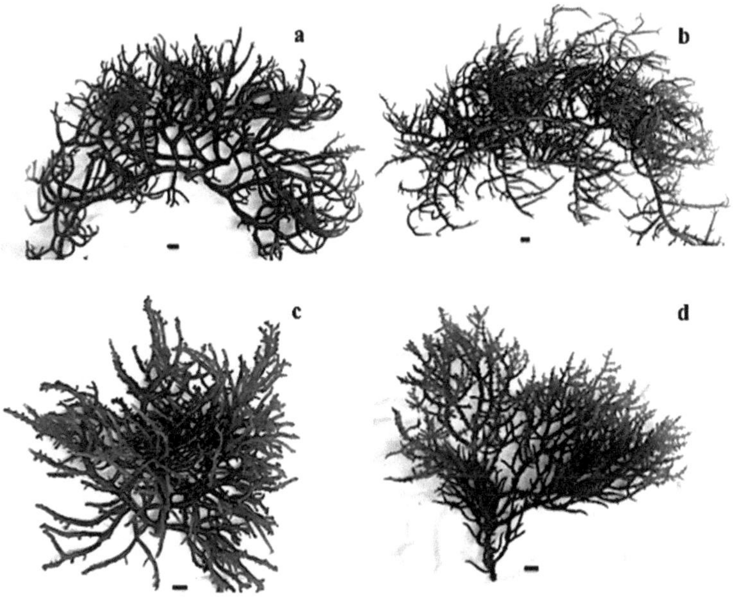

Fig. 4.12 (**a–c**) Some cultivars of *K. alvarezii*; (**d**) *E. denticulatum* in the Philippines (Photos courtesy of AQ Hurtado)

Fig. 4.13 (**a–f**) Cultivars of *K. alvarezii*; (**f–1**) Cultivars of *K. striatus* in Indonesia (Photos courtesy of I Indrayani)

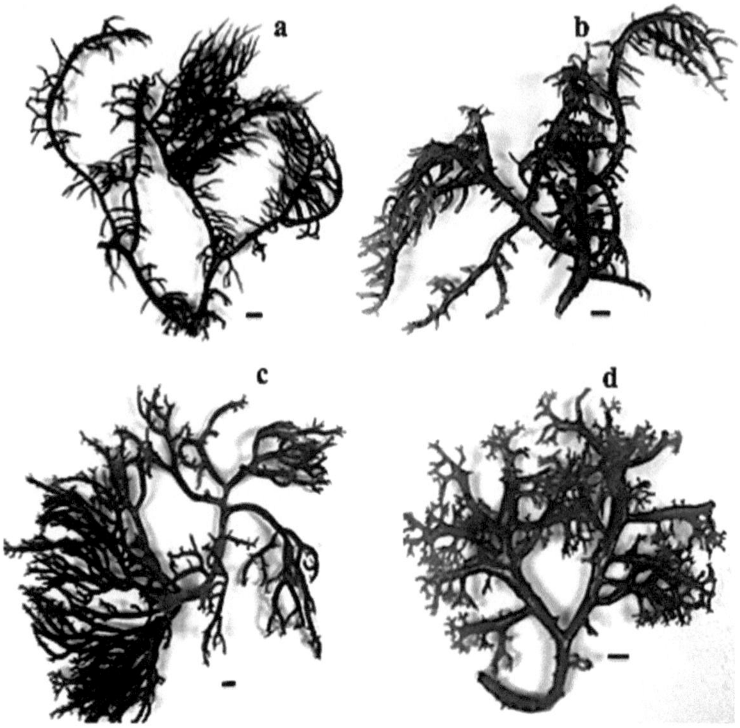

Fig. 4.14 (**a–c**) Cultivars of *K. alvarezii*; (**d**) *K. striatus* in Sabah, Malaysia (Photos courtesy of MKM Ali)

Fig. 4.15 *K. striatus* grown in tubular net in Tanzania (Photos courtesy of FE Msuya)

Floating rafts are progressively being adapted to each region of cultivation. Different materials and designs have been tested around the world. In Santa Catarina State, Brazil, the first rafts used for experimental cultivation were constructed from PVC piping. However, it was found that the piping needed to be replaced frequently because of damage from strong currents which increased production costs. Alternatives to PVC piping were tested, with double long-lines around 18.7 L buoys and wooden spacers between the long-lines and recently specifically designed buoys were currently under test. These buoys had a flat cylindrical shape: a length of 2.09 m, 0.31 m wide and 0.17 m height, with a volume of 90 L volume and 9 kg weight. These buoys were developed to withstand heavy currents and to increase raft stability. As a result of the format, there was no need to use wood spacers between the long-lines, since the buoy itself provided this function. Also, with this type of buoy and rigging, integration with mollusc farming and future mechanization should be easier (Hayashi et al. 2011, 2016; Santos et al. 2016).

Fig. 4.16 Cultivars of *K. alvarezii* in South America (Photos courtesy of HG Goes)

b

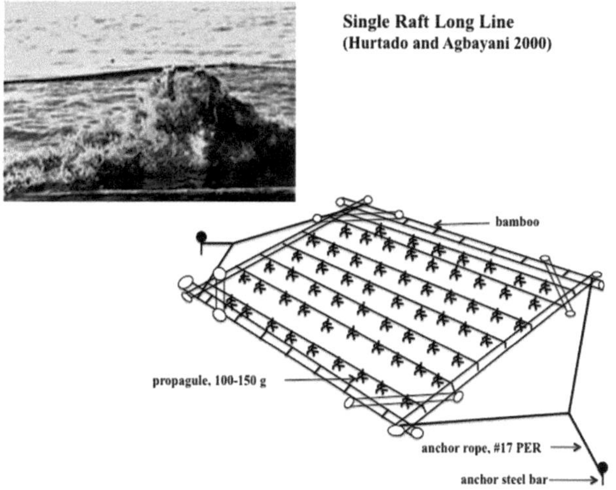

Fig. 4.17b Single raft long-line cultivation technique

a

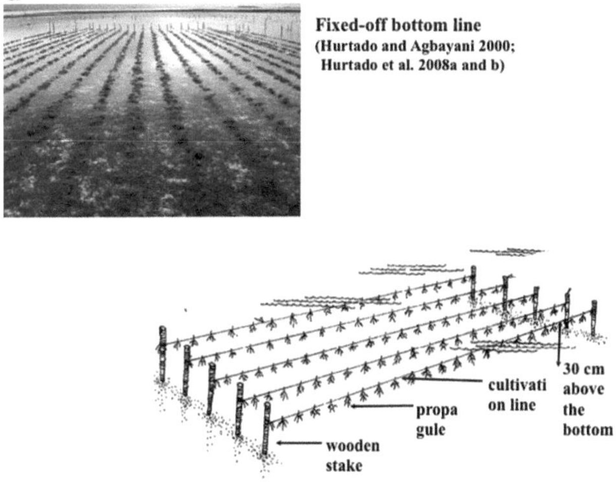

Fig. 4.17a Fixed-off bottom cultivation technique

c

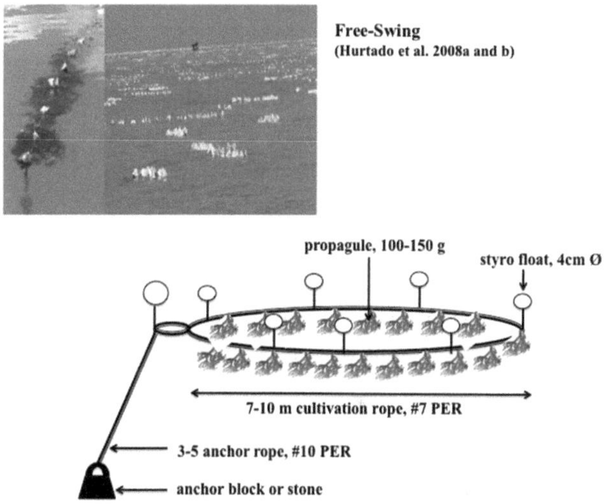

Fig. 4.17c Free-swing technique

In India, CSIR-CSMCRI, in association with CSIR – Structural Research Engineering Centre, were undertaking trials using an octagonal raft design with various degrees of inclination. The octagonal design provides a modular structure that is expandable, easy to assemble and anchor, whilst additionally providing for the free flow of seawater which replenished the nutrient supply to the growing seedlings within the raft area. The good maintenance of the rafts, as well as removal drift seaweed and silt from the seedlings, was found to be more efficient within these structures, rather than the conventional, flat rafts. The new robust and sturdy raft design and floatation system was also more suitable for deep-sea farming sites.

4.5 Volume and Value of Production to Include Export Volumes

Four countries are responsible for almost all of the world's production of eucheumatoids (tropical to sub-tropically farmed carrageenophytes). At the time of writing (April, 2017) the main producers were: Indonesia, the Philippines, Tanzania and Malaysia in decreasing order. Brazil may be an interesting addition to this list, please see elsewhere in this book). Due to unavailability of reliable information on production from the "minor producing countries", the major producing countries are presented in Table 4.4.

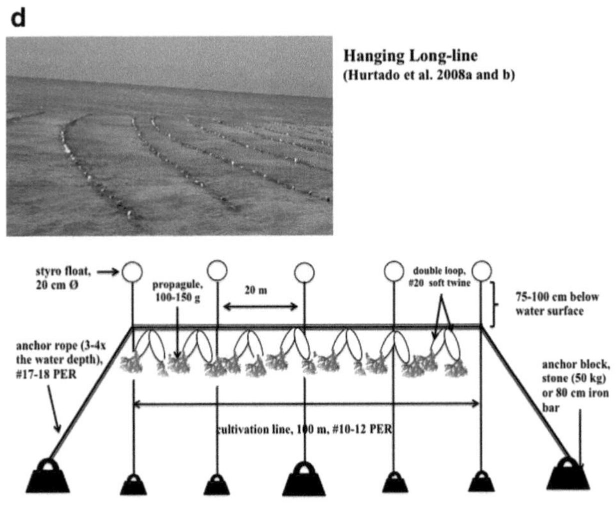

Fig. 4.17d Hanging long-line technique

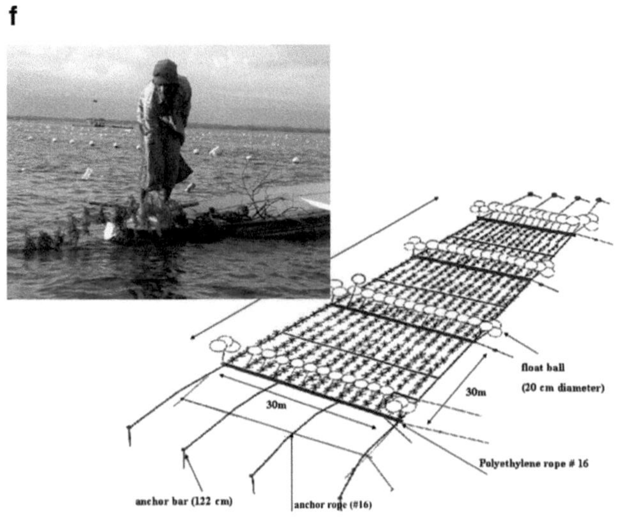

Fig. 4.17f Spider web technique

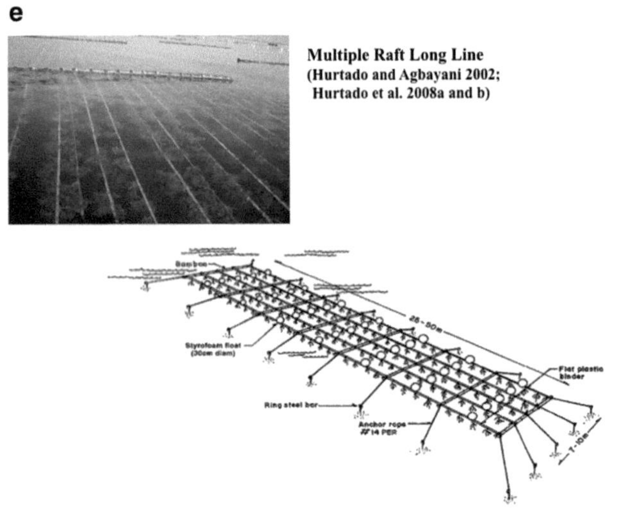

Fig. 4.17e Multiple raft long-line technique

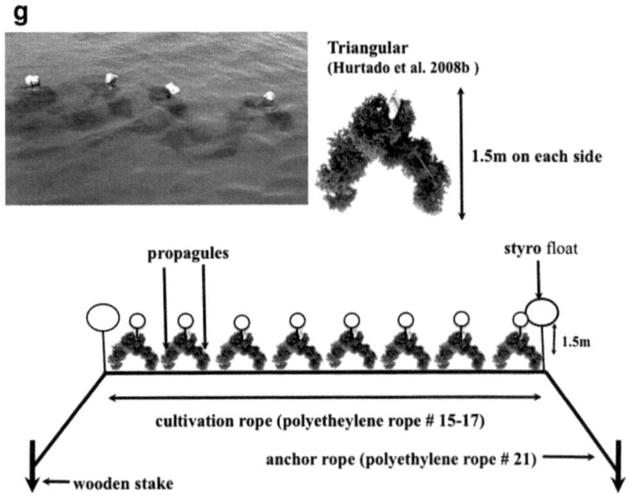

Fig. 4.17g Traingular technique

4.6 Problems and Strategies for Conservation and Sustainability

As with any agronomical practice, problems with pest species and their control are to be expected. However, several problems were becoming increasingly common to all countries which produced commercial or experimental volumes of "cotonii" and "spinosum" species, often farmed as "monocrops", without rotation (Table 4.5). Technical problems related to climate change and increasing SST indeed need attention, mainly the incidence of epi- and endophytic organisms, which have no apparent solutions even now. Low productivity of highly cloned strains, excessive grazing, coral reef encroachment and the use of mangrove trees to make stakes are all threats to sustainability (Hurtado et al. 2015). Figure 4.18 are some examples of macro-epiphytism.

Leonardi et al. (2006) classified five (5) different types of infection based on the host-epiphyte relationships at the anatomical level (Fig. 4.19). Amongst the five types of "infestation", *Ceramium* and *Neosiphonia-Polysiphonia* epiphytes were the most harmful to the host. There are reports that *Neosiphonia* spp. (Critchley et al. 2004; Hurtado et al. 2006; Hurtado and Critchley 2006; Vairappan 2006, Vairappan et al. 2008; Pang et al. 2015; Ateweberhan et al. 2015) and *Polysiphonia* (Tsiresy et al. 2016) caused significant biomass (and income) losses due to over-growth, and thallus fragmentation. Thus, a high incidence of this endophyte on

h

Fig. 4.17h (**a** and **b**) Tubular net (Photos courtesy of FE Msuya)

i

Fig. 4.17i Hanging basket technique

Kappaphycus thalli posed a huge problem in sourcing clean propagules for the next growing cycle.

Chapter 12 details linkages between farmers and buyers; the lack of communication between producers and traders compromised the marketing of the dried seaweed. Consequently, farm-gate prices have varied considerably with no established criteria. The apparent excess of middlemen and shortages of low-income laborers also contributed to this problem. Weak linkages between and amongst academics, government, stakeholders and producers/processors were responsible for delays in progress for the introduction of new technologies and sustainability (Hurtado 2016).

New strategies and renewed vigor are required in order to solve these problems. Action plans which are being considered and implemented are described in Table 4.6. One of the most important points is the need to strengthen the necessary relationships amongst academics, stakeholders (processors), producers (farmers) and government. Promotion of national and international networks with collaboration between centers of production, not as competitors, but collaborators developing better means of farming/livelihoods, as a better way forward for all, will help provide solutions for specific local and increasingly global problems. Adoption of the principles of polyculture (IMTA) would diversify production, as well as providing producers economic independency of just one production crop. Also, the seaweed biomass produced can play an important role as an extractive, biofilter in such systems, thereby minimizing the impact of aquacultural effluents in nature. Finding other value additions to the target species for cultivation can also improve the economic chain of these seaweeds (Please refer to Chap. 12 on value-chains elsewhere in this book).

In order to ensure a sustainable seaweed aquaculture industry, Best Aquaculture Practices (BAqP) must be adopted and fully implemented. The following guiding principles must be fully adopted: (1) use only of a fast growing strain, with the most appropriate culture technique, per season, in a given area, (2) use of more disease-resistant strains, (3) strict compliance to a 45–60 day growing period, to prevent the harvest of immature seaweeds with low carrageenan yields and quality (i.e. lower molecular weight), (4) use of environmentally-friendly materials for construction of supporting structures and (5) ecologically-sound farming areas, i.e. away from seagrass beds and coral reefs.

4.7 Future Prospects of the Industry

The future of the industry for the cultivation of tropical carrageenophytes is full of many challenges. In some places, seedlings have already shown signs of loss of strain vigour, after multiple years of continuous vegetative propagation from the same clone, thus resulting in decreasing productivity. New plants from spores and/or developed from micropropagation could possibly be introduced (see more in the Reddy et al. chapter of this book) using techniques which have been used with success in other seaweeds such as: *Undaria, Saccharina* and *Pyropia* (Chen 2013). Fallowing or resting the areas after 2–3 years of use must be considered to for adoption in order to invigorate or replenish the areas with new nutrients. Development and production of new strains, resistant to disease and epiphytes, are required (but without genetic manipulation by gene insertion) must also be seriously considered in order to obtain improved productivity and production. Investment in the techniques and adoption of

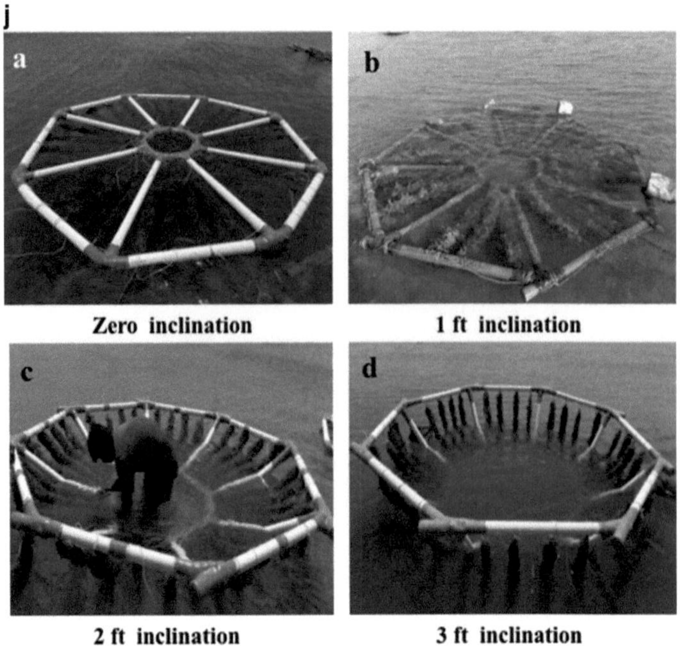

Fig. 4.17j (a–d) Polygonal raft technique (Photos courtesy of K Eswaran)

Table 4.3 Different species of eucheumatoid seaweeds cultivated around the world, each corresponding farming technique is indicated

Country	Speceis farmed	Shallow waters (1-3m LLT)		Deep waters (>5m LLT)						
		FOB	SFR	FS	HLL	MRLL	SW	TNB	B/NB	OR
Brazil	Ka, Ks		✪					✪		
Cambodia	Ka	✪								
China	Ka, Ks, Ed	✪								
Colombia	Ka		✪							
Equador	Ka		✪							
Fiji	Ka, Ks	✪	✪		✪					
Cambodia	Ka, Ks, Ed				✪					
India	Ka, Ks		✪						✪	✪
Indonesia	Ka, Ks, Ed	✪			✪					
Kenya	Ka, Ed	✪								
Kiribati	Ka	✪	✪		✪					
Madagascar	Ka, Ed	✪								
Malaysia	Ka, Ks, Ed	✪	✪		✪				✪	
Micronesia	Ka	✪								
Mozambique	Ka, Ed	✪	✪							
Myanmar	Ka		✪		✪					
Panama	Ka				✪					
Papua New Guniea	Ka	✪			✪					
Philippines	Ka, Ks, Ed	✪	✪	✪	✪	✪	✪			
Solomon Is.	Ka	✪								
Sri Lanka	Ka		✪							
Venezuela	Ka, Ed	✪	✪							
Vietnam	Ka, Ks, Ed	✪			✪				✪	
Tanzania/Zanzibar	Ka, Ks, Ed	✪			✪			✪		

Ka Kappaphycus alvarezii; *Ks Kappaphycus striatus*; *Ed Eucheuma denticulatum*
FOB fixed-off bottom, *SFR* single floating raft, *HLL* hanging long-line, *MRLL* multiple raft long-line, *FS* free swing
TNB tubular net bag, *B/NB* basket/net bag, *OR* octagonal rafts

Table 4.4 The major producers of *Kapppaphycus* and *Eucheuma* seaweeds (as per information available, April, 2017)

	Volume (MT, dwt)					
	K. alvarezii (cottonii)			*E. denticulatum* (spinosum)		
	1999	2009	2015	1999	2009	2015
Indonesia[a]	27,000	85,000	110,000	14,000	17,000	25,000
Philippines[a]	112,000	61,000	60,000	7,000	5,500	6,000
Tanzania/Zanzibar[a]				7,000	10,400	14,000
Malaysia[b]	3,700	5,000	8,000	ND	ND	ND
Total	142,700	151,000	178,000	28,000	32,900	45,000

[a]Porse and Rudolph 2016
[b]Azhar 2016/DOF-Sabah

Table 4.5 Challenges and constraints facing the current eucheumatoid industry

Technical	'Ice-ice' and epi-endo epiphytism incidence at elevated seawater temperatures
	Excessive grazing (siganid fish)
	Weak or poor acceptance of seaweed farmers to new farming innovations
Social	Non-acceptance of communities to start or continue to farm due to low Farm-gate price of dried seaweed
	Presence of many 'tiers' in the trading chain
	Weak linkages between and amongst academia-research centers-private sector
Ecological-environmental	Possible encroachment on to coral reefs
	Use of non-biodegradable planting material such as the soft plastic tie-tie
	Non-sustainable use of mangrove trees as stakes for lines
	Conflict of water usages due to poor castal resource management
Financial	Insufficient government support required to expand cultivation areas or to rehabilitate after cyclones or other natural, weather-related phenomena

mechanization of production and harvesting are required to reduce drudgery and manual labor and would improve productivity on a daily basis. Polyculture or Integrated Multi-Trophic Aquaculture (IMTA) principles can be adapted in the tropics and sub-tropics in line with sustainable farming of *Kappaphycus/Eucheuma* and at the same time to have an ecologically-balanced marine environment. Adoption of IMTA principles and methods could help mitigate the adverse effects or impacts of climate change caused by El Niño, La Niña, super typhoons, cyclones or flash floods.

Idealistically, an holistic seaweed program by the different governments of various countries in collaboration with academia, the private sector and stakeholders would guarantee a responsible, sustainable implementation of seaweed farming production (Cottier-Cook et al. 2016). Best management practices need to be established, implemented and furthermore taught and disseminated for each location. The establishment of centres of excellence in seaweed agronomy would help in this regard. In places where *Kappaphycus* and *Eucheuma* are, or will be introduced, quarantine procedures and environmental monitoring must be monitored and adhered to strictly in order to reduce the possibilities for any adverse impacts resulting from the introduction of non-native species to the new environment.

4.8 Conclusions

It is clear that the global success of eucheumatoid farming has been related to the joint-efforts and collaboration between and amongst the key stakeholders from the seaweed farmers, traders/exporter, processors, academia and government to the end-users. Several strains of *K. alvarezii* and *K. striatus* have been used by farmers for quite some time. Plasticity of these two species led to much confusion amongst seaweed farmers, traders and processors, however, the recent findings on the genetic, molecular characteristics of the two species in question shed more light on their correct taxonomic identification (Please see Tan et al. chapter of this book). Collectively, marine plants, such as cultivated seaweeds, will continue to sequester more CO_2 than terrestrial plants from the atmosphere (even if the capture is for a relatively short period of time over the seaweed growth and processing cycle). Seaweed forests and enhanced, future farming activities will assist in the mitigation of global warming (Please read more in the Largo et al. chapter of this book).

Modifications of farming techniques and especially integration of some levels of mechanization have been developed over the years, yet with much of the innovation arising from the seaweed farmers themselves, complimented with technical assistance from academia. However, further efforts are required in order to optimize the cultivation systems and provide post-harvest management strategies which maintain the qualities of the biomass through mechanization and smart farm management (Please see the Ali et al. chapter of this book). Emerging applications for all, or part, of the *Kappaphycus* and *Eucheuma*biomass as sources of biofuel, biostimulant and for example, ingredients in cosmetics are

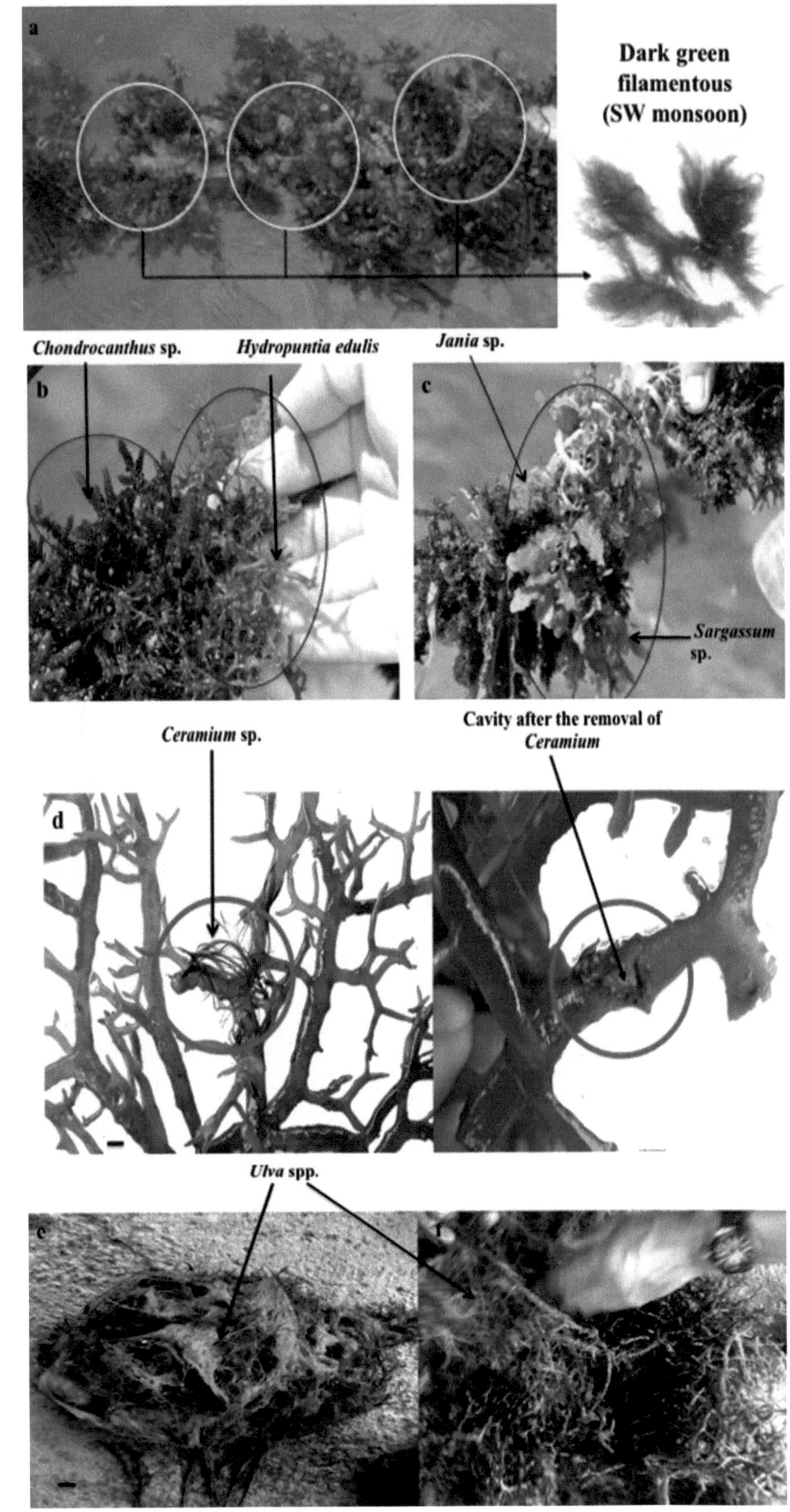

Fig. 4.18 Macro-epiphytes (**a**) filamentous blue-green algae, (**b**–**c**) red macro-epiphytes, (**d**) *Ceramium* sp., (**e**–**f**) *Ulva* spp. (Photos courtesy of AQ Huratdo)

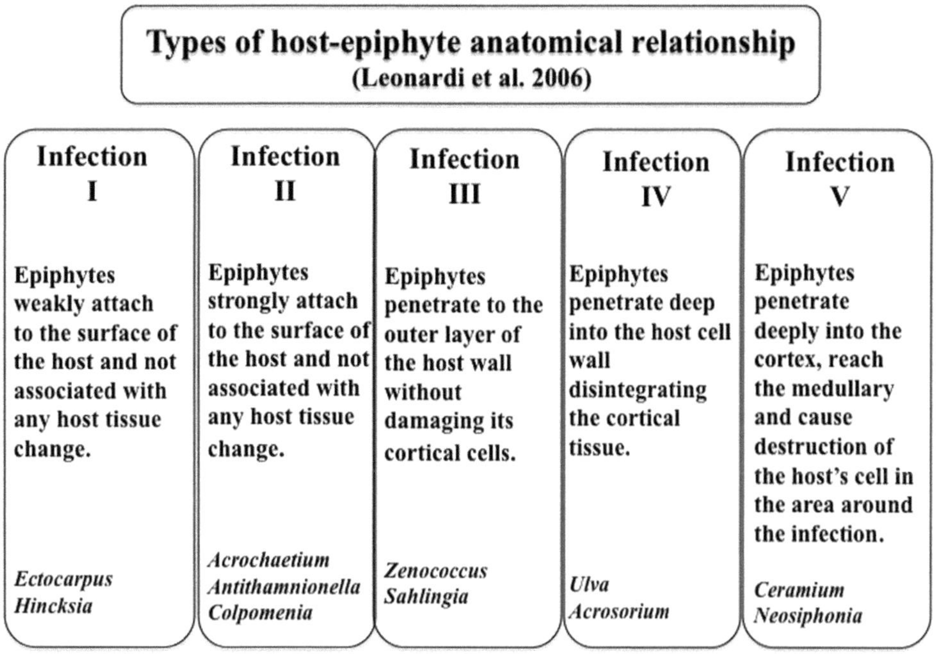

Fig. 4.19 Types of host-epiphyte anatomical relationship (Leonardi et al. 2006)

Table 4.6 Conservation, sustainability strategies and action plans which should be adopted more extensively in eucheumatoid farming

Conservation and sustainability strategies	Action plans
Capacity enhancement of human resources	Active enhancement of public promotion and environmental education through regular training/workshops/seminars
	Cross-country/area visits to successful seaweed areas/farmers and the sharing of knowledge and best practices
	National and international collaboration and networking
	Improved scientific knowledge and strong cooperation with local and international societies, stakeholders working on the conservation of marine resources
Diversified livelihoods	Introduction of invertebrate aquaculture and sea-ranching options such as sea-urchins, sea-cucumbers and abalone in conjunction with and other high value animals – gradually replacing fisheries/capture from natural populations, which should be conserved
	Cultivation of other carrageenophytes with potential for bioactive, biofuel, pharmaceutical, cosmetic and nutraceutical applications
Sound ecosystem-based management	Adoption of Best Aquaculture Practices (BAqPs) –acquired by collation of knowledge and technology transfer
	Sufficient buffer space between lines and farms to allow free movement of water
	Reduction of the number of farms in densely cultivated areas in order to achieve maximum, sustainable carrying capacity
	Use of appropriate cultivation methods suited to the environmental conditions of a given area (this requires that farmers be given choices and guidance when establishing or improving their operations)
	Use of bio-degradable planting materials
	Proper coastal zoning of aquaculture activities (this requires that managers understand the social and commercial benefits as well as the ecosystem services provided by responsible, sustainable seaweed cultivation)
	Adaption of policies for "exclusion zones" such that seaweed farms could not be sited near or on top of coral reefs or in Marine Protected Areas (MPAs)
	Prevention of marine pollution originating from anthropogenic sources such as run-off from coastal and inland domestic, agricultural and industrial effluents

(continued)

Table 4.6 (continued)

Conservation and sustainability strategies	Action plans
Resilience to climate change	Shift of farming practices:
	From near-shore, shallow coastal zone to off-shore or deep ocean
	From fixed off- bottom to floating systems
	From the use of species prone to disease to the selection of new strains with enhanced disease resistance (coupled with good productivity and carrageenan quality properties)
	From vegetative propagules to spore-generated or micropropagated propagules
	From monocrop to integrated aquaculture which may include several trophic levels
	Addition of nutrients to the cultured species in nutrient-depleted areas to boost growth and resistance to biotic and abiotic stresses
	Optimization of farming activities during favorable months of growth
	Reduction of environmental stress
Secured sustainability	Cultivation and harvest
	Cultivation of other macroalgae each with multiple potential for use in pharmaceuticals, nutraceuticals, cosmetics, food (sea vegatable), feeds and fuels
	Introduction of IMTA principles to open-water, seaweed agronomy in the tropical to sub-tropical regions of the world
	Introduction of mechanization from the seeding of ropes to harvesting to increase productivity per unit effort
	Documentation of the potential of using seaweed as a biofilter for re-capturing nutrients from the marine environment
	Thorough evaluation of the life cycle of seaweed cultivation and its role to be played in atmospheric carbon capture
	Documentation of the effects on the surrounding marine ecosystems (animals, plants etc.) of off-shore, large-scale seaweed cultivation operations

fast gaining acceptance in the market. The chapters of Ortiz et al. (processing) and Neish and Suryanarayan (value-chain analysis) and adoption of the MUZE or biorefinery approach to biomass processing and value-stream generation, if commercially adopted, could provide significant benefits to the process stream. These non-rheological, applications are aside from the traditional food applications and a renewed research focus on new, value-added products is essential.

The future of the global eucheumatoid industry is both changing and challenging, but the potential to cultivate both genera of *Kappaphycus* and *Eucheuma* in tropical and subtropical countries is large. New research efforts will bring more light to the constant improvements required for both the production and applications and their benefits to humans-see https://www.sams.ac.uk/news/sams-news-globalseaweed-launch.html.

References

Adams T, Foscarini R (1990) Proceedings of the regional workshop on seaweed culture and marketing. South Pacific Aquaculture Development Project. Field Document 1990/2, 86 p

Algae World News (2015) www.news.algaeworld.org

Amat JP (1981) Etude et Culture de l'algue Rouge *Eucheuma spinosum* dans le Golfe de Tadjourah. Thesis, l'Institut National de Toulouse, 125 pp

Araujo PG (2013) Avaliação do potencial invasor de *Kappaphycus alvarezii* (Rhodophyta, Gigartinales) no litoral da Paraíba, Brasil. PhD thesis, Universidade Federal de Pernambuco, Brazil (in Portuguese)

Areces AJ, Céspedes N (1992) Potencialidad productiva de algunas carragenófitas del Indopacífico en aguas del Caribe. Boletín de la Red Latinoamericana de Acuicultura 6(2):13–16

Areces Mallea AJ, Alvarez Villanueva FC, Bernardi J, Cabrera R (2014) Ecological risk assessment of the introduction of exotic carrageenophytes in the tropical Western Atlantic. J Appl Phycol 26(5):2055–2063

Ask, EI, Batibasaga A, Zertuche-Gonzalez JA, De San M (2003a) Three decades of *Kappaphycus alvarezii* (Rhodophyta) introduction to non-endemic locations. Proc Int Seaweed Symp 17:49–57

Ask EI, Ledua E, Batibasaga A, Mario S (2003b) Developing the cottonii (*Kappaphycus alvarezii*) cultivation industry in the Fiji Islands. Proc Int Seaweed Symp 17:81–85

Ateweberhan M, Rougier A, Rakotomahazo C (2015) Influence of environmental factors and farming technique on growth and health of farmed *Kappaphycus alvarezii* (cottonii) in south-west Madagascar. J Appl Phycol 27:923–934

Azhnar K (2016) Seaweed development and production of seaweeds in Sabah, Malaysia. Paper presented at the Seaweed Workshop, University of Malaysia at Terengganu, 26 Feb 2016

Barbaroux O, Perez R (1983) Recherches sur les possibilites d'exploitation et of. Culture de l'algue rouge *Eucheuma spinosum* aux Antilles Institut Scientifique Et Tecnique des Peches Maritimes-Laboratoire D'algologie Appliquee. Rapport N° 5 de février 1981 à juillet 1982

Barbaroux O, Perez R, Dreno JP (1984) Possibilité d'exploitation et de cultureaux Antilles d'*Eucheuma spinosum*. Science et Pêche, Bull. Inst. Pêches marit, 348, p 20

Barrios J, Bolaños J, López R (2007) Blanqueamiento de arrecifes coralinos por la invasión de *Kappaphycus alvarezii* (Rhodophyta)

en Isla Cubagua, Estado Nueva Esparta, Venezuela. Boletin del Instituto Oceanografico de Venezuela 46:147–152

Batista G, Connor J (1990) Native uses of seaweeds in the Republic of Panama. Annotated bibliography of the seaweeds used for food in the West Indies. OECS Fisheries Report No 3.7–8. Caribbean Natural Resource Institute (CANARI) St. Lucia (Eds)

Batista de Vega G (2006) The seaweed resources of the Caribbean. In: Critchley A, Ohno M, Largo DB (eds) DVDROM World seaweeds resources, an authoritative reference system. UNESCO Publishing/ETI Bioinformatics, University of Amsterdam, ISBN 90 75000 80 4

Batista de Vega G (2009) Cultivo Ecosostenible de *Kappaphycus alvarezii* en Panamá. Tesis PhD en el Departamento de Biología, Fisiología y Biotecnología Vegetal Marina, en La Universidad Las Palmas de Gran Canaria, España

Batista de Vega G, Ceballos J, Anzalone A, Digman Michelle A, Gratton E (2016) A laser-scanning confocal microscopy study of carrageenan in red algae from seaweed farms near the Caribbean entrance of the Panama Canal. J Appl Phycol. doi:10.1007/s10811-016-0914-4

Bergschmidt H (1997) Seaweed production in Kiribati: a new cash crop. Pac Bull 10:11–12

Bixler HJ, Porse H (2011) A decade of change in the seaweed hydrocolloids industry. J Appl Phycol 23:321–335

Brown NA (1999) Finding their legs for seawater: women and seamoss mariculture in Praslin, St. Lucia. CANARI Technical Report No. 257:16 pp

Bulboa CR, Paula EJ (2005) Introduction of non-native species of *Kappaphycus* (Rhodophyta, Gigartinales) in subtropical waters: comparative analysis of growth rates of *Kappaphycus alvarezii* and *Kappaphycus striatum in vitro* and in the sea in southeastern Brazil. Phycol Res 53:183–188

Bulboa CR, Paula EJ, Chow F (2007) Laboratory germination and sea out-planting of tetraspore progeny from *Kappaphycus striatum* (Rhodophyta) in subtropical waters of Brazil. J Appl Phycol 19:357–363

Bureau of Agricultural Statistics (2001) Fisheries statistics of the Philippines 1997–2001. Manila Philippines

Castelar B, Reis RP, Bastos M (2009a) Contribuição ao Protocolo de Monitoramento Ambiental da Maricultura de *Kappaphycus alvarezii* (Doty) Doty ex P.C. Silva (Areschougiaceae – Rhodophyta) na Baía de Sepetiba, RJ, Brasil. Acta Bot Brasilica 23:613–617

Castelar B, Reis RP, Moura AL, Kirk R (2009b) Invasive potential of *Kappaphycus alvarezii* off the south coast of Rio de Janeiro state, Brazil: a contribution to environmentally secure cultivation in the tropics. Bot Mar 52:283–289

Castelar B, Siqueira MF, Sanchez-Tapia A, Reis RP (2015) Risk analysis using species distribution modeling to support public policies for the alien alga *Kappaphycus alvarezii* aquaculture in Brazil. Aquaculture 446:217–226

Chen J (2013) Seaweed culture in China. Paper presented at FAO Seaweed Workshop 3–6 October 2013, Jakarta, Indonesia

Cheney DP (1988) The genus *Eucheuma* J. Agardh in Florida and the Caribbean. In: Abbott IA (ed) Taxonomy of economic seaweeds with reference to some Pacific and Caribbean species, vol 2. La Jolla, California Sea Grant College Program, University of California, pp 209–219

Contreras A, Rosenthal W (2003) Data analysis of the water temperature on Bahia Las Minas, Panama Costal Resarch Institute Forschungszentrum (GKSS), 1–19

Cottier-Cook EJ, Nagabhatla N, Badis Y, Campbell M, Chopin T, Dai W, Fang J, He P, Hewitt C, Kim GH, Huo Y, Jiang Z, Kema G, Li X, Liu F, Liu H, Liu Y, Lu Q, Luo Q, Mao Y, Msuya FE, Rebours C, Shen H, Stentiford GD, Yarish C, Wu H, Yang X, Zhang J, Zhou Y, Gachon CMM (2016) Safeguarding the future of the global seaweed aquaculture industry. United Nations University (INWEH) and Scottish Association for Marine Science Policy Brief. ISBN 978-92-808-6080-1. 12pp

Critchley AT, Largo DB, Wee W, Bleicher-L'Honneur G, Hurtado A, Shubert (2004) A preliminary summary of *Kappaphycus* farming and the impacts of epiphytes. Jap J Phycol 52: 231–232

Dawes CJ (1989) Temperature acclimation in cultured *Eucheuma isiforme* from Florida and *E. alvarezii* from the Philippines. J Appl Phycol 1:59–65

Dawes CJ, Mathieson AC, Cheney DP (1974) Ecological studies of Floridean *Eucheuma* (Rhodophyta, Gigartinales): I. Seasonal growth and reproduction. Bull Mar Sci 24:235–273

Dawes CJ, Stanley NF, Stancioff DJ (1977) Seasonal reproductive aspects, chemistry and iota carrageenan from floridian *Eucheuma* (Rhodophyta, Gigartinales). Bot Mar 2:137–147

Domínguez-May R, Gasca-Leyva E, Robledo D (2015) Harvest optimization and risk analysis for the mariculture of *Kappaphycus alvarezii* (Rhodophyta, Solieriaceae) reviews in aquaculture. doi:10.1111/raq.12134

Doty MS (1973) Farming the red seaweed, *Eucheuma*, for carrageenans. Micronesica 9:59–73

Doty MS, Alvarez VB (1981) *Eucheuma* productivity. Int Seaweed Symp 8:688–691

Duedall ID, Maul GA (2005) Demography of coastal populations In: Schwartz ML (ed) Encyclopedia of coastal science. Springer, Dordrecht, pp 368–374

Dung PH, Nhan PT (2001) Some results of study on experimental culture of Elkhorn Sea moss *Kappaphycus alvarezii* in Cat Bay, Hai Phong. Proceedings, marine fisheries research. Tuyen tap cac cong trinh nghien cuu nghe ca bien 2:537–547

Eklöf JS, Msuya FE, Lyimo TJ, Buriyo AS (2012) Seaweed farming in Chwaka Bay: a sustainable alternative in aquaculture? In: de la Torre-Castro M, Lyimo TJ (eds) People, nature and research in Chwaka Bay. WIOMSA, Zanzibar, pp 213–233

Eldredge LG (1994) Introductions of commercially significant aquatic organisms to the Pacific Islands. Perspectives in aquatic species management in the Pacific islands Volume 1. Inshore Fisheries Research Project Technical Document No 7, South Pacific Commission, New Caledonia

Eranza DR, Bahron A, Alin J (2015) Sustaining seaweed farming in Malaysia. Int J Bus Mgt 3:201–205

Eswaran K, Ghosh PK, Mairh OP (2002) Experimental field cultivation of *Kappaphycus alvarezii* (Doty) Doty ex Silva at Mandapam region. Seaweed Res Util 24:67–72

FAO, Food and Agriculture Organization of the United Nations (2016) The state of world fisheries and aquaculture. Rome, Italy, p 190

Freile-Pelegrín Y, Robledo D (2006) Carrageenan of *Eucheuma isiforme* (Solieriaceae, Rhodophyta) from Yucatán, Mexico. II. Seasonal variations in carrageenan and biochemical characteristics. Bot Mar 49:72–78

Goes HG, Reis RP (2011) An initial comparison of tubular netting versus tie-tie methods of cultivation for *Kappaphycus alvarezii* (Rhodophyta, Solieriaceae) on the south coast of Rio de Janeiro State, Brazil. J Appl Phycol 23:607–613

Goes HG, Reis RP (2012) Temporal variation of the growth, carrageenan yield and quality of *Kappaphycus alvarezii* (Rhodophyta, Gigartinales) cultivated at Sepetiba bay, southern Brazilian coast. J Appl Phycol 24:173–180

Hambrey Consulting in association with Nautilus Consultants (2011a) Opportunities for the development of the Pacific Mariculture Section, Country Report, Marshall Islands (www.hambreyconsulting.co.uk), 16 pp

Hambrey Consulting in association with Nautilus Consultants (2011b) Opportunities for the development of the Pacific Mariculture Section, Country Report, Papua New Guinea (www.hambreyconsulting.co.uk), 14 pp

Hambrey Consulting in association with Nautilus Consultants (2011c) Opportunities for the development of the Pacific Mariculture Section, Country Report, Solomon Islands (www.hambreyconsulting.co.uk), 15 pp

Hay ME (1981) Herbivory, algal distribution and the maintenance of between habitat diversity on a tropical fringing reef. Am Nat 118:520–540

Hayashi L, Oliveira EC, Bleicher-Lhonneur G, Boulenguer P, Pereira RTL, Seckendorff R, Shimoda VT, Leflamand A, Vallée P, Critchley AT (2007) The effects of selected cultivation conditions on the carrageenan characteristics of *Kappaphycus alvarezii* (Rhodophyta, Solieriaceae) in Ubatuba Bay. São Paulo Braz J Appl Phycol 19:505–511

Hayashi L, Hurtado AQ, Msuya FE, Bleicher-Lhonneur G, Critchley AT (2010) A review of *Kappaphycus* farming: prospects and constraints. In: Israel A, Einav R, Seckbach J (eds) Seaweeds and their role in globally changing environments, cellular origin, life in extreme habitats and astrobiology 15, pp 251–283. ISBN 978-90-481-8568-9, Springer Science, London

Hayashi L, Santos AA, Faria GSM, Nunes BG, Souza MS, Fonseca ALD, Barreto PLM, Oliveira EC, Bouzon ZL (2011) *Kappaphycus alvarezii* (Rhodophyta, Areschougiaceae) cultivated in subtropical waters in Southern Brazil. J Appl Phycol 23:337–343

Hayashi L, Bulboa C, Kradolfer P, Soriano G, Robledo D (2014) Cultivation of red seaweeds: a Latin American perspective. J Appl Phycol 26:719–727

Hayashi L, Novaes A, Santos A (2016) Design and prototype of Kappaphycus alvarezii mechanical harvester. In: 22 international seaweed symposium, 19–24 June 2016, Copenhagen, Denmark (Book of Abstracts)

Hurtado AQ (2016) Genetic resources of farmed seaweeds. Final report submitted to Food and Agriculture Organization (FAO), Rome, Italy, 15 July 2016

Hurtado AQ, Critchley AT (2006) Seaweed industry of the Philippines and the problem of epiphytism in *Kappaphycus* farming. In: Phang SM, Critchley AT, Ang P Jr (eds) Advances in seaweed cultivation and utilization in Asia. University of Malaya Maritime Research Centre, Kuala Lumpur, pp 21–38

Hurtado AQ, Critchley AT, Trespoey A, Lhonneur GB (2006) Occurrence of *Polysiphonia* epiphytes in *Kappaphycus* farms at Calaguas Is., Camarines Norte, Philippines. J Appl Phycol 18:301–306

Hurtado AQ, Critchley AT, Bleicher-L'honneur G (2008) *Kappaphycus* 'cottonii' farming, rev edn. Degussa Texturant Systems, 26p

Hurtado AQ, Gerung GS, Yasir S, Critchley AT (2014) Cultivation of tropical red seaweeds in the BIMP-EAGA region. J Appl Phycol 26:702–718

Hurtado AQ, Reis RP, Loureiro RR, Critcheley AT (2015) *Kappaphycus* (Rhodophyta) cultivation: problems and the impacts of Acadian Marine plant Extract powder, pp 251–299. In: Pereira L, Neto JM (2014) Marine algae: biodiversity, taxonomy, environmental assessment, and biotechnology, 381p

Hussin H, Khoso A (2017) Seaweed cultivation and coastal communities in Malaysia: an Overview. Asian Fish Sci 30:87–100

Jara SZ, Riffo MM (2016) Bioeconomic analysis of small-scale cultures of *Kappaphycus alvarezii* (Doty) Doty in India. J Appl Phycol 28:1133–1143

Komarudin U, Hussein R (2015) The assessment of the seaweed farming sector in Kiribati (a technical, environmental, socio-economic and value chain evaluation). Final Report to FAO, Rome, 34 pp

Krishnan M, Narayanakumar R (2010) Structure, conduct and performance of value chain in seaweed farming in India. Agric Econ Res Rev 23:505–514

Kronen M (2013) Social and economic dimensions of carrageenan seaweed farming in Solomon Islands. In: Valderrama D, Cai J, Hishamunda N, Ridler N (eds) Social and economic dimensions of carrageenan seaweed farming, pp 147–161. Fisheries and Aquaculture Technical Paper No. 580, Rome, FAO, 204 pp

Lal A, Vuki V (2010) The historical development of seaweed farming, including roles of men and women, and prospects for its future development in Fiji. SPC Women in Fish Info Bull 21:11–16

Lang O (2015) Current status of sustainable aquaculture in Cambodia. In: Romana-Eguia MRR, Parado-Estepa FD, Salayo ND, Lebata-Ramos MJH (eds) Resource enhancement and sustainable aquaculture practices in Southeast Asia: challenges in responsible production of aquatic species: proceedings of the international workshop on resource enhancement and sustainable aquaculture practices in Southeast Asia 2014 (RESA), pp 27–40. Aquaculture Department of Southeast Asian Fisheries Development Center, Tigbauan, Iloilo, Philippines

Le Q (2011) Poverty reduction from seaweed *Kappaphycus alvarezii* cultivation. Vietfish Int 8(3) May–Juny issue

Leon RER, Tirado DAM (2011) Aspectos técnicos y económicos para el establecimiento comercial del maricultivo de algas en Colombia: Experiencias en la Península de La Guajira. In: Ambiente Y Desarrolo, v. XV, n. 28, pp 123–144

Leonardi PI, Miravalles AB, Faugeron S, Flores V, Beltran J, Correa J (2006) Diversity, phenomenology and epidemiology of epiphytism in farmed *Gracilaria chilensis* (Rhodophyta) in northern Chile. Eur J Phycol 41:247–257

Lirasan T, Twide P (1993) Farming *Eucheuma* in Zanzibar, Tanzania. Hydrobiologia 260(261):353–355

Luxton D (2001) Trial seaweed farming in the northern Cook Islands. Report prepared for the Ministry of Marine Resources, Cook Islands Government

Luxton DM, Luxton PM (1999) Development of commercial production in Line Islands, Central Pacific. Hydrobiologia 398(399):477–486

Luxton DM, Robertson M, Kindley MJ (1987) Farming of *Eucheuma* in the South Pacific islands of Fiji. Hydrobiologia 151(152):359–362

McLachlan J (1992) Seaweed aquaculture research and development study. South Pacific, University of South Pacific Report, Suva, Fiji, 78 p

Msuya FE (1995) Feasibility study for starting seaweed farming in Lindi and Mtwara Regions, Tanzania. Consultancy report submitted to the Regional Integrated Project Support (RIPS) Programme (Phase I), Mtwara, Tanzania, Institute of Marine Sciences, Zanzibar, Tanzania. 61 pp

Msuya FE (1996) Seaweed farming in Lindi and Mtwara regions, Phase-Two, Implementation and Expansion. Consultancy report for Rural Integrated Project Support (RIPS) Programme, Mtwara, Tanzania, IMS 1997/01, 71 pp

Msuya FE (2010) Development of seaweed cultivation in Tanzania: the role of the University of Dar es Salaam and other institutions. In: Aquaculture compendium. CAB International, Wallingford, UK

Msuya FE (2013) Social and economic dimensions of carrageenan seaweed farming in the United Republic of Tanzania. In: Valderrama D, Cai J, Hishamunda N, Ridler N (eds) Social and economic dimensions of carrageenan seaweed farming, pp 115–146. Fisheries and Aquaculture Technical Paper No. 580. Rome, FAO, 204 pp

Msuya FE, Buriyo A, Omar I, Pascal B, Narrain K, Ravina JJM, Mrabu E, Wakibia JG (2014) Cultivation and utilisation of red seaweeds in the western Indian Ocean (WIO) region. J Appl Phycol 26:699–705

Muñoz J, Freile-Pelegrín Y, Robledo D (2004) Mariculture of *Kappaphycus alvarezii* (Rhodophyta, Solieriaceae) color strains in tropical waters of Yucatán, México. Aquaculture 239:161–177

Nang HQ (2005) Results of study on production of *Eucheuma*, *Kappaphycus alvarezii* in Vietnam and development orientations. Fish Rev 3:17–21

Nang HQ, Dinh NH (1998) The seaweed resources of Vietnam. In: Critchley AT, Ohno M (eds) Seaweed resources of the world. Japan International Cooperation Agency (JICA), Yokosuka, pp 62–69

Neish IC (2013) Social and economic dimensions of carrageenan seaweed farming in Indonesia. In: Valderrama D, Cai J, Hishamunda N, Reidler N (eds) Social and economic dimensions of carrageenan seaweed farming, pp 61–89. Fisheries and Aquaculture Technical Paper No. 580. FAO, Rome, Italy

Nelson SG (1988) Development of phycocolloid-related industries in Oceania. South Pacific Commission workshop on Pacific inshore fisheries resources. Nouméa, New Caledonia, 14–25 March 1988. SPC/Inshore.Fish.Res./WP12, 14 p

Ohno M, Nang HQ, Dinh NH, Triet VD (1995) On the growth of the cultivated *Kappaphycus alvarezii* in Vietnam. Jpn J Phycol (Sorui) 43:19–22

Ohno M, Nang HQ, Hirase S (1996) Cultivation and carrageenan yield and quality of *Kappaphycus alvarezii* in the waters of Vietnam. J Appl Phycol 8:431–437

Oliveira EC (2006) Seaweed resources of Brazil. In: Critchley AT, Ohno M, Largo DB (eds) World seaweed resources-an authoritative reference system. ETI BioInformatcs, Amsterdam. DVD-ROM

Oliveira EC, Silva BNT, Amancio CE (2009) Capítulo 5 – Fitobentos (Macroalgas). In: Rubens ML (ed) Informe sobre as espécies exóticas invasoras marinhas no Brasil/Ministério do Meio Ambiente, Brasília: MMA/SBF, Série Biodiversidade 33, pp 107–142

Ortega MM, Godínez JL, Garduño G (2001) Catálogo de algas bénticas de las costas mexicanas del Gofo de México y el Mar Caribe. Cuadernos 34, Universidad Nacional Autónoma de México, México. 594 pp

Pang T, Liu J, Liu Q, Li Q, Li J (2015) Observations on pests and diseases affecting a eucheumatoid farm in China. J Appl Phycol 27:1975–1984

Parker HS (1974) The culture of the red algal genus *Eucheuma* in the Philippines. Aquaculture 3:425–439

Paula EJ, Pereira RTL, Ohno M (1999) Strain selection in *Kappaphycus alvarezii* var. *alvarezii* (Solieriaceae, Rhodophyta) using tetraspore progeny. J Appl Phycol 11:111–121

Paula EJ, Pereira RTL, Ohno M (2002) Growth rate of the carragenophyte *Kappaphycus alvarezii* (Rhodophyta, Gigartinales) introduced in subtropical waters of São Paulo State, Brazil. Phycol Res 50:1–9

Perez R, Kaas R, Campello F, Arbault S, Barbaroux O (1992) La culture des algues marines dans le monde Institut français de recherche pour l'exploitation de la mer Ifremer. ISBN 2-905434-41-4

Perez JE, Alfonsi C, Nirchi M, Muñoz C, Gómez JA (2003) The introduction of exotic species in aquaculture: a solution or part of the problem? Interciencia 18:234–238

Perez-Enríquez R (1996a) Growth of *Eucheuma isiforme* (C. Agardh) J. Agardh on experimental rafts off the coast of Yucatan State, Mexico. J Appl Phycol 8:27–28

Perez-Enríquez R (1996b) Summer biomass density and weight assessment of the red seaweed *Eucheuma isiforme* (Rhodophyta, Gigartinales) at Dzilam Yucatan state, Mexico. Bot Mar 39:251–253

Pickering T (2005) Advances in seaweed aquaculture among Pacific Island countries (updated October 2005). Presentation to the Secretariat of the Pacific Community (SPC) sub-regional sea-weed meeting, 25–27 October 2005. Nadi, Fiji. www.spc.int/aquaculture

Porse H, Rudolph B (2017) The seaweed hydrocolloid industry: 2016 updates, requirements, and outlook. J Appl Phycol. doi:10.1007/s10811-017-1144-0

Prakash J (1990) Fiji, Fiji. In: Adams T, Foscarini R (eds) Proceedings of the regional workshop on seaweed culture and marketing. Suva, pp 14–17

Ram V (1991) The seaweed industry in Fiji, with special reference to *Eucheuma* (Rhodophyta). University of the South Pacific Marine Studies Programme Technical Report 1991 (2)

Randriambola T, Rafalimanana T (2005) L'algoculture ou la culture d'algues rouges du genre *Eucheuma*. Premier atelier national sur l'aménagement de la pêcherie de céphalopodes, 28–29 juillet 2005. Toliara, Madagascar 8 pp

Reis RP, Pereira RR, Góes HG (2015) The efficiency of tubular netting method of cultivation for *Kappaphycus alvarezii* (Rhodophyta, Gigartinales) on the southeastern Brazilian coast. J Appl Phycol 27:421–426

Reis RP, Castelar B, Santos AA (2016) Why is algaculture still incipient in Brazil? J Appl Phycol. doi:10.1007/s10811-016-0890-8

Ricohermoso MA, Deveau LE (1979) Review of comercial propagation of *Eucheuma* (Florideophyceae) clones in the Philippines. Int Seaweed Symp 9:525–531

Rincones RER, Tirado DAM (2011) Aspectos técnicos y económicos para el establecimiento comercial del maricultivo de algas en Colombia: Experiencias en la Península de La Guajira. Ambiente y Desarollo 28:123–144

Rincones R, Cruz T, Montúfar M (2016) Preliminary assessment of *Kappaphycus alvarezii* cultivation in Ecuador. Int Seaweed Symp (Book of Abstract) 22

Robledo D (2006) Seaweed resources of Mexico. In: Critchley A, Ohno M, Largo D (eds) World seaweed resources. DVD-ROM multimedia interactive software – UNESCO Publishing/ETI Bioinformatics, University of Amsterdam, ISBN 90 75000 80 4

Robledo D, Freile-Pelegrín Y (2011) Prospects for the cultivation of economically important carrageenophytes in Southeast Mexico. J Appl Phycol 23:415–419

Robledo D, Gasca-Leyva E, Fraga J (2013) Social and economic dimensions of carrageenan seaweed farming in Mexico. In: Valderrama D, Cai J, Hishamunda N, Ridler N (eds) Social and economic dimensions of carrageenan seaweed farming, Fisheries and Aquaculture Technical Paper No 580. Rome, FAO, pp 185–204

Sade A, Ali I, Ariff MRM (2006) The seaweed industry in Sabah, East Malaysia. Jati 11:97–107

Salas S, Chuenpagdee R, Charles A, Seijo JC (2011) Coastal fisheries of Latin America and the Caribbean. Food and Agriculture Organization of the United Nations (FAO). Rome. 430pp

Samarakoon JI (2009) Mangroves for the future. National Strategy and action plan. An Ecosystem-Based Integrated Coastal Management in Sri Lanka. In: IUCN, (International Union for Conservation of Nature) Publication, pp 18–20, Colombo, Sri Lanka

Santos AA (2014) Potencial de cultivo da macroalga *Kappaphycus alvarezii* no litoral. PhD's thesis, Universidade Federal de Santa Catarina, Brazil. (in Portuguese)

Santos AA, Novaes ALT, Hayashi L (2016) Flutuadores desenvolvidos para o cultivo da macroalga *Kappaphycus alvarezii* em Florianópolis, Santa Catarina. Abstract in: Congresso Brasileiro de Aquicultura e Biologia Aquática, Belo Horizonte, Brasil

SEDAC – Socio Economic Data and Aplications Center (2016) http//sedac.ciesin.columbia.edu. Accessed 27 Aug 2016

Sepúlveda M (2016) O potencial da maricultura – Oferta mundial da macroalga *Kappaphycus alvarezii* cresce para atender a demanda por carragena. Panorama da Aquicultura 26:38–45

Serpa-Madrigal A, Areces AJ (1998) Influencia de algunos factores fitotécnicos en la actividad de los herbívoros sobre el cultivo de *Kappaphycus* Doty (Gigartinales: Rhodophyta) en el ecosistema arrecifal caribeño. Bol Inst Oceanogr Venez Univ Oriente 37:63–68

Serpa-Madrigal A, Areces AJ, Cano M, Bustamante G (1997) Depredación sobre las carragenófitas comerciales *Kappaphycus alvarezii* y *K. striatum*. Rev Inv Mar (Cuba) 18:65–69

Shanmugam M, Pahalawattaarachchi V, Sivaram K (2016) Farming of seaweed *Kappaphycus alvarezii* in Sri Lanka – a status note and strategy for farm expansion. Int Conf Fish Aqua 3:24–25 (Abstract Only)

Silva BNT, Amancio CE, Oliveira Filho EC (2010) Exotic marine macroalgae on the Brazilian coast: a revision. Oecologia Australis 14(2):403–414

Smith MT (1991) Seaweed farming trials, 1988–1991. Final report of project. Solomon Islands Government document # SOL 205/277/01

Smith AH (1992) Seaweed cultivation in the West Indies. In: Mshigeni K, Bolton J, Critchley A, Kiangi G (eds) Sustainable seaweed resource development in Sub-Saharan Africa. University of Namibia, Windhoek, pp 337–351

Smith AH (1997) Seamoss cultivation in the West Indies. CANARI Guidelines Series 23 pp

Smith AH (1998) Seaweed resources of the Caribbean. In: Critchley AH, Ohno YM (eds) Seaweed resources of the world. Japan International Cooperation Agency, Yokosuka, pp 324–330

Smith AH, Rincones RE (2006) Seaweed resources of the Caribbean In: Critchley AT, Ohno M, Largo DB (eds) World seaweed resources: an authoritative reference system [DVD-ROM]. DVD-ROM multimedia interactive software – UNESCO Publishing/ETI Bioinformatics, University of Amsterdam, ISBN 90 75000 80 4

Solly RK, Booth W (1977) Seaweed colloids from the South Pacific. South Pacific Commission, ninth regional technical meeting on fisheries, Noumea, New Caledonia, 24–28th January 1977, SPC/Fisheries 9/WP 10

South GR (1993) Seaweeds. In: Wright A, Hill L (eds) Nearshore marine resources of the South Pacific. Institute of Pacific Studies, Suva, pp 685–690

Sullivan JB, Ru Wd, Lu W, Chunhua C, Changsheng C, Rawson M, Chopin T, Yarish C (2001) Developemnt of a carrying capacity model for aquaculture in Xincum Bay, Hainan Province P. R. China. In: Proceedings of the 12th Biennial Costal Zone conference, Clevelad Ohio, July 15–19, 2001

The Cambodia Today (2005) Growing seaweed sector brings cash, scandal. www.cambodiadaily.com

The Phnom Penh Post (2008) Port threatens seaweed industry. www.phnompenhpost.com

Tiroba G, McHugh DJ (2006) Solomon Islands – country report. In: McHugh DJ (ed) The seaweed industry in the Pacific islands, pp 47–53. Canberra, Australian Government, Australian Centre for International Agricultural Research. (also available at http://aciar.gov.au/publication/WP61)

Trono GC (1977) *Eucheuma* farming in the Philippines. Lecture delivered during the in-service field training course in seaweed research for BFAR personnel. University of the Philippines, June 1977

Tsiresy G, Preux J, Lavitra T, Dubois P, Lepoint G, Eeckhaut I (2016) Phenology of farmed seaweed *Kappaphycus alvarezii* infestation by the parasitic epiphyte *Polysiphonia* sp. in Madagascar. J Appl Phycol. doi:10.1007/s10811-016-0813-8

Tsutsui I, Nang HQ, Dinh NH, Arai S, Yoshida T (2005) The common marine plants of southern Vietnam. Japan Seaweed Association, Tokyo, 250 p

Uan J (1990) Kiribati. In: Proceedings of the regional workshop on seaweed culture and marketing. South Pacific Aquaculture Development Project. Field Document 1990/2, 10–15

URT-United Republic of Tanzania (2005) Seaweed development strategic plan. Ministry of Natural Resources and Tourism, 47pp

Vairappan CS (2006) Seasonal occurrences of epiphytic algae on the commercially cultivated red alga *Kappaphycus alvarezii* (Solieriaceae,Gigartinales, Rhodophyta). J Appl Phycol 18:611–617

Vairappan CS, Chung CS, Hurtado AQ, Msuya FE, Bleicher-Lhonneur G, Critchley AT (2008) Distribution and symptoms of epiphyte infection in major carrageenophyte-producing farms. J Appl Phycol 20:477–483

Valderrama D, Cai J, Hishamunda N, Ridler N, Neish IC, Hurtado AQ, Msuya FE, Krishnan M, Narayanakumar R, Kronen M, Robledo D, Gasca-Leyva E, Fraga J (2015) The economics of *Kappaphycus* seaweed cultivation in developing countries: a comparative analysis of farming systems. Aquac Econ Manag 19(2):251–277

Why S (1985) *Eucheuma* seaweed farming in Kiribati, central Pacific (1983). Overseas Development Administration, Research Project R3775, 91 p

Wu C, Li J, Xia E, Peng Z, Tan S, Li J, Wen Z, Huang X, Cai Z, Chen G (1989) On the transplantation and cultivation of *Kappaphycus alvarezii* in China. Chin J Oceanol Limnol 7:327–334

Micro-propagation of *Kappaphycus* and *Eucheuma*: Trends and Prospects

5

C.R.K. Reddy, Nair S. Yokoya, Wilson Thau Lym Yong, Maria Rovilla J. Luhan, and Anicia Q. Hurtado

Abstract

Among the red algae, *Kappaphycus* and *Eucheuma* are the two most commercially important carrageenophytes farmed extensively in Southeast Asian countries; they rank top in production in terms of volume of raw material produced. The farming of carrageenophytes has emerged as a successful enterprise and provides a promising, alternative livelihood option for low-income, coastal communities in a number of countries. In 2014, carrageenophyte production worldwide, surpassed other red seaweeds and was placed at the top of the production rankings with 10.99 million MT wet weight (77% of total production of farmed red seaweed) and Indonesia being a major producer. However, over time, the productivity of the crop (carrageenophytes) has declined in some regions due to sourcing of seedlings from single, selected genetic stocks considered to have higher yield potential which resulted in strain fatigue, or loss of vigour. The incidences of disease and epiphytic infestations are on rise in recent times which have severely affected biomass production, as well as the yield and product quality (carrageenan). In order to circumvent the crop productivity issues arising from clonal propagation, the raising of planting materials from spores, derived through the red algal sexual reproductive cycle, has been initiated to support the sustainability of selected, farmed carrageenophytes. Alternatively, *in vitro* tissue culture techniques have also been explored to not only rejuvenate the vigour of seedlings (i.e. the out-planting material) but also to seek seedlings resilient to stress, disease and epiphytes to act as an invigorated mother stock. These efforts have succeeded to a great extent in the development of appropriate techniques for explant culture, callus induction, callus sub-culture and regeneration to micro-propagules with improved traits. The present chapter briefly summarizes the developments and success achieved in micro-propagation of *Kappaphycus* and *Eucheuma* and also provides pointers to both gaps and priority areas for future research required for the advancement of sustainable farming of these carrageenophytes.

C.R.K. Reddy (✉)
Division of Marine Biotechnology and Ecology,
CSIR-Central Salt and Marine Chemicals Research Institute,
Bhavnagar 364002, India
e-mail: crk@csmcri.res.in

N.S. Yokoya
Institute of Botany, Environmental Secretary of São Paulo State,
P.O. Box 68041, 04045-972 São Paulo, Brazil

W.T.L. Yong
Biotechnology Research Institute, Universiti Malaysia Sabah,
Jalan UMS, 88400 Kota Kinabalu, Sabah, Malaysia

M.R.J. Luhan
Aquaculture Department, Southeast Asian Fisheries Development Center (SEAFDEC), Tigbauan, Iloilo 5021, Philippines

A.Q. Hurtado
Integrated Services for the Development of Aquaculture and Fisheries (ISDA) Inc., McArthur Highway, Tabuc Suba, Jaro, 5000 Iloilo City, Philippines

A.Q. Hurtado et al. (eds.), *Tropical Seaweed Farming Trends, Problems and Opportunities*, Developments in Applied Phycology 9, DOI 10.1007/978-3-319-63498-2_5

5.1 Introduction

Kappaphycus and *Eucheuma* are the two important carrageenophyte genera that are extensively, commercially farmed in tropical and subtropical waters. In 2014, carrageenophyte production worldwide surpassed other red seaweeds and remained at the top of the production rankings with 10.99 million MT wet weight (77% of total production of farmed red seaweed with a value over USD 1.1 billion), and Indonesia being a major producer (Nayar and Bott 2014; FAO 2016). Carrageenophyte farming has indeed developed into a successful enterprise in a number of tropical (and subtropical) developing countries and has emerged as a promising alternative livelihood option for low-income, coastal communities (Valderrama et al. 2013). The economic value of carrageenophytes mainly stems from their cell wall polysaccharides including several types of carrageenan which are used as ingredients in a number of processed food products in dairy, meat, food, cosmetics, personal care and also some applications in the pharmaceutical industry; usage continues to grows by 4–5% per annum (Bixler and Porse 2011). Another emerging application of *Kappaphycus* biomass is the production of plant growth biostimulants, as an integrated biorefinery product, manufactured as a part of kappa-carrageenan extraction, through a unique liquefaction process using fresh seaweed directly harvested from the sea (Ghosh et al. 2015; Singh et al. 2016).

Dwindling supplies of cultivated carrageenophytes, more particularly from the Philippines (for various reasons – see elsewhere in this book), coupled with the constantly growing global market demands for carrageenan, provided impetus to seek out new farming grounds in other parts of the world, with similar climatic zones (see chapter by Hayashi et al.). This initiative of crop dissemination (not unlike many terrestrial crops spreading globally over centuries) led to the introduction of *Kappaphycus* and *Eucheuma* in 32 countries, in a relatively short time-frame, in order to sustain commercial production around the world (Zemke-White and Smith 2006).

Commercial farming of carrageenophytes traditionally employs vegetative thalli as asexual propagules broken into convenient segments and termed "seedlings" by the industry (Hurtado et al. 2006). The practice of propagule selection in *Kappaphycus* and *Eucheuma* involves setting aside the fastest growing plants from one season to be utilised as "seed" for the successive season – unfortunately when practiced over and over again, this led to decline in "seedling vigour" and inevitably a decline in a crop production (Dawes et al. 1993). The persistent outbreaks of disease such as: 'ice-ice' and the over-growth of harmful epiphytes and endophytes, the successes of which have been attributed to farm management practices as well as (at least in part) to environmental stresses (Largo et al. 1995) arising from climate change, have also further seriously affected the carrageenophyte farming industry. As a consequence of the downturn and loss of income, there have been serious social and trade-related issues leading to social unrest and unethical practices (Valderrama et al. 2013). Consequently, the importance for the selection and establishment of new varieties, strains and cultivars, together with their preservation using tissue culture and genetic manipulation techniques, has been a real necessity for survival. These techniques provide an immense potential to increase not only crop production but also carrageenan yield and quality and hence value addition to the farmers. In addition, this pioneering work, as applied to seaweeds with marine agronomic value, will also provide further positive applications in marine macroalgal biotechnology as a whole.

A number of studies have dealt with the necessary adaptations and application of pre-existing (as previously used in revolutionary roles for terrestrial crop production), tissue culture techniques to that of economically important seaweeds. As a result, a great deal of success with respect to callus induction and thallus regeneration were reported for a diverse range of seaweeds (Polne-Fuller and Gibor 1987; Butler and Evans 1990; Aguirre-Lipperheide et al. 1995; Reddy et al. 2008, 2010). Although the initial studies were centred on the development of various basic techniques in seaweed tissue culture, their application for *in vitro* selection and propagation of desired strains (Dawes et al. 1993; Kirihara et al. 1997) and production of secondary metabolites (Lawlor et al. 1990; Maliakal et al. 2001; Rorrer and Cheney 2004) has also been accomplished. The present chapter briefly summarizes the developments and successes achieved in micro-propagation of *Kappaphycus* and *Eucheuma* and points to both gaps and thrust areas required for future research in order to further continue and advance sustainable, extensive and future intensive, farming of carrageenophytes.

5.2 Spore Production and Their Developmental Morphology

Trono and Ohno (1989), and Trono and Lluisma (1992) had already reported dramatic decreases in the hardiness and carrageenan quality of the then cultivated varieties of carrageenophytes. One factor which may have contributed to this phenomenon first reported nearly 30 years ago, could have been the common practice of farmers using simple, vegetative cuttings from their pre-existing stocks as further propagules for the succeeding generations of cropping. It was a type of "strain selection" from the "best looking, fastest growing" plants at the earliest sites where cultivation was initiated. Unfortunately, other than growth and colour, there were no other traits selected for such as yield and properties of the carrageenans (Hurtado and Agbayani 2000). This continuous re-transplantation from a single parent stock is likely to have eroded some important genetic traits that resulted in the lower quality of the present stocks (though no studies using molecular markers are known to have taken place). Periodic replacement of propagules with new stocks (various strains, colour and physical morphotypes) as sources of invigorated plantlets was necessary to prevent the deteriorating agronomic traits of the seedlings used at that time. Hence, initiated by the farmers, the industry began to look for alternative approaches to vegetative cuttings in order to produce the new plantlets with vigour and value for a changing sustainable, carrageenophyte seaweed industry.

The earliest studies made extensive efforts to produce seedlings from reproductive, as well as somatic sources. For instance, mature carposporophytes of *K. alvarezii* were shown to produce viable spores *in vitro* (Azanza-Corrales et al. 1992; Azanza and Aliaza 1999; Azanza-Corrales and Ask 2003). Successful culture of micro-propagules of *K. alvarezii* was already undertaken at the laboratory scale (Dawes and Koch 1991; Dawes et al. 1993, 1994). The explants of seaweeds were used for tissue culture and achieved great success (Reddy et al. 2003; Munoz et al. 2006; Hurtado and Biter 2007; Hurtado et al. 2009). Tetrasporophyte progenies were shed (released) in the laboratory and grown outdoors (Bulboa et al. 2007, 2008; Bulboa and de Paula 2005).

The use of spores in seaweed farming has been demonstrated successfully for *Porphyra (Pyropia)*, *Gracilaria*, *Laminaria (Saccharina)* and *Undaria*. In *Kappaphycus*, the convenient use of spores is still at its earliest stage of development as the technique applied to produce plantlets for culture results in the higher production costs of propagules. Of greater importance will be the determination of the quality of seedlings raised from the spores for their growth, carrageenan yield and quality, disease susceptibility etc.. A similar concern is applicable to seedlings generated through *in vitro* tissue culture or any other means, in order to ascertain beneficial features of seedlings prior to their distribution to farmers. Such endeavours are necessary to further understanding of the species in order to achieve enhanced efficiencies of biomass production. *Kappaphycus* has a *Polysiphonia*-type triphasic life cycle (Dawes 1993). Female gametophytes have a carpogonial carpogonial branch while male gametophytes have a superficial spermatangia. In the water column, the spermatia released by the spermatangia find their way to the protruding carpogonial branches on the surface of the thallus and lead the spermatia to the carpogonium for fertilization from which carpospores will develop. The diploid carpospores are a product of syngamy of non-flagellate, spermatia and oogonia; they are contained within cystocarpic bodies and grow on the haploid, female gametophytes. The carposporophytes are attached and acquire nourishment from the female gametophytes. Mature carpospores will be shed and find a substratum up on which to settle and germinate. The germinated carposporeling (2N), remain vegetative until it develops spores which mature into tetrasporophytes. Tetrasporophytes develop tetrasporangial bodies and are the site of meiosis to release haploid tetraspores, which will develop to gametophytes (N). A further complicating factor is that the male and female gametophytes and the tetrasporophytes are isomorphic and are therefore not distinguishable at their early stages of somatic growth. This is yet another major track essential to basic and applied research endeavours that would lead to identify each genotype at the earliest stages of development necessary for improvement of cultivars.

In "wild populations", adult *Kappaphycus,* with mature carposporophytes, are usually found in areas where cover of the seagrasses *Thallasia* and *Cymodocea* is dense. The carposporophytes are also found associated with other seaweeds growing on dead coral and rock substrata. Often they are undistinguishable due to a covering of fine silt. A very keen eye is therefore required to scrutinize seaweed patches in order to collect mature, reproductive *Kappaphycus,* cystocarpic individuals. Cystocarpic *Kappaphycus* could be present year-round, but since mature specimens are rare to find in the field, few individuals are ever collected. Hence there are relatively small amounts of biomass to work with at any one time using this method of selection and propagation. The processing of seaweed tissue in the laboratory involves the use of sterile materials. Seawater is filtered using a 0.45 μm GFC Whatman filter and other tools such as brush, plankton nets, bottle caps, forceps, Petri plates and autoclaved needles. Glassware must be heat sterilized at 100 °C in an oven. Other materials that cannot be autoclaved should be placed under a UV light inside a laminar flow bench. Directly from the point of collection, live seaweed material is placed in a container with sterile seawater (SSW). This tissue is cleaned using a soft brush and rinsed repeatedly to remove adhering debris. Portions of thalli with cystocarps are excised and placed on to a glass slide in a Petri dish with SSW (34‰) and placed at ambient temperature (28 °C) in the dark. The excised cystocarps, thus incubated in Petri plates, showed shedding of spores immediately. Carposporophytic *Kappaphycus* plants will also shed spores *in vitro* by manipulation of culture conditions. "Shocks" caused by increasing or decreasing temperature or light can induce the release of spores. Continuous exposure to light and longer incubation in the dark also induces the shedding of spores in mature carposporophytes (Azanza and Aliaza 1999).

Simple, filtered seawater is a good medium to initiate the release of spores. A substratum is needed for the spores to settle upon. Glass slides are preferred for easy handling, cleaning and replenishing the culture medium. Spores that are released early disperse the furthest distance from the cystocarps on the substrata provided and can be most easily

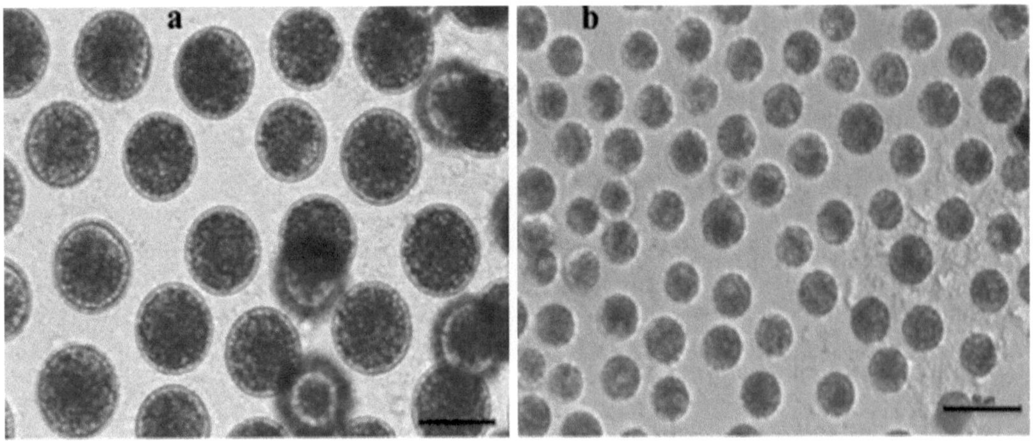

Fig. 5.1 (**a**) Diploid spores and (**b**) haploid spores (bar = 50 μm) (Photos courtesy of MRJ Luhan)

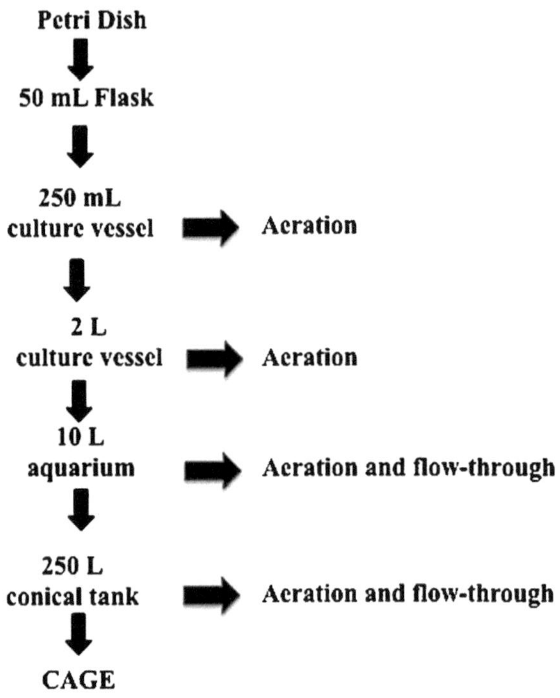

Fig. 5.2 Multi-step culture of *Kappaphycus* in the land-based nursery

collected and/or separated. Those spores released later will be trapped in the natural mucilage of the thalli. The majority of released spores are embedded in the mucilage which does dissolve, but leaves many individuals coalesced. Some spores will be suspended in the medium but settle within 24–48 h after shedding. Spores may settle individually or two or more may coalesce and develop into a juvenile plantlet. There are approximately 200 spores released per mature cystocarp. The mean diameter of diploid and haploid spores is 44.9 ± 2.5 μm and 19.2 ± 1.9 μm, respectively. Diploid spores (Fig. 5.1a) are relatively larger than the haploid spores (Fig. 5.1b). Hyaline hairs appear 6–8 days after settlement and are visible until the sporelings develop a dome shape or shoot primordia. Contamination in culture will appear after 24 h. After 48 h the culture vessel should be changed and a fresh medium added. This process is repeated regularly in order to reduce contamination and over-growth of contaminants such as diatoms, ciliates and bacteria. Initially the growth of sporelings is very slow but can be boosted by gradually increasing the culture medium volume and supply of nutrients in pace with the growth of the sporelings (Luhan and Sollesta 2010).

The first stage of laboratory culture is to grow the new sporelings on glass slides placed onto the bottom of the Petri dish (Fig. 5.2). Following the settlement of spores and their attachment on to the glass slides, the slides are then placed inside a wide-mouth, capped bottle (50 mL capacity) filled with sterile seawater enriched with modified Grund medium (Brown et al. 1977) and kept in an incubator with a fluorescent lamp, with an irradiance of 50 μmol m^{-2} s^{-1}, 13:11 light:dark photoperiod and at 25 °C. An irradiance between 50–175 μmol m^{-2} s^{-1} should be maintained during the culture. Slides are oriented at 45°. Small volumes of medium in bottles are used during the initial culture period with tiny sporelings (<3 mm). The culture medium should be replenished on a daily basis. Sporelings are then scraped off the glass when they reach 3 mm and are transferred to a container with 50 mL medium and cultured in suspension without aeration. As the culture progresses, larger flasks (250 mL) are used and aeration provided to tumble the water and simulate natural conditions. The use of bubblers prove to be ideal for the mixing of water in order to prevent shading and ensure the efficient distribution of nutrients for the sporelings.

Sporelings harvested from the 2L flasks are then transferred to a 10L aquarium outdoors. Filtered seawater is used to grow the plantlets in the aquarium. The water, in the aquarium is vigorously aerated with flow-through, filtered seawater. Fertilization is made twice a week by adding

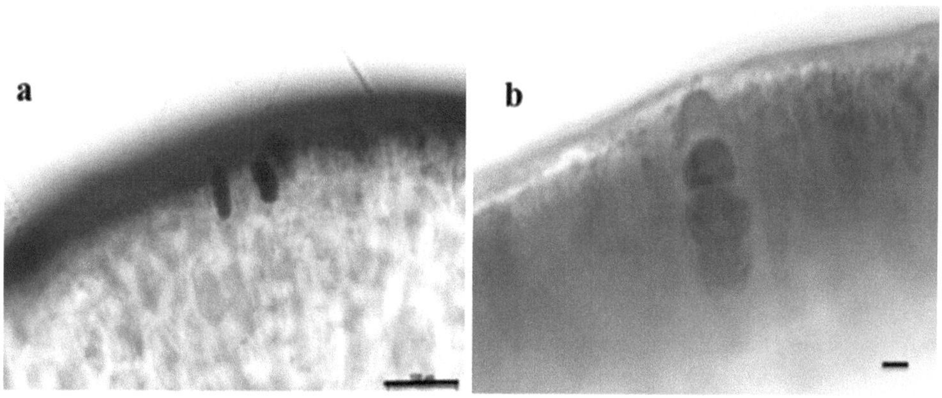

Fig. 5.3 (**a** and **b**) Cross-section of tetrasporophytes (bar = 100 μm) (Photos courtesy of MRJ Luhan)

modified Grund medium. The water supply is shut off before the application of fertilizer. The water supply is reopened after 2 h. Irradiance (172–1200 μmols m^{-2} s^{-1}), salinity (32–34 ppt) and temperature (28–30 °C) are to be maintained under ambient outdoor conditions. Plantlets are grown for 28 days in the aquaria. Plantlets are stocked at 50 individuals per aquarium. Sporelings from the 10 L aquaria are further grown in 250 L outdoor tanks with 200 L seawater in a flow-through system, with 400 individuals, at a flow rate of 1L min^{-1} with vigorous aeration. A flow-through water system brings constant water movement that further simulates natural conditions, preventing self-shading and providing additional nutrients favouring good growth. The tank culture is the last step in the land-based nursery.

Five transfers/steps are made until the sporelings reach approximately 50–60 mm in length. Generally, a marked increase in growth is noted when the sporelings are transferred from a smaller to larger culture containers. The additional extra space and nutrient availability explains the increase in growth at each change. Initially the growth of sporelings is very slow. It will take 150 days from spores before the sporelings are transferred to 250 mL flasks which is between 5 and 8 mm. The growth of sporelings in 250 mL (approx. 0.04 g), 2 L (approx. 0.6 g), aquaria (approx. 0.28 g) and tanks (approx. 15 g) are 4.9%, 9.6%, −2.7% and 4.6% day^{-1}, respectively. Negative growth was observed in the aquaria which could be due to static conditions (i.e. the absence of flow-through) in the first few days after stocking in outdoor conditions. 'Ice-ice' was observed after 5 days of culture and recovered after 12 days when the water was continuously changed or flow-through system started.

The diploid *Kappaphycus* sporelings will mature and can be transferred to produce mature tetrasporophytes *ex situ* in cages. Similarly, mature tetrasporophytes will shed their own (haploid) spores *in vitro* when environmental parameters are manipulated. Again, exposure to longer periods of light, or dark, can be used to induce the release of spores. Tetrasporangia of *Kappaphycus* are zonate bodies (Fig. 5.3). The spores can be found mostly solitary on the glass substratum.

Not all species (strains/variants and morphotypes) of *Kappaphycus* thrive well in tanks. However, a tank-culture phase is essential if a land-based nursery is to provide a continuous supply of healthy seedlings for out-planting by the farmers especially during those situations where there are shortages of seedlings arising from high mortalities at the cultivation grounds (this is not unlike the general practice for a number of aquaculture species nor indeed the concept of specialist (expert centres) which look after breeding and improvements of major terrestrial crops such as banana, rice, potato, corn, etc.). In countries where several typhoons may occur each year (see Chapter by Largo et al.), a land-based nursery is essential over a sea-based nursery, since total loss of stocks and mortalities due to typhoons can be avoided. The land-based system is a secure repository which also will be increasingly important for chain of custody assurances required by high value applications of biomass, i.e. pharmaceutical applications where there must be strict traceability back to the original germplasm. In many instances, destruction of seaweed stocks has been observed in grow-out farms after typhoons, even contributing to a decline in Philippines' production and the surge of Indonesia as the number one producer (since the coastlines of that country effectively do not suffer the devastating consequences of typhoons and storm surges). It is expensive to maintain land-based nurseries, but this may be undertaken by institutions with existing facilities or perhaps funded by the companies buying biomass for processing.

5.3 Clonal Propagation

Most of the carrageenophytes cultivars which are farmed at present, in a circumferential band around the oceans bounded by the tropics (see the maps in, Chap. 3) have been sourced from limited seed stocks with proven productivity and high yield of product. The widespread proliferation of such stocks across different geographical regions led to very large populations. However, these stocks need to be reinvigorated and improved in order to achieve overall sustainability in the carageenophyte farming industry.

All of the studies dealing with clonal selection have successfully utilised the regeneration potentials for isolating superior clones from wild stocks. The species studied include: *Chondrus* (Cheney et al. 1981), *Gigartina* (Sylvester and Waaland 1983), *Gracilaria* (Patwary and van der Meer 1982, 1983) and *Kappaphycus* (Doty and Alvarez 1973). From the very inception, the seaweed selection approach has been from an empirical perspective for a single trait but later studies have combined superior growth with superior polysaccharide yields. Unfortunately, these studies, apart from confirming the faster growth in a particular location, did not progress further to an understanding of the genetics of these traits. This has led to the development of modern molecular biology techniques in order to transform and improve the cultivars (Wang et al. 2010). Recently Handayani et al. (2014) attempted to produce transgenic *Kappaphycus* resistant to bacterial infections by cloning the lysozyme (*Lys*) gene from chickens through an *Agrobacterium*-mediated gene transfer.

Recently, Fadilah et al. (2016) employed recurrent selection methods to select high growth rate germplasm from *K. alvarezii*, from a population having a DGR >10%. Incidentally, the selected germplasm also had the favourable characteristics of higher clump weight, average daily growth rate, longer thallus potential, with a greater number of branches with shorter internodes, as compared to the control plants. Also, total sugar content and the ratio of kinetin and indole-3-acetic acid were reportedly higher in the selected germplasm than that of the control control. The growth rate of this germplasm was assessed for four generations and found that more than 90% showed aconsistent DGR of 10%. Although this approach, in general, has been hailed as successful (van der Meer and Patwary 1983; van der Meer 1986), it entails continuous, cumbersome monitoring and isolation of clones with superior quality for a given trait and finally the costly maintenance of the selected strains. In another study, Titlyanov and Titlyanova (2006) and Titlyanov et al. (2006) reported methods for the mass production of *Gelidium* for land-based cultivation, using fragments and cell aggregates of apical meristems. Furthermore, freeze-thawing of apical meristem tissues enabled the production of plantlets which produced rhizoids that could be used for cultivation in the sea. These methods effectively maximized the number of propagules per donor plant and further facilitated mass production of seed stock. However, the selection of the "elite germplasm" through clonal propagation is a continuous process and a substantial amount of harvest is required to be utilized as initial seed material from which the subsequent selections are to be made, which are then require to be validated in subsequent cultivation cycles. This largely forms part of "domestication of species" to bring wild germplasm under a farming line. Further, isolation of useful germplasm lines for cultivation through this process is less likely to be successful as genetic variant cells are small in number and are masked by more common non-variant cells (Garcia-Reina et al. 1991).

5.4 Seaweed Extracts as Sources of Bio-Stimulant Used in Generating Microplantlets of *Kappaphycus*

5.4.1 Brief Description of AMPEP (Acadian Marine Plant Extract Powder) as a Bio-Stimulant for a red Seaweed Propagation and Enhanced Vigour

AMPEP (Acadian Marine Plant Extract Powder) is an acronym proposed and popularised by Hurtado (pers comm). Although based on an existing commercial agrochemical, extract, the product is not currently promoted. It is however a brown seaweed-based bio-stimulant (bioeffector) extracted from *Ascophyllum nodosum*. This brown seaweed grows luxuriantly along coasts of the North Atlantic, notably Canada, France, Iceland, Ireland, Norway and the United Kingdom (the limits to distribution are north from New York in the US to Portugal) (Ugarte and Sharp 2012).

Ascophyllum nodosum is an important commercial seaweed in Canada (Ugarte et al. 2006). It is a dominant, perennial seaweed and forms extensive beds where conditions are suitable (Ugarte and Sharp 2001). In the past, *A. nodosum* was used as a source of biomass for the extraction of alginate and production of "kelp meal" (unspecified origin, dried and milled seaweed flakes and/or powders). Other uses of this resource include manufacture of extracts (various methods of extraction) to produce a bio-stimulant i.e. a liquid concentrate and dried seaweed powder extract for agricultural and horticultural uses. The production of this agrochemical extract is an important economic activity of the Maritime Provinces, Canada, and similarly extracts are manufactured in Ireland, Norway and US.

AMPEP is brownish-black in colour, crystalline in nature with a typical marine odour, 100% soluble in water with a pH 10.0–10.5. Its total nitrogen (N), available phosphoric acid (P_2O_5) and soluble potash (K_2O) contents are 0.8–1.5%, 1–2% and 17–22%, respectively. Its total amino acid content is 4.4% plus other elements such as sulphur, magnesium, calcium, sodium range from 0.2–5%; boron, iron, manganese, copper, zinc range from 1–250 ppm. This seaweed extract is a proven source of bio-stimulant (bioeffector) properties (Khan et al. 2009; Craigie 2011).

Earlier studies reported on the significant role of *A. nodosum* extract (ANE) on the improvement of disease tolerance in some agricultural crops, as reported by Jayaraj et al. 2008, 2011; Fan et al. 2011, to name a few. AMPEP has been used not only in the micro-propagation of *Kappaphycus* plantlets,

Table 5.1 Varieties of *Kappaphycus* treated in various concentrations of AMPEP and AMPEP + PGR produced shoot primordia at different rates (Hurtado et al. 2009)

Treatment	Concentration	Shoot primordia (days post-culture initiation)			
	Mg L^{-1}	Kapilaran (KAP)	Tambalang (PUR)	Adik-Adik (AA)	Green sacol (GS)
AMPEP	0.001	–	–	25	–
	0.01	–	–	39	–
	0.1	72	–	39	–
	1.0	–	–	25	–
	2.0	–	–	19	–
	3.0	–	21	17	–
	4.0	60	21	17	–
	5.0	60	21	17	–
AMPEP + PGR	0.001	–	41	17	–
	0.01	–	22	–	–
	0.1	–	22	17	–
	1.0	49	22	21	25
	2.0	49	22	19	–
	3.0	46	21	17	–
	4.0	46	21	17	–
	5.0	–	31	17	–
ESS/2 (control)		–	78	–	–

Note: (–) indicates absence of cell stage

in the Philippines (Hurtado et al. 2009; Yunque et al. 2011) but also to enhance the growth rate of field cultivated *Kappaphycus* (Philippines, Hurtado et al. 2012) and to mitigate the incidence of epiphytes (in Brazil, Loureiro et al. 2010, 2012) and the harmful endophyte *Neosiphonia* (Borlongan et al. 2011), improved carrageenan yield (Loureiro et al. 2014) and the impact of epibiont settlement on *Kappaphycus* (Marriog et al. 2016; see chapter by Loureiro et al.).

The role of AMPEP as an efficient medium for direct shoot formation and development in the micro-propagation of *Kappaphycus* plantlets is briefly discussed below. These results are likely to be of great interest in the pursuit of developing plantlets at the commercial scale. AMPEP studies with *Kappaphycus* have indeed provided a lot of proof of concept information, but there is still much further research required, not least in the efficacy of various types of other extracts from the other seaweeds along with understanding the underlying mechanisms imparting the positive responses observed.

5.4.2 Use of AMPEP + Plant Growth Regualtors (PGRs)

When three varieties of *Kappaphycus alvarezii*:viz. (Kapilaran, KAP), Tambalang purple (PUR), Adik-adik (AA) and one variety of *Kappaphycus striatum* var. sacol (green sacol; GS) were tested with different concentrations of AMPEP + PGR (=Phenyl Acetic Acid (PAA) + Zeatin at 1 mg L^{-1}), each variety responded differently in the number of days taken for shoot formation, as shown in the Table 5.1. PUR and AA initiated shoot formation with the use of AMPEP only at higher concentrations (i.e. 3–5 mg L^{-1}) after a shorter period. Only PUR responded positively to ESS/2 for shoot initiation.

5.4.3 Optimization of AMPEP as a Culture Medium

The use of different low concentrations of AMPEP with, or without, PGR led to the enhanced development of shoots in six varieties of *Kappaphycus*. The earliest shoot emergence was observed to occur at low concentrations of AMPEP (e.g. 0.001–1.0 mg L^{-1}) with PGR, or at high concentrations of AMPEP (3–4 mg L^{-1}) without PGRs (Table 5.2). Only the control treatment of PUR formed shoots after 78 days of culture in ESS/2. The rate of production of new and improved *Kappaphycus* explants for a commercial nursery stock was improved through the use of AMPEP with optimized culture media pH, temperature and density conditions.

Explants of TGR cultured at all AMPEP concentrations with, or without PGR, produced the first shoots within the same time frame (i.e. 19 days post-culture). An optimum concentration of 3.0 mg L^{-1} AMPEP and 0.001 mg L^{-1} AMPEP + PGR was determined for TGR, based on the generated shoot length (Fig. 5.4). Significant differences ($P < 0.01$) were observed between the lengths of shoots under different treatments. However, no definite pattern of response to AMPEP with, or without PGR, was observed (Fig. 5.5).

Table 5.2 Summary of media composition that produced the earliest shoot primordia in the six *Kappaphycus* varieties tested, as compared to the control medium (ESS/2 + PGRs) (Yunque et al. 2011)

	Variety	First occurrence of shoot primordia (no. of days)	AMPEP concentration	ESS/2
K. alvarezii	Tambalang purple (PUR)	21	3 mg L^{-1}; 0.1 mg L^{-1} + PGR	78 days
	Kapilaran brown (KAP)	49	1 mg L^{-1} + PGR	–
	Vanguard brown (VAN)	22	0.1 mg L^{-1} + PGR	–
	Adik-Adik brown (AA)	17	3 mg L^{-1}; 0.001 mg L^{-1} + PGR	–
	Tungawan green (TGR)	19	3 mg L^{-1}; 0.001 mg L^{-1} + PGR	–
K. striatum	Sacol green (GS)	25	1 mg L^{-1} + PGR	–

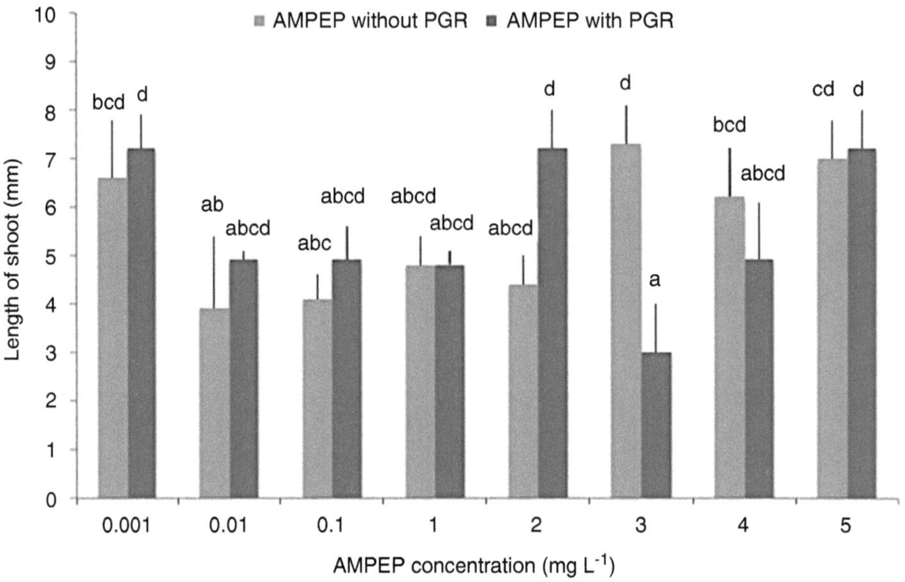

Fig. 5.4 Average length (±SD) of shoots developed from TGR grown in various concentrations of AMPEP with and without PGRs after 19 days of culture. Means of the same letter are not statistically different at the p = 0.01 level of significance (Yunque et al. 2011)

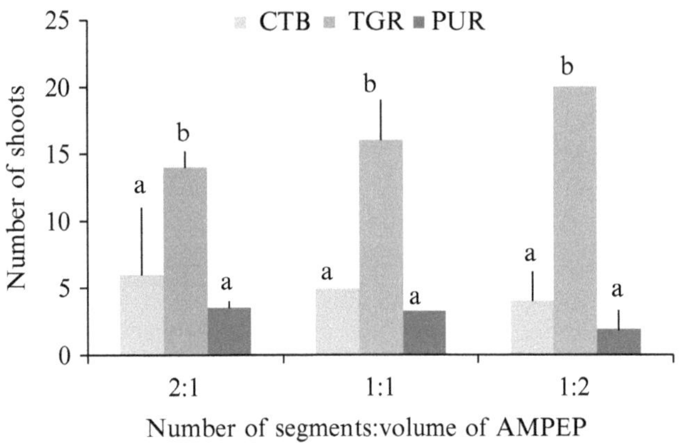

Fig. 5.5 Average density (±SD) of segments with shoots developed from CTB, TGR, and PUR. Means having the same letter are not statistically different at the p = 0.01 level of significance (CTB curly *top brown*, TGR Tambalang *Green*, PUR *Purple*) (Yunque et al. 2011)

The culture medium supplemented with optimum medium concentrations of AMPEP, in the presence of plant growth regulators (PGR), were used *vis-a-vis*: (1) pH– temperature combinations (2) explant density and volume of medium in order to improve the production of *Kappaphycus* plantlets in tissue culture.

Various pH–temperature combinations elicited shoot growth in GS, PUR and BRN. After 54 days of incubation, the highest number of shoots formed with the least mortality, were pH 8.7 and 9.7 at 25 °C for GS (11 = 46%), pH 6.7 at 25 °C for PUR (15 = 63%) and pH 6.7 and 7.7 at 20 °C for BRN (5 = 21%). All varieties responded positively to almost all pH–temperature combinations, except GS, which did not produce shoots at any pH value at 20 °C (Table 5.3).

Explant number versus the culture medium volume combination was also tested. Formation of *Kappaphycus* shoot primordia was observed as early as day 9 for TGR and PUR and after day 21 for Tungawan CTB. The highest average number of new shoots formed was determined at 2:1 density of 1.0 mg L^{-1} AMPEP + PGR for both brown and purple color morphotypes (35% and 100%, respectively). The green morphotype produced the most number of shoots at 1:2 density of the same culture media (16.7%). Remarkably, no mortality was recorded in either brown or green morphotypes. Mortality was only observed in PUR cultures with the highest value of 53.3% at the 2:1 density. Significant differences ($P < 0.01$) were observed between the strains and volume of AMPEP:density treatments.

5.4.4 Use of AMPEP + Spindle Inhibitors + PGRs

AMPEP K^+ is a bio-stimulant produced from *A. nodosum*, however, the K^+ content is higher than in the regular AMPEP (Tibubos et al. 2016). When AMPEP K^+ was used, singly or in combination with spindle inhibitors (colchicine and oryzalin) + PGR, the production of 1–4 direct axes (shoots) of *K. alvarezii* were recorded. This confirmed the observations of Hayashi et al. (2008) and Neves et al. (2015) for *K. alvarezii*. A highly significant difference ($p < 0.01$) in shoot length was observed when *K. alvarezii* was compared with those treated with AMPEP K^+ + PGR (Control B), indicating the efficacy of the PGR addition in shoot formation and elongation. No significant differences ($p > 0.05$) in shoot lengths were observed when *K. alvarezii* shoots were treated with AMPEP K^+ (Control A) vs AMPEP K^+ + colchicine + PGR at all levels of concentrations, which simply indicated that colchicine demonstrated the same effect as the Control A in shoot formation and elongation. However, highly significant differences ($p > 0.01$) were observed when *K. alvarezii* was treated with AMPEK K^+ + oryzalin + PGR vs. Control A. This indicated a strong efficacy of oryzalin to aid in shoot formation and elongation. When AMPEK K^+ + oryzalin + PGR and AMPEP K^+ + colchicine + PGR were compared in relation to their shoot elongation, no significant differences ($p > 0.05$) were observed at all concentrations, which simply indicated that either of these two types of spindle inhibitor could be useful in controlling shoot formation and elongation in *K. alvarezii* (Figs. 5.6a–e, 5.7a–e, 5.8a–o and 5.9a–o).

Table 5.3 Effect of pH and temperature combinations on number of explants with shoots developed from GS, PUR, and BRN after 54 days incubation, in 1.0 mg L^{-1} AMPEP + PGR enriched medium (n = 24) (Yunque et al. 2011)

Variety	Temperature (°C)	pH			
		6.7	7.7	8.7	9.7
Sacol green (GS)	20	0	0	0	0
	23	5	2	4	3
	25	10	8	11	11
Tambalang purple (PUR)	20	6	4	6	4
	23	8	6	10	5
	25	15	13	11	4
Tambalang reddish brown (BRN)	20	5	5	4	3
	23	2	2	0	0
	25	3	5	0	3

5.5 Tissue Culture and Micropropagule Production

Several protocols for callus induction, callus sub-culture and thallus regeneration have been reported for a wide range of seaweeds (Polne-Fuller and Gibor 1984; Aguirre-Lipperheide et al. 1995; Kumar et al. 2007; Reddy et al. 2003). The earliest tissue culture attempts for *Eucheuma* and *Kappaphycus* species were those of Polne-Fuller and Gibor (1987) followed by Dawes and Koch (1991). The methods employed in seaweed tissue culture generally involve the preparation of axenic explants; their culture in optimized, enriched growth media supplemented with plant growth regulators; callus induction; callus sub-culture and *de novo* thallus regeneration (Bradley 1991; Reddy et al. 2003). The first report for the production of callus from the cortical tissues of *Eucheuma uncinatum* (Polne-Fuller and Gibor 1987), followed by medullary tissue of *Kappaphycus alvarezii* (Dawes and Koch

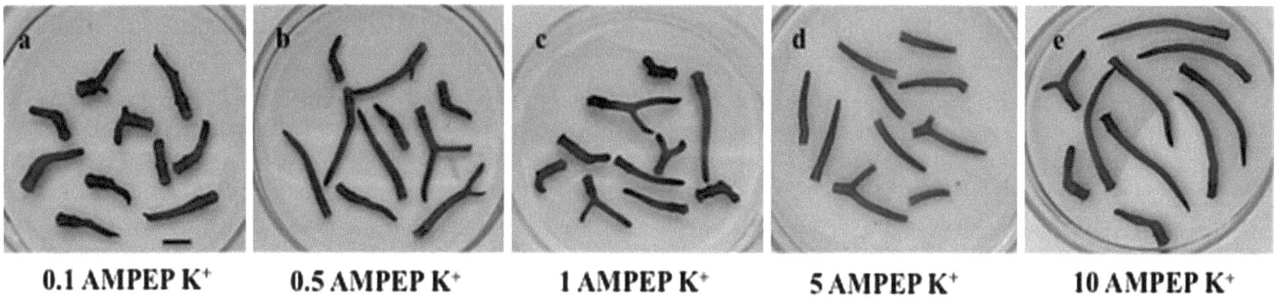

Fig. 5.6 (a–e) Length of shoots (mm) developed after 90 days using different concentrations (mg L^{-1}) of AMPEP K^+ (bar = 1 cm)

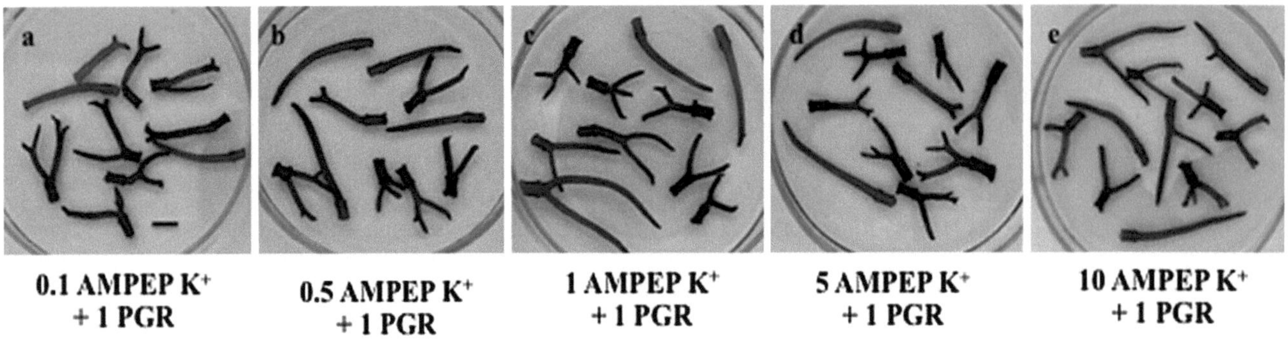

Fig. 5.7 (**a**–**e**) Length of shoots (mm) developed after 90 days using different concentrations (mg L^{-1}) of AMPEP K^+ + 1 mg L^{-1} PGR (IAA + Kinetin) (bar = 1 cm) (Photos by AQ Hurtado)

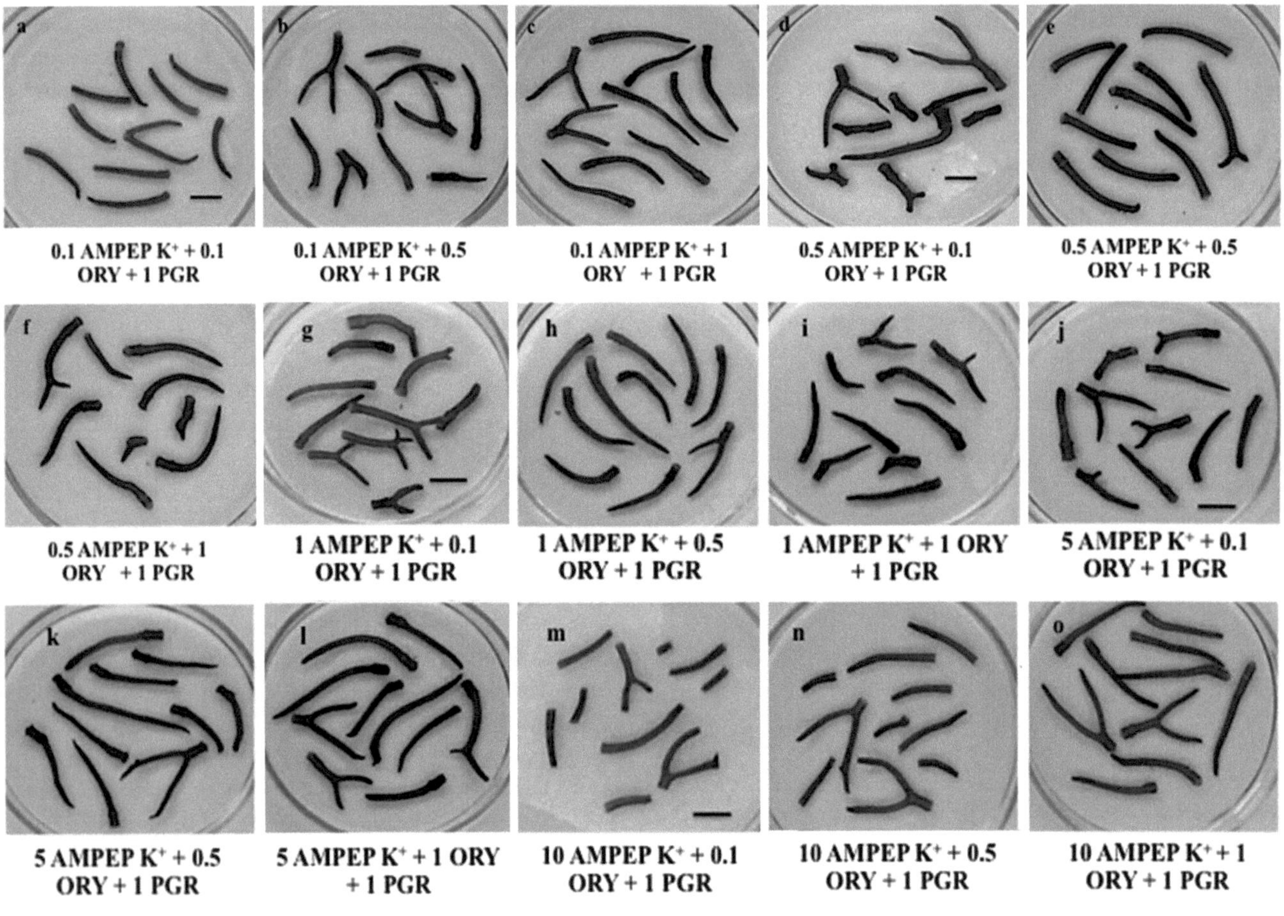

Fig. 5.8 (**a**–**o**) Length of shoots (mm) developed after 90 days using different concentrations (mg L^{-1}) of AMPEP K^+ + Oryzalin (mg L^{-1}) + 1 mg L^{-1} PGR (IAA + Kinetin) (bar = 1 cm) (Photos by AQ Hurtado)

1991) stimulated researchers to develop further successful tissue culture protocols for *Eucheuma* and *Kappaphycus* species encouraged by their economic importance. A number of studies subsequently showed regeneration of micropropagules directly from explants of *E. serra* (Sahoo et al. 2002), *K. striatus* (Yunque et al. 2011), *K. alvarezii* (Neves et al. 2015; Yong et al. 2014b) and other red seaweeds, as reviewed by Reddy et al. (2008) and Baweja et al. (2009). Seaweed breeders used micro-propagation technology as a means of maintenance and clonal propagation of seedstock for the mariculture of economically important seaweeds including *Kappaphycus* and *Eucheuma* (Hurtado et al. 2015).

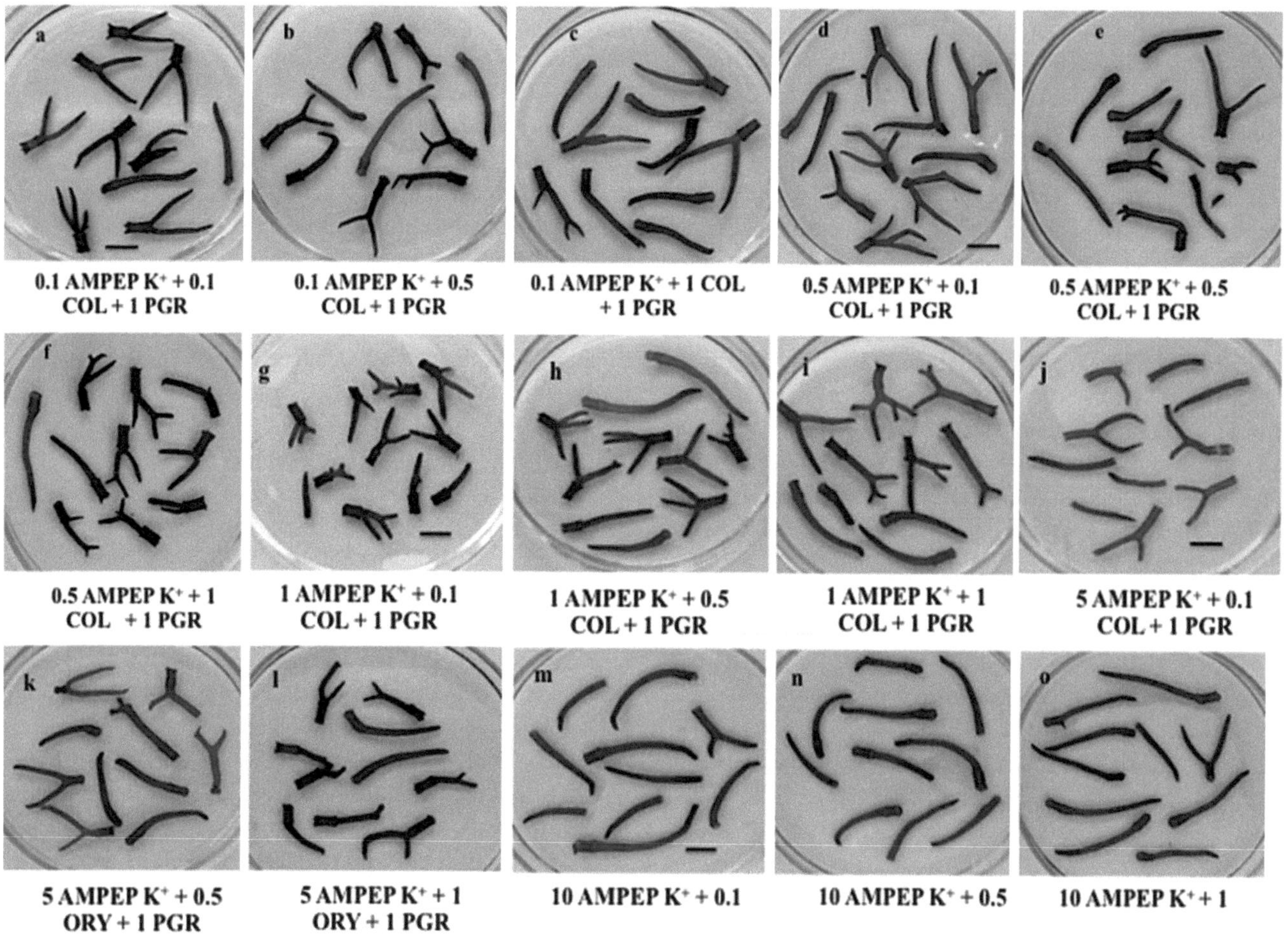

Fig. 5.9 (**a–o**) Length of shoots (mm) developed after 90 days using different concentrations (mg L^{-1}) of AMPEP K^+ + Colchicine (mg L^{-1}) + 1 mg L^{-1} PGR (IAA + Kinetin) (bar = 1 cm) (Photos by AQ Hurtado)

While tissue culture offers an exciting avenue for the selection of high quality germplasms and improvement of productivity, achievement of definite goals in seaweed tissue culture is often accompanied by the adoption of appropriate techniques and culture methods. In general, callus induction and somatic embryogenesis are used for the selection and propagation of desired strains, with economic importance, as reported by Reddy et al. (2003) and Yeong et al. (2014) in *K. alvarezii*. Somaclonal variation induced from undifferentiated callus with a high degree of genetic (including genomic, chromosome and epigenetic) variability provides a significant opportunity for the selection of improved phenotypes with desired heritable changes from regenerated cultures (Garcia-Reina et al. 1991).

While achieving rapid propagation of a superior strain, maintaining the genotype, micro-propagation via direct regeneration is an advantage over typical callus culture in order to produce "true-to-type" materials devoid of somaclonal variation which can arise from callus de-differentiation (Yong et al. 2014b). Establishment of seaweed culture through the development of adventitious rhizoids from small thallus fragments has been recommended to exploit the regeneration potential of explants for the production of uniform seedlings in high quantity, from a small number of donor plants (Garcia-Reina et al. 1991; Yunque et al. 2011). With the cutting-edge production technology, as compared to conventional vegetative propagation, tissue culture and micro-propagation have provided tools for an improved strategy to enhance the production of commercially viable *Kappaphycus* and *Eucheuma* seaweeds in order to fulfil the growing global demands of their raw materials (Hurtado et al. 2015; Yong et al. 2014b; Yunque et al. 2011). Muñoz et al. (2006) and Yong et al. (2014b) published the detailed design and application of an air-lift photobioreactor which could achieve optimal growth of *K. alvarezii* controlling all of the important abiotic growth factors such as light delivery, mixing time and supplementation of plant growth regulators. Successful implementation of bioreactors in future seaweed cultivation enterprises will be of benefit to control and intensify the biomass production while circumventing the barriers of seasonality and environmental variability (Reddy et al. 2008; Yong et al. 2014b). Further investigations are aimed at cutting down the capital and operational costs of bioreactors. Additionally, bioprocess engineering of cell and tissue cultures created a potential platform for the controlled production of high value chemicals such as pharmaceuticals and nutraceuticals from marine seaweeds (Rorrer and Cheney 2004), including *Eucheuma* and *Kappaphycus* which could provide rich sources of valuable, bioactive compounds (Farah Diyana et al. 2015; Holdt and Kraan 2011; Kumar et al. 2008; Lau et al. 2014; Matanjun et al. 2008; Raman and Doble 2014; Reddy et al. 2008). However, the manufacturing capacity, chain of custody and security of supply are found lacking and required for safer and more efficacious biopharmaceuticals production from valuable species. A listing of *Eucheuma* and *Kappaphycus* species from which different types of cell and tissue cultures have been reported to date is provided in Table 5.4.

5.5.1 Axenic Culture Development

The steps adopted for preparing axenic, viable explants from seaweeds are quite similar to those techniques followed for land plant tissues, and are briefly described in the following.

The development of axenic culture techniques is an indispensable feature for successful tissue culture. However, it is difficult to achieve due to the absence of a cuticle on the algal surface, which would normally protect land plant tissues from damage caused by sterilizing chemicals. Therefore, it is very important to develop a protocol that selectively eliminates the contaminating bacteria, fungi and other microorganisms without affecting the viability of the donor plant tissues.

The initial explants to establish axenic tissue culture of *Kappaphycus* species are generally taken from either apical segments or thallus sections isolated from epiphyte-free, thallus or unialgal cultures (Table 5.5). Basically, the protocol for preparing axenic explants employs two steps, i.e. surface sterilization and incubation in a culture medium supplemented with a broad spectrum antibiotics and antifungal agents. The former employs the treatment of explants - first with detergent and then povidone iodine (Reddy et al. 2003; Yunque et al. 2011; Sulistiani et al. 2012), or sodium hypochlorite (Hayashi et al. 2008; Zitta et al. 2013). The latter involves the incubation in broad spectrum antibiotics with an antifungal agent, as summarized in Table 5.5, for eliminating microbial contaminants from both epiphytic and endophytic origin. In contrast, Dawes and Koch (1991) succeeded in producing axenic explants with antibiotic treatment alone from *Eucheuma* and *Kappaphycus*, and the most effective antibiotic soak was 30 min in polymixin B sulphate (300 mg 100 mL^{-1}).

The procedure to obtain axenic material of *K. alvarezii* described by Reddy et al. (2003) was the following: apical thallus fragments of approximately 5 cm in length were manually cleaned with a brush in filtered seawater with 0.5% liquid detergent (Charmy green, Japan) in sterile seawater for 10 min. Explants were then treated with 2% povidone iodine (available iodine 0.5% w/v) in sterile seawater for 3 min. Finally the explants were incubated in sterile PES medium (Provasoli 1968) containing 3% filter-sterilized broad spectrum antibiotic mixture (Polne-Fuller and Gibor 1984) with nystatin (fungicide) for 2 days (Table 5.5).

Sulistiani et al. (2012) employed a similar protocol, as described by Reddy et al. (2003). Apical segments (4–5 cm) were soaked in Tween® solution (5 drops in 200 mL) for 10 min, after which they were rinsed twice in 0.5% povidone iodine for 3 min. After, explants were rinsed 3–4 times with sterilized seawater, dried with sterilized tissue paper, and immersed in Conwy medium (Liao et al. 1983) containing 3% antibiotic mixture of Polne-Fuller and Gibor (1984).

Yunque et al. (2011) studied seven varieties of *K. alvarezii* and one variety of *K. striatus* (var. Sacol Green), and used the following protocol: apical segments (2 cm) were brushed with 0.05% povidone iodine, cleaned by shaking three times in 50 mL centrifuge tubes with fine glass beads and sterile filtered seawater. Cleaned segments were incubated in 9.1% E3 anti-bacterial solution (Bradley et al. 1988) in sterile, filtered seawater.

Hayashi et al. (2008) incubated explants in antibiotic and anti-fungal solutions. Apical branches of *K. alvarezii* (5 cm length) were selected from unialgal cultures of tetrasporophytes (red, brown and green strains) and a gametophyte strain (named as "Edison de Paula", EP). These branches were incubated with seawater enriched with a half strength Von Stosch solution (Oliveira et al. 1995) with anti-fungal and antibiotic solution (composed of 10,000 units Penicillin G, 10 mg Streptomycin sulphate and 25 μg Amphotericin B per liter, and 0.1 μg mL^{-1} Nystatin, Sigma) for 2 days. After, explants were washed with a solution of sterile seawater with 0.05% sodium hypochlorite and 2% organic detergent (Amway, USA) for 20 s, followed by successive rinsing with sterile seawater inside a laminar flow chamber.

Zitta et al. (2013) reported on the callus ontogeny of the brown tetrasporophyte of *K. alvarezii* and used a similar protocol to obtain axenic explants, as described by Hayashi et al. (2008), with the exception that explants were washed four times in a solution of sodium hypochlorite 0.05% and four times in 2% organic Triton X-100 detergent (Sigma) for 5 s at each washing.

Table 5.4 Types of cell and tissue culture for *Eucheuma* and *Kappaphycus* species with related culture media and exogenous factors

Species	Tissue	Culture	Medium	Irradiance (μmol m^{-2} s^{-1})	Photoperiod (L:D)	Temperature (°C)	References
E. uncinatum & *E. alvarezii*	Explants	Callus formation & regeneration	PES (solid)	40–60	12:12	18	Polne-Fuller and Gibor (1987)
E. denticulatum	Branches	Direct regeneration & callus formation	ESS (solid)	20–160	12:12	18–26	Dawes and Koch (1991)
E. denticulatum	Branches	Clonal propagation & callus formation	ESS + CW (solid)	25–50	NA	23–25	Dawes et al. (1993)
E. serra	Thallus	Direct regeneration	ASW (liquid)	7–50	12:12	12–25	Sahoo et al. (2002)
E. denticulatum	Explants	Callus formation & regeneration	ESS/2 (liquid)	5–80	16:8	25	Hurtado and Cheney (2003)
K. alvarezii	Branches	Direct regeneration & callus formation	ESS (solid)	20–160	12:12	18–26	Dawes and Koch (1991)
K. alvarezii	Branches	Clonal propagation & callus formation	ESS + CW (solid)	25–50	NA	23–25	Dawes et al. (1993)
K. alvarezii	Callus	Somatic embryogenesis & regeneration	PES (liquid)	5–70	12:12	22 (±1)	Reddy et al. (2003)
K. alvarezii	Thallus	Protoplast isolation & germination	F/2 and F/4 (liquid)	20–160	12:12	20–30	Salvador and Serrano (2005)
K. alvarezii	Explants	Callus formation & regeneration	PES (solid and liquid)	5–90	12:12	26 (±1)	Muñoz et al. (2006)
K. alvarezii	Explants	Callus formation & regeneration	ESS/2 (solid and liquid)	10–15	13:11	23–25	Hurtado and biter (2007)
K. alvarezii	Explants	Callus formation & regeneration	VS 50 (solid)	40 (±10)	14:10	23 (±2)	Hayashi et al. (2008)
K. alvarezii, K. striatum	Explants	Shoot formation	ESS/2 (liquid)	5–45	13:11	23	Hurtado et al. (2009)
K. striatum	Spores	Sporeling development	MGM (liquid)	50	13:11	25	Luhan and Sollesta (2010)
K. alvarezii, K. striatum	Explants	Shoot formation	ESS/2 (liquid)	10–45	13:11	20–25	Yunque et al. (2011)
K. alvarezii	Thallus	Callus formation & regeneration	PES (solid)	1500 (lux)	12:12	22–25	Sulistiani et al. (2012)
K. alvarezii	Thallus	Callus formation & regeneration	PES (solid and liquid)	25	12:12	25	Yeong et al. (2014)
K. alvarezii	Thallus	Direct regeneration	PES (liquid)	100 (±10)	12:12	25 (±1)	Yong et al. (2014a, b)
K. alvarezii	Explants	Protoplast isolation & germination	PES (liquid)	1500–2000 (lux)	12:12	15–40	Zhang et al. (2014)
K. alvarezii	Thallus	Direct regeneration	VS 50 (liquid)	15–50	12:12	25 (±1)	Neves et al. 2015

PES = Provasoli enriched seawater (Provasoli 1968), ESS and ESS/2 (half strength) = Enriched sterile seawater (Saga 1986), CW = coconut water, ASW = autoclaved seawater, F/2 (half strength) and F/4 (quarter strength) = F medium (Guillard and Ryther 1962), VS 50 = 50% von Stosch's solution (Guiry and Cunningham 1984), MGM = modified Grund medium (McLachlan 1973), NA = not available

5.5.2 Callus Morphogenesis

Dawes and Koch (1991) observed a higher rate of "bubbly type" of callus in green and brown forms of *K. alvarezii*, which was formed on the cut ends which were exposed to the air. These then formed an undifferentiated brown to beige mass of cells which originated from the medullary and inner cortical cells. Additionally, crystalline filaments grew from the cortical region of the explants. On the other hand, the majority of *E. denticulatum* explants formed a filamentous type of callus. Similarly, Reddy et al. (2003) observed filamentous outgrowths originating from both medullary and cortical regions of the cut surface of *K. alvarezii* explants, which gave rise to pigmented, uniseriate-branched filaments, which after 2 months' growth became a bright whitish, cap-like structure. Callus formation on cut surfaces, is indeed common in seaweed tissue culture (e.g. Polne-Fuller and Gibor 1987; Huang and Fujita 1996; Reddy et al. 2008), and is induced by the wounding process.

Apical calluses developed from the intact apical tips of *K. alvarezii* explants which originated from the apical region as a result of wound healing (Reddy et al. 2003). Apical calluses

Table 5.5 Protocols employed for obtain axenic explants from *Kappaphycus* and *Eucheuma*

Explants	Culture medium	Antibiotic and/or antimycotic solution	Surface sterilization solution	References
Thallus section (5 mm) from epiphyte-free main axes of *K. alvarezii* and *E. denticulatum*	ESS enriched seawater (Saga 1986)	Polymyxin B sulphate (3000 units L^{-1}) for up to 3 h	–	Dawes and Koch (1991)
Apical fragments (5 cm) of *K. alvarezii*	PES medium (Provasoli 1968)	Explants were incubated in 3% filter-sterilized spectrum antibiotic mixture (Polne-Fuller and Gibor 1984) for 2 days	Firstly, thallus was brushed in filtered seawater and then in 0.5% liquid detergent in sterile seawater for 10 min, and treated with 2% povidone iodine (available iodine 0.5% w/v) in sterilized seawater for 3 min	Reddy et al. (2003)
Apical branches (5 cm) of brown, red and green tetrasporophytes, and gametophyte ("Edison de Paula" strain) of *K. alvarezii*	Von Stosch (half strength, Oliveira et al. 1995)	Firstly, explants were incubated in antibiotic and antifungal solution (composed by 10,000 units Penicillin G, 10 mg Streptomycin sulphate, and 25 μg Amphotericin B per liter, and 0.1 μg mL^{-1} Nystatin) for 2 days	Explants were washed with sterilized seawater with 0.05% sodium hypochlorite and 2% organic detergent for 20 s	Hayashi et al. (2008)
Apical segments (2 cm) of seven varieties of *K. alvarezii* and one variety of *K. striatum* (var. sacol green)	Sterile filtered seawater	Cleaned segments were incubated in 9.1% E3 anti-bacterial solution (Bradley et al. 1988) in sterilized filtered seawater.	Firstly, apical segments were brushed with 0.05% povidone iodine, cleaned by shaking three times in 50 mL centrifuge tubes with fine glass beads in sterilized filtered seawater	Yunque et al. (2011)
Apical segments (4–5 cm) of *K. alvarezii*	Conwy medium (Liao et al. 1983)	Explants were immersed in medium with 3% antibiotic mixture (1 g Penicillin G, 2 g Streptomycin sulphate, 1 g Kanamycin, 25 mg Nystatin, 200 mg Neomycin in 100 mL distilled water)	Firstly, explants were soaked in Tween solution (5 drops in 200 mL) for 10 min, rinsed twice in 0.5% povidone iodine for 3 min, and then washed 3–4 times with sterilized seawater	Sulistiani et al. (2012)
Apical branches (5 cm) of brown tetrasporophytes of *K. alvarezii*	Von Stosch (half strength, Oliveira et al. 1995)	Firstly, explants were incubated in 2 mL L^{-1} antibiotic and anti-fungal solution (composed of 10,000 units mL^{-1} Penicillin G, 10 mg mL^{-1} Streptomycin sulphate and 25 μg mL^{-1} Amphotericin B, Sigma) and 0.1 μg mL^{-1} nystatin (Sigma) for 2 days	Explants were washed four times in solution of sodium hypochylorite 0.05% and four times in 2% organic Triton X-100 detergent (Sigma) for 5 s each rinsing	Zitta et al. (2013)

were also formed in *Solieria filiformis* (Robledo and Garcia-Reina 1993), *Meristotheca papulosa* (Huang and Fujita 1997), *Gracilariopsis tenuifrons* (Yokoya 2000), *Hypnea musciformis* (Yokoya et al. 2003), *Gracilaria perplexa* and *G. tenuistipitata* (Yokoya et al. 2004) and *G. domingensis* (Ramlov et al. 2013).

Hayashi et al. (2008) observed two types of callus in four colour strains of *K. alvarezii*: (a) filamentous callus developed on the explants surface exposed to the air which originated from divisions of the cortical and medullary cells, and (b) compact callus, growing on the explant surface which was in direct contact with the culture medium, also formed by the cortical and medullary cells. The "Edison de Paula" (EP) strain produced filamentous cells growing from the surface of the explant in contact with the culture medium, similarly of those described by Dawes and Koch (1991). In another instance, Sulistiani et al. (2012) observed three types of callus in explants as cultured in PES liquid medium supplemented with PGRs: (a) white compact callus, (b) white filamentous callus and (c) greenish/brownish callus. Compact (as opposed to diffuse) calluses have also been reported in other red algae, e.g. *Gracilaria domingensis* (Ramlov et al. 2013).

The ontogeny of callus development in the brown tetrasporophyte of *K. alvarezii* was analysed using light and transmission electron microscopy (TEM) by Zitta et al. (2013). Under the light microscope, callus filaments were observed to have elongated and clavate-shaped cells. Filamentous calluses composed of elongated filaments were also observed in *Agardhiella subulata* (Bradley and Cheney 1990), *Grateloupia dichotoma* (Yokoya and Handro 1996), *Solieria filiformis* (Yokoya and Handro 2002) and *Hypnea musciformis* (Kumar et al. 2007). Under TEM analysis, the most notable feature was the thickening of the cell wall during the de-differentiation process of initially of apical cells and later, in all cells of the callus (Zitta et al. 2013). The increased thickness of the cell wall could be a defensive response against desiccation since callus was growing on a solid culture medium and exposed to the air. In addition, it was observed that the cell walls had two different types of distribution of their microfibrils, a phenomenon also reported for callus cells originating from *Gracilariopsis tenuifrons* (Bouzon et al. 2010). Other ultrastructural changes in callus cells were seen to be the proliferation of convoluted membranes, increased numbers of mitochondria and modifications to normal chloroplast ultrastructure, these observations were taken to indicate cell de-differentiation. The cytoplasm showed a large amount of starch grains, and the chloroplasts had an altered state with plasto-globules and disorganized thylakoids, as well as the presence of several convoluted membranes and vacuoles (Zitta et al. 2013).

5.5.3 Factors Affecting Callus Induction and Growth

Different types and concentrations of auxins and cytokinins have been added to culture media for callus induction and growth in *Kappaphycus* and *Eucheuma* species (see Table 5.6 for a summary). However, exogenous applications of NAA and BAP did not increase callus induction or the growth in *K. alvarezii* (Reddy et al. 2003). Similar results were reported by Sulistiani et al. (2012) using combinations of BAP + NAA and BAP + IAA. Dawes and Koch (1991) observed that the largest calluses (up to 5 mm in height) in *K. alvarezii* explants were induced by treatments with NAA + 2iP, NAA+ BAP, and PAA + 2iP at concentrations of (1 + 1) mg L^{-1}.

The type of culture medium used was also reported to affect callus formation in explants of *K. alvarezii* and *E. denticulatum*, and ESS-enriched seawater was the suitable medium for both species (Dawes and Koch 1991). Reddy et al. (2003) observed a percentage callus formation greater than 80% in explants cultured in PES medium with 1.5% of bacto-agar, without the addition of PGRs. Hayashi et al. (2008) observed the highest percentage of callus formation (100%) in the brown strain of *K. alvarezii* cultured in F/2 50 medium (seawater enriched with 50% of Guillard and Ryther solution), while the "Edison de Paula" strain showed higher levels of callus induction in explants cultured in VS 50 (i.e. seawater enriched with 50% of von Stosch solution) and F/2 50 medium. However, explants cultured in PES medium showed relatively faster callus induction after 18 days, while in CW medium callus formation was observed after 22 days (Sulistiani et al. 2012).

5.5.4 Factors Affecting Regeneration and Micro-propagation

Micro-propagation, or *in vitro* clonal propagation, allows for the production of a large number of individuals within a short period and also selection of higher yielding strains that could be used as future seedlings for seaweed cultivation (Yokoya and Yoneshigue-Valentin 2011). Micro-propagation of *Kappaphycus* and *Eucheuma* species can be achieved either indirectly from regeneration of thalli from callus cells or by directly regenerating thalli from the explant tissue by passing the callus-induction phase. Polne-Fuller and Gibor (1987) reported the successful regeneration of new thalli in *Eucheuma uncinatum* and *K. alvarezii*. Micro-propagation of *E. denticulatum* and *K. alvarezii* were stimulated by the addition of several combinations of auxins and cytokinins (see Table 5.6, Dawes and Koch 1991). Similar results were reported by Muñoz et al. (2006) and PGRs stimulated regeneration from callus in *K. alvarezii*. On the other hand, Hayashi et al. (2008) observed that explants from brown, red and green tetrasporophytes and the "Edison de Paula" strain were able to be directly regenerated; indirect regeneration was very rare (only 1%), but when it occurred, that particular explant produced approximately 90 micro-propagules.

A novel method for clonal propagation using somatic embryos from uniseriate, pigmented, filamentous callus of *K. alvarezii* was described by Reddy et al. (2003); here micropropagule production was improved by the addition of NAA and a mixture of NAA and BAP in to the culture medium. Also, the propagules obtained from pigmented callus of *Kappaphycus* showed an enhanced growth rate of as much as 1.5–1.8 times the rate of farmed plants which were propagated through vegetative (splitting) means, without compromising on the yield of the carrageenan after cultivation in the sea. This method could be very useful as a potential tool for mass clonal propagation of desired seed stock of *Kappaphycus*, and *Eucheuma* species for commercial farming.

5.5.5 Endogenous Concentrations of PGRs in *Kappaphycus alvarezii*

Endogenous PGRs (auxins, cytokinins, abscisic acid, gibberellins, oxylipins (jasmonic acid), polyamines are reported as detected in a number of algae belonging to different evolutionary lineages (Stirk et al. 2003, 2009; Tarakhosvskaya et al. 2007; Yokoya et al. 2010; Mikani et al. 2015). Auxin and cytokinin-like activities have been reported in a number of different commercial seaweed extracts (e.g. Stirk and van Staden 1997), which are useful as biostimulants in agriculture and horticulture (e.g. Stirk and van Staden 2006; Khan

Table 5.6 Plant growth regulators (PGRs) used for callus induction and growth, regeneration and micro-propagation of *Kappaphycus* and *Eucheuma* species

PGR type	Full name/abbreviation	Concentrations (mg L^{-1})	Culture medium	References
Auxins	Indole 3-acetic acid (IAA)	1, 5, 10	ESS	Dawes and Koch (1991)
		0.5, 1.0, 5.0	VS50	Hayashi et al. (2008)
		2.5, 5.0	PES & CW	Sulistiani et al. (2012)
	1-Naphthalene acetic acid (NAA)	0.1, 1.0	PES	Reddy et al. (2003)
		0.1–10	ESS	Dawes and Koch (1991)
		0.5, 1.0	PES & CW	Sulistiani et al. (2012)
	Indole 3-butyric acid (IBA)	1, 5, 10	ESS	Dawes and Koch (1991)
	Phenyl acetic acid (PAA)	0.1–10	ESS	Dawes and Koch (1991)
		1	ESS/2	Hurtado and Cheney (2003)
		1	AMPEP	Hurtado et al. (2009)
		1	ESS/2	Hurtado et al. (2009)
		1	AMPEP	Yunque et al. (2011)
	2,4-Dichlorophenoxy-acetic acid (2,4-D)	0.5, 5.0	VS50	Hayashi et al. (2008)
Aromatic cytokinins	6-benzylaminopurine (BAP)	0.1, 1.0	PES	Reddy et al. (2003)
		0.1–10.0	ESS	Dawes and Koch (1991)
		0.5, 1.0, 5.0	VS50	Hayashi et al. (2008)
		0.5, 1.0	PES & CW	Sulistiani et al. (2012)
	Kinetin (K)	1, 5, 10	ESS	Dawes and Koch (1991)
Isoprenoid cytokinins	Zeatin (Z)	1, 5, 10	ESS	Dawes and Koch (1991)
		1	ESS/2	Hurtado and Cheney (2003)
		1	AMPEP	Hurtado et al. (2009)
		1	AMPEP, ESS/2	Yunque et al. (2011)
	N^6-[-2-isopentenyl] adenine (2iP)	1, 5, 10	ESS	Dawes and Koch (1991)

CW – Conwy medium (Liao et al. 1983), ESS – ESS enriched seawater (Saga 1986), PES – Seawater enriched with Provasoli solution (Provasoli 1968), VS50 – Seawater with half-strength of von Stosch solution (Oliveira et al. 1995)

et al. 2009). Similarly, the sap from freshly harvested *K. alvarezii* has been reported to be a potently active, foliar spray, which could be related to the presence of the cytokinins (kinetin and zeatin), the auxin (indole 3-acetic acid), and the gibberellic acid, GA_3 reported to be present (Prasad et al. 2010). These PGRs were quantified by tandem mass spectrometry of various organic extracts, and the endogenous concentration of IAA in the sap was estimated to be 23.36 ± 0.15 ppm, while concentrations of kinetin and zeatin were 7.94 ± 0.30 and 23.97 ± 0.47 ppm, respectively; concentrations of GA_3 was 27.87 ± 0.14 ppm (Prasad et al. 2010). It is important to highlight that the first report on the occurrence of endogenous kinetin in algae was for *K. alvarezii* (Prasad et al. 2010).

5.5.6 Quality Assessment of Tissue Culture Progeny

In order to achieve the goals of cell and tissue culture in supporting the growth and value addition to the seaweed farming industry, continuous quality assessment of tissue culture progeny is mandatory to ensure the production of high quality seedlings. Stepwise acclimatization of tissue-cultured seedlings from *in vitro* conditions to partial *in vitro* tank culture, prior to transfer to an *ex situ* nursery or glasshouse is strongly recommended in order to improve the growth performance and survival capacity of the seedlings (Yong et al. 2015a). Tissue-cultured *K. alvarezii* thalli have been reported to exhibit higher growth rates and resistance to epiphytes and 'ice-ice' disease, as compared to same age, farm-propagated seedlings after out-planting to the open sea (Reddy et al. 2003; Yong et al. 2015a). Hurtado et al. (2015) reported on the production of new and improved *Kappaphycus* thalli grown under optimized growth conditions with the addition of a biostimulant to maintain the high-yielding and rapidly growing seed-stock for a commercial nursery. Moreover, comparison of the tissue culture vs farm-propagated thalli of *K. alvarezii* revealed economically important quality benefits: e.g. carrageenan yield and quality (Reddy et al. 2003; Yong et al. 2014a), higher nutritional value (Yong et al. 2015b) and less heavy metal contamination (Yong et al. 2017). This suggested that the tissue culture strains were a superior quality to serve as planting materials for the future of sustainable seaweed farming and use of the biomass for higher value applications.

5.6 Conclusions

To strengthen and indeed advance the *Kappaphycus* farming industry, it is fundamentally important to achieve success in the development of *in vitro* tissue culture technology for the furtherance of sustainable, commercial farming. The foregoing account has clearly demonstrated not only the success achieved in development of basic techniques for routine tissue culture of *Kappaphycus* and *Eucheuma,* but also underlined the benefits derived from the application of these techniques in the production of clones, their propagation and maintenance. In parallel, these efforts have also created a wide array of choices for methods, approaches and strategies for the consistent production of propagules directly from explants thereby bypassing callus induction. The propagules differentiated from callus were also shown to have improved growth characteristics suggesting the possibility of generating improved variants. Such variants will be of great benefit, especially for those countries where carrageenophytes are farmed from a single, exotic genotype. There has also been significant advancement in establishing commercial nurseries which are now considered as very important for maintaining and further boosting the activity of commercial carrageenophyte farming. The rate of production of new and improved *Kappaphycus* explants for a commercial nursery stock has been substantially improved through the novel use of AMPEP with optimized culture media, pH, temperature and density of thalli. The use of AMPEP alone and/or in combination with PGR in culture media for propagation of micro-plantlets using tissue culture techniques has been highly successful in the production of seedstock for nursery and out-planting purposes. Combinations of AMPEP K^+ with PGR + colchicine or oryzalin provided better formulations for robust seedlings and shoot production in *K. alvarezii.* Likewise, AMPEP K^+ + PGR could be used to induce direct, erect shoots for the micro-propagation of *K. alvarezii* plantlets under laboratory conditions.

It is also crucial to establish homozygous lines of economically important seaweeds for agronomically important traits such as growth, phycocolloid, disease and epiphyte resistance, stress tolerance etc. Establishment of such lines is of fundamental importance, and can form an invaluable resource for attempting breeding aimed at developing robust hybrid varieties for commercial farming of carrageenophytes, more particularly in those countries where *Kappaphycus* grows naturally. Nevertheless, the benefits of cell suspension culture have not been realized largely due to absence of friable callus formation in seaweeds, a phenomenon which is quite common for terrestrial plants. This could be possible, if the role of biostimulants in cell division and development at the molecular level is better understood.

Acknowledgment The authors would like to profoundly thank their respective heads of the organization for extending their support, encouragement and facilities while preparing this article.

References

Aguirre-Lipperheide M, Estrada-Rodríguez FJ, Evans LV (1995) Facts, problems and needs in seaweed tissue culture: an appraisal. J Phycol 31:677–688

Azanza RV, Aliaza TT (1999) *In vitro* carpospore release and germination in *Kappaphycus alvarezii* (Doty) Doty from Tawi-Tawi, Philippines. Bot Mar 42:281–284

Azanza-Corrales R, Ask E (2003) *Kappaphycus alvarezii* (Doty) Doty carposporeling germination growth and development in the laboratory. In: Chapman A, Andersen AJ, Vreeeland V, Davison I (eds) Proc Int Seaweed Symp 17:95–99

Azanza-Corrales R, Mamauag SS, Alfiler E, Orolfo MJ (1992) Reproduction in *Eucheuma denticulatum* (Burman) Collins and Hervey and *Kappaphycus alvarezii* (Doty) Doty farmed in Danajon Reef, Philippines. Aquaculture 103:29–34

Baweja P, Sahoo D, García-Jiménez P, Robaina RR (2009) Seaweed tissue culture as applied to biotechnology: problems, achievements and prospects. Phycol Res 57:45–58

Bixler HJ, Porse H (2011) A decade of change in seaweed hydrocolloids industry. J Appl Phycol 23:321–335

Borlongan IA, Tibubos KR, Yunque DAT, Hurtado AQ, Critchley AT (2011) Impact of AMPEP on the growth and occurrence of epiphytic *Neosiphonia* infestations on two varieties of commercially cultivated *Kappaphycus alvarezii*, grown at different depths in the Philippines. J Appl Phycol 23:615–621

Bouzon ZL, Schmidt EC, Almeida AC, Yokoya NS, Oliveira MC, Chow F (2010) Cytochemical characterization and ultrastructural organization in calluses of the agarophyte *Gracilariopsis tenuifrons* (Gracilariales, Rhodophyta). Micron 42:80–86

Bradley P (1991) Plant hormones do have a role in controlling growth and development of algae. J Phycol 27:317–321

Bradley PM, Cheney DP (1990) Some effects of plant regulators on tissue cultures of the marine red alga *Agardhiella subulata* (Gigartinales, Rhodophyta). Hydrobiologia 204(205):353–360

Bradley PM, Cheney DP, Saga N (1988) One step antibiotic disk method for obtaining axenic cultures of multicellular marine algae. Plant Cell Tissue Organ Cult 12:55–60

Brown V, Ducker SC, Rowan KS (1977) The effect of orthophosphate concentration on the growth of articulated coralline algae (Rhodophyta). Phycologia 16:125–131

Bulboa CR, de Paula EJ (2005) Introduction of non-native species of *Kappaphycus* (Rhodophyta, Gigartinales) in subtropical waters: comparative analysis of growth rates of *Kappaphycus alvarezii* and *Kappaphycus striatum in vitro* and in the sea in south-eastern Brazil. Phycol Res 53:183–188

Bulboa C, de Paula E, Chow F (2007) Laboratory germination and sea out-planting of tetraspore progeny from *Kappaphycus striatum* (Rhodophyta) in subtropical waters of Brazil. J Appl Phycol 19:357–363

Bulboa C, de Paula EJ, Chow F (2008) Germination and survival of tetraspores of *Kappaphycus alvarezii* var. *alvarezii* (Solieriaceae, Rhodophyta) introduced in subtropical waters of Brazil. Phycol Res 56:39–45

Butler DM, Evans LV (1990) Cell and tissue culture of macroalgae. In: Akatsuka I (ed) Introduction to applied phycology. SPB Academic Publishing, The Hague, pp 629–645

Cheney DP, Mathiesen A, Schubert D (1981) The application of genetic improvement techniques to seaweed cultivation: I. Strain selection in the carrageenophyte *Chondrus crispus*. Proc Int Seaweed Symp 10:559–567

Craigie JS (2011) Seaweed extracts stimuli in plant science and agriculture. J Appl Phycol 23:371–393

Dawes CJ (1993) Marine botany. Wiley Interscience Publication. John Wiley & Sons, New York, p 478

Dawes C, Koch E (1991) Branch, micropropagule and tissue culture of the red algae *Eucheuma denticulatum* and *Kappaphycus alvarezii* farmed in the Philippines. J Appl Phycol 3:247–257

Dawes CJ, Trono GC Jr, Lluisma AO (1993) Clonal propagation of *Eucheuma denticulatum* and *Kappaphycus alvarezii* for Philippine seaweed farms. Hydrobiologia 260(261):379–383

Dawes CJ, Lluisma AO, Trono GC (1994) Laboratory and field growth studies of commercial strains of *Eucheuma denticulatum* and *Kappaphycus alvarezii* in the Philippines. J Appl Phycol 6:21–24

Doty MS, Alvarez V (1973) Seaweed farms: a new approach for U.S. industry. Mar Technol Soc Annu Conf Pap 9:701–708

Fadilah S, Alimuddin A, Masak PR, Santoso J, Parenrengi A (2016) Growth, morphology and growth related hormone level in *Kappaphycus* produced by mass selection in Gorontalo waters, Indonesia. Hayati J Biosci. http://dx.doi.org/10.1016/j.hjb.2015.09.004

Fan D, Hodges DM, Zhang JZ, Kirby CW, Ji XH, SLocke SJ, Critchley AT, Prithiviraj B (2011) Commercial extract of the brown seaweed *Ascophyllum nodosum* enhances phenolic antioxidant content of spinach (*Spinacia aloracea* L.) which protects *Caenorhabditis elegans* against oxidative and thermal stress. Food Chem 124:195–120

FAO (2016) The state of world fisheries and aquaculture 2016. Contributing to food security and nutrition for all. FAO, Rome, 200 pp

Farah Diyana A, Abdullah A, Shahrul Hisham ZA, Chan KM (2015) Antioxidant activity of red algae *Kappaphycus alvarezii* and *Kappaphycus striatum*. Food Res Int 22:1977–1984

Garcia-Reina G, Gómez-Pinchetti JL, Robledo DR, Sosa P (1991) Actual, potential and speculative applications of seaweed cellular biotechnology: some specific comments on *Gelidium*. Hydrobiologia 221:181–194

Ghosh A, Vijay Anand KG, Seth A (2015) Life cycle impact assessment of seaweed based biostimulant production from onshore cultivated *Kappaphycus alvarezii* (Doty) Doty ex Silva—is it environmentally sustainable? Algal Res 12:513–521

Guillard RR, Ryther JH (1962) Studies of marine planktonic diatoms: I. *Cyclotella nana* Hustedt, and *Detonula confervacea* (Cleve) Gran. Can J Microbiol 8:229–239

Guiry MD, Cunningham EM (1984) Photoperiodic and temperature responses in the reproduction of north-eastern Atlantic *Gigartina acicularis* (Rhodophyta: Gigartinales). Phycologia 23:357–367

Handayani T, Alimuddin WU, Suryati E, Parenrengi A (2014) Binary vector construction and *Agrobacterium tumefaciens*-mediated transformation of lysozyme gene in seaweed *Kappaphycus alvarezii*. Biotropia 21:80–90

Hayashi L, Yokoya NS, Kikuchi DM, Oliveira EC (2008) Callus induction and micro-propagation improved by colchicine and phytoregulators in *Kappaphycus alvarezii* (Rhodophyta, Solieriaceae). J Appl Phycol 20:653–659

Holdt SL, Kraan S (2011) Bioactive compounds in seaweed: functional food applications and legislation. J Appl Phycol 23:543–597

Huang W, Fujita Y (1996) Callus induction and thallus regeneration in some species of red algae. Phycol Res 45:105–111

Huang W, Fujita Y (1997) Callus induction and thallus regeneration of the red alga *Meristotheca papulosa* (Rhodophyta, Gigartinales). Bot Mar 40:55–61

Hurtado AQ, Agbayani R (2000) The farming of seaweed *Kappaphycus*. SEAFDEC AQD, Tigbauan, Iloilo, Philippines, p 25

Hurtado AQ, Biter A (2007) Plantlet regeneration of *Kappaphycus alvarezii* var. adik-adik by tissue culture. J Appl Phycol 19:783–786

Hurtado AQ, Cheney DP (2003) Propagule production of *Eucheuma denticulatum* (Burman) Collins et Harvey by tissue culture. Bot Mar 46:338–341

Hurtado AQ, Critchley A, Trespoey A, L'honneur-Bleicher G (2006) Occurrence of *Polysiphonia* epiphytes in *Kappaphycus* farms at Calaguas Is., Camarines Norte, Phillippines. J Appl Phycol 18:301–306

Hurtado AQ, Yunque DA, Tibubos K, Critchley AT (2009) Use of Acadian marine plant extract powder from *Ascophyllum nodosum* in tissue culture of *Kappaphycus* varieties. J Appl Phycol 21:633–639

Hurtado AQ, Joe M, Sanares RC, Fan PB, Critchley AT (2012) Investigation of the application of Acadian Marine Plant Extract Powder (AMPEP) to enhance the growth, phenolic content, free radical scavenging, and iron chelating activities of *Kappaphycus* Doty (Solieriaceae, Gigartinales, Rhodophyta). J Appl Phycol 24:601–611

Hurtado AQ, Neish IC, Critchley AT (2015) Developments in production technology of *Kappaphycus* in the Philippines: more than four decades of farming. J Appl Phycol 27:1945–1961

Jayaraj JA, Wan M, Rahman PZK (2008) Seaweed extract reduces foliar fungal disease on carrot. Crop Protect 10:1360–1366

Jayaraj J, Norrie J, Punja ZK (2011) Commercial extract from brown seaweed *Ascophyllum nodosum* reduces fungal disease in greenhouse cucumber. J Appl Phycol 23:353–361

Khan W, Rayirath UP, Subramanian S, Jithesh MN, Rayorath P, Hodges DM, Critchley AT, Craigie JS, Norrie J, Prithiviraj B (2009) Seaweed extracts as biostimulants of plant growth and development. J Plant Growth Regul 28:386–399

Kirihara S, Fujikawa Y, Notoya M (1997) Axenic tissue culture of *Sargassum confusum* C. Agardh (Phaeophyta) as a source of seeds for artificial marine forests. J Mar Biotech 5:142–146

Kumar GR, Reddy CRK, Jha B (2007) Callus induction and thallus regeneration from callus of phycocolloid yielding seaweeds from the Indian coast. J Appl Phycol 19:15–25

Kumar KS, Ganesan K, Rao PVS (2008) Antioxidant potential of solvent extracts of *Kappaphycus alvarezii* (Doty) Doty – an edible seaweed. Food Chem 107:289–295

Largo DB, Fukami K, Nishijima T (1995) Occasional pathogenic bacteria promoting 'ice-ice' disease in the carrageenan-producing red algae *Kappaphycus alvarezii* and *Eucheuma denticulatum* (Solieriaceae, Gigartinales, Rhodophyta). J Appl Phycol 7:545–554

Lau TY, Vittal DF, Chew CSY, Yong WTL (2014) Antiproliferative potential of extracts from *Kappaphycus* seaweeds on HeLa cancer cell lines. Sains Malaysiana 43:1895–1900

Lawlor HJ, Borowitzka MA, McComb JA (1990) Tissue culture of brown seaweeds. Aust J Biotechnol 4:260–264

Liao IC, Su HM, Lin JH (1983) Larval foods for penaeid prawn. In: McVey JP (ed) CRC handbook of Mariculture volume 1: crustacean aquaculture. CRC Press Inc., Boca Raton, pp 29–60

Loureiro RR, Reis RP, Critchley AT (2010) *In vitro* cultivation of three *Kappaphycus alvarezii* (Rhodophyta, Areschougiaceae) variants (green, red and brown) exposed to a commercial extract of the brown alga *Ascophyllum nodosum* (Fucaceae, Ochrophyta). J Appl Phycol 22:101–104

Loureiro RR, Reis RP, Berrogain FD, Critchley AT (2012) Extract powder from the brown alga *Ascophyllum nodosum* (Linnaeus) Le Jolis (AMPEP): a "vaccine-like" effect on *Kappaphycus alvarezii* (Doty) Doty ex P.C. Silva. J Appl Phycol 24:427–432

Loureiro RR, Reis RP, Berrogain FD, Critchley AT (2014) Effects of a commercial extract of the brown alga *Ascophyllum nodosum* on the biomass production of *Kappaphycus alvarezii* (Doty) Doty ex P. C Silva and its carrageenan yield and gel quality cultivated in Brazil. J Appl Phycol 26:763–768

Luhan MRJ, Sollesta H (2010) Growing the reproductive cells (carpospores) of the seaweed, *Kappaphycus striatum*, in the laboratory until outplanting in the field and maturation to tetrasporophyte. J Appl Phycol 22:579–585

Maliakal S, Cheney DP, Rorrer GL (2001) Halogenated monoterpene production in regenerated plantlet cultures of *Ochtodes secundiramea* (Rhodophyta, Cryptonemiales). J Phycol 37:1010–1019

Marriog RG, Loureiro RR, Reis RP (2016) The effect of *Ascophyllum nodosum* (Ochrophyta) extract powder on the epibiosis of *Kappaphycus alvarezii* (Rhodophyta) commercially cultivated on floating rafts. J Appl Phycol 28:2471–2477

Matanjun P, Mohamed S, Mustapha NM, Muhammad K, Ming CH (2008) Antioxidant activities and phenolics content of eight species of seaweeds from north Borneo. J Appl Phycol 20:367–373

McLachlan J (1973) Growth media – marine. In: Stein JR (ed) Handbook of phycological methods. Culture methods and growth measurements. Cambridge University Press, Cambridge, pp 25–51

Mikani K, Mori IC, Matsuura T, Ikeda Y, Kojima M, Sakakibara H, Hirayama T (2015) Comprehensive quantification and genome survey reveal the presence of novel phytohormone action modes in red seaweeds. J Appl Phycol 28:2539–2548

Muñoz J, Cahue-López AC, Patiño R, Robledo D (2006) Use of plant growth regulators in micro-propagation of *Kappaphycus alvarezii* (Doty) in airlift bioreactors. J Appl Phycol 18:209–218

Nayar S, Bott K (2014) Current status of global cultivated seaweed production and markets. World Aq 45:32–37

Neves FAS, Simioni C, Bouzon ZL, Hayashi L (2015) Effects of spindle inhibitors and phytoregulators on the micro-propagation of *Kappaphycus alvarezii* (Rhodophyta, Gigartinales). J Appl Phycol 27:437–445

Oliveira EC, Paula EJ, Plastino EM, Petti R (1995) Metodologias para o cultivo no axênico de macroalgas marinas *in vitro*. In: Alveal K, Ferrario ME, Oliveira EC, Sar E (eds) Manual de Métodos Ficológicos. Universidad de Concepción, Concepción, pp 429–455

Patwary MU, van der Meer JP (1982) Genetics of *Gracilaria tikvahiae*. Phenotypic and genetic characterization of some selected morphological mutants. Can J Bot 60:2556–2564

Patwary MU, van der Meer JP (1983) Growth experiments on morphological mutants of *Gracilaria tikvahiae* (Rhodophyceae). Can J Bot 6:1654–1659

Polne-Fuller M, Gibor A (1984) Developmental studies in *Porphyra*. I. Blade differentiation in *Porphyra perforata* as expressed by morphology, enzymatic digestion and protoplast regeneration. J Phycol 20:609–616

Polne-Fuller M, Gibor A (1987) Calluses and callus-like growth in seaweeds: induction and culture. Hydrobiologia 151(152):131–138

Prasad K, Das AK, Oza MD, Brahmbhatt H, Siddhanta AK, Meena R, Eswaran K, Rajyaguru MR, Ghosh PK (2010) Detection and quantification of some plant growth regulators in a seaweed-based foliar spray employing a mass spectrometric technique sans chromatographic separation. J Agric Food Chem 58:4594–4601

Provasoli L (1968) Media and prospects for the cultivation of marine algae. In: Watanabe A, Hattore A (eds) Culture and collection of algae. Japanese Society of Plant Physiologists, Tokyo, pp 63–75

Raman M, Doble M (2014) Physicochemical and structural characterisation of marine algae *Kappaphycus alvarezii* and the ability of its dietary fibres to bind mutagenic amines. J Appl Phycol 26:2183–2191

Ramlov F, Plastino EM, Yokoya NS (2013) Growth, callus formation and plant regeneration in color morphs of *Gracilaria domingensis* (Gracilariales, Rhodophyta) cultured under different irradiance and plant growth regulators. Phycologia 52:508–516

Reddy CRK, Kumar GRK, Siddhanta AK, Tewari A (2003) *In vitro* somatic embryogenesis and regeneration of somatic embryos from pigmented callus of *Kappaphycus alvarezii* (Doty) Doty (Rhodophyta, Gigartinales). J Phycol 39:610–616

Reddy CRK, Jha B, Fuhita Y, Ohno M (2008) Seaweed micropropagation techniques and their potentials: an overview. J Appl Phycol 20:609–617

Reddy CRK, Gupta V, Jha B (2010) Developments in biotechnology of red algae. In: Seckbach J, Chapman DJ (eds) Red algae in the genomic age, cellular origin, life in extreme habitats and astrobiology 13. Springer, Dordrecht, pp 307–341

Robledo DR, Garcia-Reina G (1993) Apical callus formation in *Solieria filiformis* (Gigartinales, Rhodophyta) cultured in tanks. Hydrobiologia 260(261):401–406

Rorrer GL, Cheney DP (2004) Bioprocess engineering of cell and tissue cultures for marine seaweeds. Aquac Eng 32:11–41

Saga N (1986) Pure culture of algae. In: Yamada Y, Okada Y (eds) Plant biotechnology. Tokyo Kaymkes Dojin, Tokyo, pp 55–69. (in Japanese)

Sahoo D, Ohno M, Hiraoka M (2002) Laboratory, field and deep seawater culture of *Eucheuma serra* – a high lectin yielding red alga. Algae 17:127–133

Salvador RC, Serrano AE (2005) Isolation of protoplasts from tissue fragments of Philippine cultivars of *Kappaphycus alvarezii* (Solieriaceae, Rhodophyta). J Appl Phycol 17:15–22

Singh S, Singh MK, Pal SK, Trivedi K, Yesuraj D, Singh CS, Vijay Anand KG, Chandramohan M, Patidar R, Kubavat D, Zodape ST, Ghosh A (2016) Sustainable enhancement in yield and quality of rain-fed maize through *Gracilaria edulis* and *Kappaphycus alvarezii* seaweed sap. J Appl Phycol 28:2099–2112

Stirk WA, van Staden J (1997) Screening of some South African seaweeds for cytocinin-like activity. S Afr J Bot 63:161–164

Stirk WA, van Staden J (2006) Seaweed products as biostimulants in agriculture. In: Critchley AT, Ohno M, Largo DB (eds) World seaweed resources. DVD-ROM multimedia interactive software – UNESCO Publishing/ETI Bioinformatics, University of Amsterdam, ISBN 90 75000 80 4

Stirk WA, Novák O, Strnad M, Van Staden J (2003) Cytokinins in macroalgae. J Plant Growth Regul 41:13–24

Stirk WA, Novák O, Hradecká V, Pěnčík A, Rolčík J, Strnad M, Van Staden J (2009) Endogenous cytokinins, auxins and abscisic acid in *Ulva fasciata* (Chlorophyta) and *Dictyota humifusa* (Phaeophyta): towards understanding their biosynthesis and homeostasis. Eur J Phycol 44:231–240

Sulistiani E, Soelistyowati DT, Alimuddin YSA (2012) Callus induction and filaments regeneration from callus of cottonii seaweed *Kappaphycus alvarezii* (Doty) collected from Natuna Islands, Riau Islands Province. Biotropia 19:103–114

Sylvester AW, Waaland R (1983) Cloning the red alga *Gigartina exasperata* for culture on artificial substrates. Aquaculture 31:305–318

Tarakhosvskaya ER, Maslov YI, Shishova MF (2007) Phytohormones in algae. Russian J Plant Physiol 54:163–170

Tibubos K, Hurtado AQ, Chritchley AT (2016) Direct formation of axes in new plantlets of *Kappaphycus alvarezii* (Doty) Doty, as influenced by the use of AMPEP K$^+$, spindle inhibitors and plant growth hormones. J Appl Phycol. doi:10.1007/s10811-016-0988-z

Titlyanov E, Titlyanova T (2006) Production of plantlets of the red alga of *Gelidium* genus (Rhodophyta) from thallus fragments. Russ J Mar Biol 32:284–288

Titlyanov EA, Titlyanova TV, Kadel P, Lüning K (2006) Obtaining plantlets from apical meristem of the red alga *Gelidium* sp. J Appl Phycol 18:167–174

Trono GC Jr, Lluisma A (1992) Differences in biomass production and carrageenan yields among four strains of farmed carrageenophytes in Northern Bohol, Philippines. Hydrobiologia 247:223–227

Trono GC, Ohno M (1989) Seasonality in the biomass production of *Eucheuma* strains in northern Bohol, Philippines. In: Umezaki I (ed) Scientific survey of marine algae and their resources in the Philippine Islands. Monbusho International Scientific Research Program, Japan, pp 71–80

Ugarte RA, Sharp GJ (2001) A new approach to seaweed management in eastern Canada: the case of *Ascophyllum nodosum*. Cah Biol Mar 42:63–70

Ugarte RA, Sharp GJ (2012) Management and production of the brown algae *Ascophyllum nodosum* in the Canadian Maritimes. J Appl Phycol 24:409–416

Ugarte RG, Sharp G, Moore B (2006) Changes in the brown seaweed *Ascophyllum nodosum* (L.) Le Jol. plant morphology and biomass produced by cutter rake harvests in southern New Brunswick, Canada. J Appl Phycol 18:351–359

Valderrama D, Cai J, Hishamunda N, Ridler N (2013) Social and economic dimensions of carrageenan seaweed farming. FAO Fisheries and Aquaculture Tech Paper 580. Food and Agriculture Organization of the United Nations, Rome

van der Meer JP (1986) Genetics of *Gracilaria tikvahiae* (Rhodophyceae). XI. Further characterization of a bisexual mutant. J Phycol 22:151–158

van der Meer JP, Patwary MU (1983) Genetic modification of *Gracilaria tikvahiae* (Rhodophyceae). The production and evaluation of polyploids. Aquaculture 33:311–316

Wang J, Jiang P, Cui Y, Deng X, Li F, Liu J, Qin S (2010) Genetic transformation in *Kappaphycus alvarezii* using micro-particle bombardment: a potential strategy for germplasm improvement. Aquac Int 18:1027–1034

Yeong HY, Phang SM, Reddy CRK, Khalid N (2014) Production of clonal planting materials from *Gracilaria changii* and *Kappaphycus alvarezii* through tissue culture and airlift photobioreactors. J Appl Phycol 26:729–746

Yokoya NS (2000) Apical callus formation and plant regeneration controlled by plant growth regulators on axenic culture of the red alga *Gracilariopsis tenuifrons* (Gracilariales, Rhodophyta). Phycol Res 48:133–142

Yokoya NS, Handro W (1996) Effects of auxins and cytokinin on tissue culture of *Grateloupia dichotoma* (Gigartinales, Rhodophyta). Hydrobiologia 326(327):393–400

Yokoya NS, Handro W (2002) Effects on plant growth regulators and culture medium on morphogenesis of *Solieria filiformis* (Rhodophyta) cultured *in vitro*. J Appl Phycol 14:97–102

Yokoya NS, Yoneshigue-Valentin Y (2011) Micro-propagation as a tool for sustainable utilization and conservation of populations of Rhodophyta. Braz J Pharmacogn 21:334–339

Yokoya NS, Plastino EM, Artel R (2003) Physiological responses and pigment characterization of two colour strains of the carrageenophyte *Hypnea musciformis* (Rhodophyta). In: Chapman ARO, Anderson RJ, Vreeland VJ, Davison JR (eds) Proceedings of 17th international seaweed symposium. Oxford University Press, New York, pp 425–433

Yokoya NS, West JA, Luchi AE (2004) Effects of plants growth regulators on callus formation, growth and regeneration in axenic tissue cultures of *Gracilaria tenuistipitata* and *Gracilaria perplexa* (Gracilariales, Rhodophyta). Phycol Res 52:244–254

Yokoya NS, Stirk WA, Van Staden J, Novák O, Turecková V, Pénčík A, Strnad M (2010) Endogenous cyotikinins, auxins, and abscisic acid in red algae from Brazil. J Phycol 46:1198–1205

Yong WTL, Chin JYY, Thien VY, Yasir S (2014a) Evaluation of growth rate and semi-refined carrageenan properties of tissue-cultured *Kappaphycus alvarezii* (Rhodophyta, Gigartinales). Phycol Res 62:316–321

Yong WTL, Ting SH, Yong YS, Thien VY, Wong SH, Chin WL, Rodrigues KF, Anton A (2014b) Optimization of culture conditions for the direct regeneration of *Kappaphycus alvarezii* (Rhodophyta, Solieriaceae). J Appl Phycol 26:1597–1606

Yong YS, Yong WTL, Thien VY, Ng SE, Anton A, Yassir S (2015a) Acclimatization of micropropagated *Kappaphycus alvarezii* (Doty) Doty ex Silva (Rhodophyta, Solieriaceae) in outdoor nursery system. J Appl Phycol 27:413–419

Yong YS, Yong WTL, Ng SE, Anton A, Yassir S (2015b) Chemical composition of farmed and micropropagated *Kappaphycus alvarezii* (Rhodophyta, Gigartinales), a commercially important seaweed in Malaysia. J Appl Phycol 27:1271–1275

Yong WTL, Chin JYY, Thien VY, Yasir S (2017) Heavy metal accumulation in field cultured and tissue cultured *Kappaphycus alvarezii* and *Gracilaria changii*. Int Food Res J 24:970–975

Yunque DA, Tibubos KR, Hurtado AQ, Critchley AT (2011) Optimization of culture conditions for tissue culture production of young plantlets of carrageenophyte *Kappaphycus*. J Appl Phycol 23:433–438

Zemke-White WL, Smith JE (2006) Environmental impacts of seaweed farming in the tropics CD-ROM World Seaweed Resources – an authoritative reference system Version:1

Zhang S, Liu C, Jin Y, Chi S, Tang X, Chen F, Fang X, Liu T (2014) Studies on the isolation and culture of protoplasts from *Kappaphycus alvarezii*. Acta Oceanol Sin 33:114–123

Zitta CS, Rover T, Hayashi L, Bouzon ZL (2013) Callus ontogeny of the *Kappaphycus alvarezii* (Rhodophyta, Gigartinales) brown tetrasporophyte strain. J Appl Phycol 25:615–629

Impacts of AMPEP on Epiphytes and Diseases in *Kappaphycus* and *Eucheuma* Cultivation

6

Rafael R. Loureiro, Anicia Q. Hurtado, and Alan T. Critchley

Abstract

The causes for epiphyte outbreaks known as 'goose bumps' and the 'ice-ice' malaise are generally assigned to be due to stresses resulting from rapid shifts in environmental parameters, followed by fungal, bacterial and algal infestations. These conditions are responsible for the loss of many *Kappaphycus* and *Eucheuma* 'seedlings' affecting the quantity and quality of biomass from farming operations, which mainly affect poor coastal communities which rely on macroalgal farming as their only source of cash income. "Acadian Marine Plant Extract Powder" (an acronym coined by Hurtado – AMPEP) is a commercial extract from the brown alga *Ascophyllum nodosum;* it has proven to be, amongst other macroalgal-based extracts, one of the most promising solutions to ameliorate the harmful consequences of "disease" outbreaks; mode of action is suggested as being via elicitation of the treated eucheumatoid's own defense mechanisms thereby dampening epiphyte spore germination and establishment, while improving daily growth rate and carrageenan gel yield and quality (i.e. molecular weight).

6.1 Introduction

Several macroalgal species have been traditionally exploited for their commercial benefits, from natural populations or cultivated (in coastal or on land facilities) by many coastal communities. These were mostly used for direct human consumption and animal feed, but also highly valued in traditional medicine, especially in Asian countries (Sutherland et al. 2011; Buchholz et al. 2012). The seaweed industries that started in the seventeenth and eighteenth centuries were largely for potash production which was started through burning of seaweeds in Scotland and other parts of the United Kingdom; commercial extraction of alginate was established in the late nineteenth century. Many scientific discoveries provided for the growth and needs of the developing global revolution in the commercial and industrial "seaweed industry". Painstaking, yet basic biological discoveries were absolutely necessary, such as the life cycle of *Pyropia* (*Porphyra*) by Kathleen M. Drew and methods of alginate extraction by Edward Curtis Stanford, collectively allowed a greater diversity in the number of seaweed species that could be wild harvested and/or cultivated and used for the production of significant, industrial volumes of biomass for direct consumption and/or processing/extraction of commercial products.

According to the Food and Agriculture Organization of the United Nations (FAO 2014), close to 25 million MT (dry weight) of macroalgae were produced globally through aquaculture annually, making the seaweed industry one of the most prominent industries in the world (i.e. 8.13% growth

R.R. Loureiro (✉)
Blue Marble Space Institute of Science,
1001 4th Ave, Suite 3201, Seattle, WA 98154, USA
e-mail: rafael.rloureiro@gmail.com; rafael@bmsis.org

A.Q. Hurtado
Integrated Services for the Development of Aquaculture and Fisheries (ISDA) Inc., McArthur Highway, Tabuc Suba, Jaro, 5000 Iloilo City, Philippines

A.T. Critchley
The Evangeline Trail, Paradise, Nova Scotia, Canada

A.Q. Hurtado et al. (eds.), *Tropical Seaweed Farming Trends, Problems and Opportunities*, Developments in Applied Phycology 9, DOI 10.1007/978-3-319-63498-2_6

in dry weight year^{-1}) and profitable (i.e. a net value of over six billion US$ year^{-1}). However, the exploitation of natural macroalgal beds is still and will remain a common practice.

Seaweed farming industries and managed, natural seaweed populations have contributed to significant, socio-economic benefits over the years. New industries, companies and jobs have been generated thereby providing alternative income options to coastal and often remote, rural communities in the over-fished areas of East Africa and South-East Asia in particular. Transitioning to coastal cultivation of selected species has reduced the chances of, as well as, putting a halt to the over-exploitation of natural seaweed beds in some of those locations (Rebours et al. 2014 provided a positive summary of the benefits).

While most of the dry tonnage production of seaweeds is destined for direct human consumption as food (84%) the remainder is used for the extraction of a variety of products with commercial applications for their rheological and biological properties. By-products, due to macroalgal chemical diversity and bio-prospecting potential, hold significant financial importance and represented close to 7.5% of the annual seaweed aquaculture growth in 2014 (FAO 2014). Such significant growth in the collective seaweed industry was mainly driven by the world-wide popularization of a variety of methods for seaweed aquaculture. The farming of carrageenan-bearing (carrageenophyte) seaweeds began on farms using 'seedling' stocks derived from the world's initial sources of raw materials, The Philippines.

The species and development of selected strains of carrageenophyte, which are the basis of the current industrialized carrageenan industry, are discussed elsewhere in this book.

Eucheuma and *Kappaphycus* farming in The Philippines started at the southern-most portion of the country (due to favourable farming conditions and the natural occurrence of wild stocks from which the initial cultivation germplasm was selected. The domestication process, through a series of pilot -farming sites, began in 1967 and after 8 years (1974) the first commercial quantities of "cottonii" (i.e. *Kappaphycus alvarezii*) started to be produced using extensive cultivation systems (Hurtado et al. 2014). Decades after the first domestication of cultivars from the *Kappaphycus* genus, The Philippines held the title of the world's leader in carrageenophyte farming. This was until 2008 when The Philippines was overtaken by Indonesia for total production of raw material biomass. The Philippines remains a model that several countries have followed for the production (and processing) of *Kappaphycus alvarezii* (as a source of kappa carrageenan) and *Eucheuma denticulatum* (as a source of iota carrageenan) biomass and adapted to their own socio-economic needs (see Chap. 4 on the spread of carrageenophyte cultivation and strains globally).

As of 2017, the carrageenan industry is largely dominated by the cultivation of *Kappaphycus* and *Eucheuma* and has become an alternative or leading source of cash-based incomes for many coastal families worldwide. From East Africa to southern China, from the Caribbean countries to South America, the societal benefits from the popularization of macroalgal aquaculture, even though it is a small fraction of what is produced in The Philippines and Indonesia, has proven to be a very promising way to help under-privileged sectors of the local populace in those countries (Valderama et al. 2013).

6.2 Macroalgal Diseases and Infestation Outbreaks

Just like any land-based crop, the cultivation of macroalgae is known to be subject to disease and infestations (e.g. 'ice-ice' and *Neosiphonia* sp.) with remarkable similarities regardless of geographic location and the distance between farming sites (locally and globally; Gachon et al. 2010; Hafting et al. 2015). This "global crop" concept does not exempt *Kappaphycus* and *Eucheuma* farms from such issues, since they share the same key aspects which are suitable for farming and also have a very restricted genetic pool from many to years of vegetative, seedling propagation and the inherent similarities of the farming sites, selected to accommodate the physiology of these genera (Hafting et al. 2015).

The impacts of disease and infestation outbreaks have generated disastrous consequences, not only for the industry, but also the many families which rely on seaweed aquaculture as a stable source of income. To date, very few solutions can be taken to contain or mitigate damaging outbreaks and, most of these solutions, involve the complete loss of the crop and then total replacement with fresh seedlings, something that for many family-based farms results in loss of their income, since many do not receive any governmental or private company support.

6.3 Environmental Stress, Malaise and 'Ice-Ice'

Since the early years of cultivation of *Kappaphycus alvarezii* in the Philippines, a phenomenon known as 'ice-ice' (Fig. 6.1) has been reported at farming sites. The characteristic bleaching to white of the thalli gives 'ice-ice' its name; it is recognized as an indication of abiotic, physiological stress (e.g. due to fluctuations in salinity and temperature which seem to be the main causative agents) rather than being initially induced by a pathogen. The bleaching effect is mainly caused by a defense response mechanism of the seaweed triggered by the presence of volatile, halogenated organic compounds. The subsequent oxidative burst, driven by the stress factors, releases H_2O_2 (hydrogen peroxide) which explains the white thalli after extended exposure (Pedersén et al. 1996a, b; Loureiro et al. 2012).

Even though members from the Cytophaga-Flavobacterium-Bacteroids group and two marine fungal genera (e.g., *Aspergillus* sp. and *Phoma* sp.) have been isolated from affected thalli (Largo et al. 1995a, b; Butardo et al. 2003; Solis et al. 2010; Achmad et al. 2016), these contaminations seem to share no link to the causative events of 'ice-ice'. The microbial "attack" is a secondary phase affecting the apical macroalgal tissues after being weakened by the physiological stresses. This, secondary effect is obvious in the form of complete necrosis of the thalli due to both bacteria and fungi, which were isolated from the affected thalli. These microbial agents decompose and consume cellulose; the fungi have been shown to have a preference for carrageenan as its main carbon source (Solis et al. 2010).

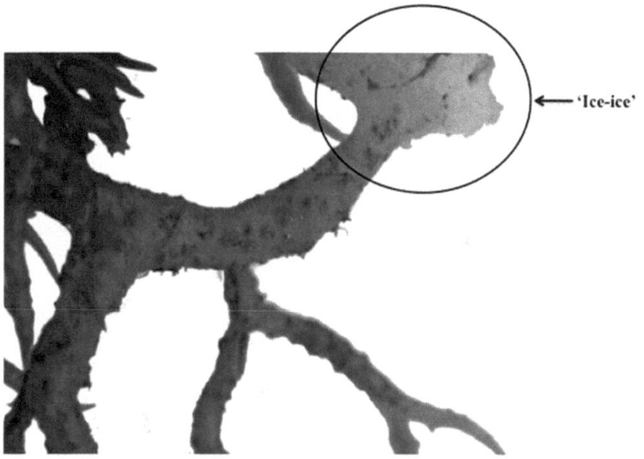

Fig. 6.1 'Ice-ice' infected *Kappaphycus alvarezii* (Photo courtesy of AQ Hurtado)

6.3.1 'Ice-Ice' Infestation

'Goose bumps' which appear as "black pimples" (Fig. 6.2a) on the surface of *Kappaphycus* are actually the settling sites for spores from the filamentous red alga *Neosiphonia*. The germinating spores penetrate to the cortical and medullary tissues of the host seaweed. The spores will later develop into an Endophytic Filamentous Alga infestation (EFA; Fig. 6.2b). Infestations in *Kappaphycus* spp. cultivation sites were first recorded in 1994 at Sitangkai, Tawi-Tawi, Philippines; it was first erroneously ascribed as a *Polysiphonia* sp. epiphyte outbreak (due to a mis-identification), only later to be confirmed as *Neosiphonia apiculata*. Other species of *Neosiphonia* have since been implicated. *Neosiphonia* is a genus with species which are common epiphytes of *Sargassum* spp. and could have been transferred by drift *Sargassum* being entangled in cultivation ropes and the hitch-hiking epiphyte "jumping ship" from one host to the new carrageenophyte host, where it was very successful at attachment and subsequent reproduction through copious spore production under favorable conditions (Vairappan 2006; Vairappan et al. 2008). These outbreaks where seemingly deemed harmless and initially disregarded initially by the processing industry (the purchasers of the raw biomass). However, in 2000 a major outbreak of epiphytic *Neosiphonia* occurred on Calaguas Is., Camarines Norte (Critchley et al. 2004; Hurtado 2005). The result of this and other outbreaks that followed were recognized as being disastrous to the coastal farmers to the point that some had to entirely relocate their farming sites and re-seed their entire site with then unaffected 'seedlings' in order to avoid further contamination (Hurtado et al. 2014).

Fig. 6.2 (**a** and **b**) 'Goose bumps' and *Neosiphonia* sp. infected *Kappaphycus alvarezii* (bar = 1 cm) (Photos courtesy of AQ Hurtado)

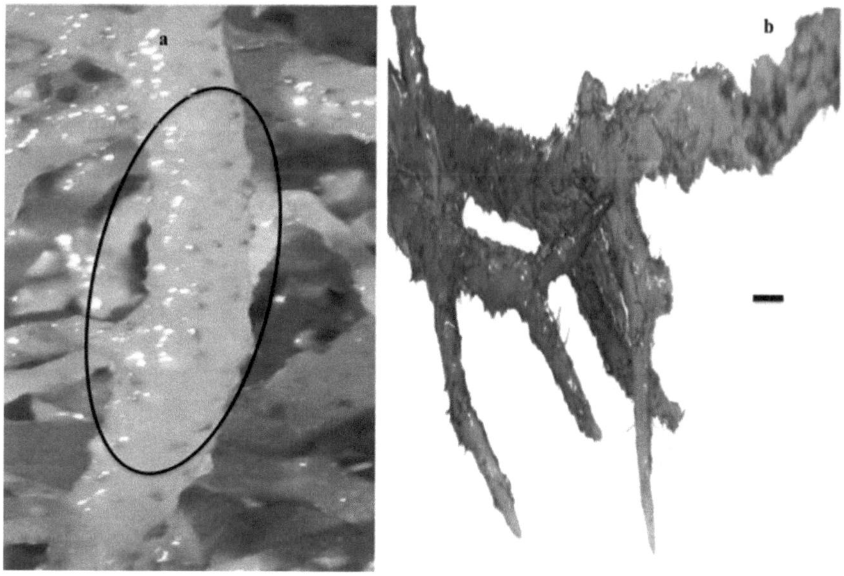

The reasons that *Kappaphycus* species were susceptible to epiphyte infestations are still unclear. However, increasing abiotic stresses are likely to have played a role in overpowering the natural, seaweed defense mechanisms over a long period of time, thereby giving epiphytes an enhanced chance of successful settlement and development on the host's thalli; much of the time, an increase in salinity and temperature were seen to be followed by an increase in epiphyte coverage (Vairappan 2006; Vairappan et al. 2008). It is not understood why *Neosiphonia* was the successful genus, other than a pool of reproductive material was likely to have been encountered in cultivation areas with floating *Sargassum* mats, however the epiphyte seems to be far more prolific on compromised *Kappaphycus* than its original, brown algal host. What followed was a similar effect on thalli that had been affected by 'ice-ice', where the individuals became contaminated by fungi and bacteria causing tissue necrosis around the areas damaged by the epiphytes. Shortages of the essential 'seedling' stocks led to the unregulated distribution of infected biomass, which unfortunately precipitated unprecedented and widespread dispersal of the "epiphyte problem" (Hurtado et al. 2014; Loureiro et al. 2014a). Sadly these initial reports were largely dismissed by the carrageenan industry at that time.

6.4 The Use of AMPEP (Acadian Marine Explant Powder) in Macroalgal Cultivation and Its Beneficial Effects on Epiphytes and Diseases

Unprocessed seaweed biomass and macroalgal extracts of macroalgae have been used as agricultural inputs (soil and health enhancers) as well as growth promoters (biostimulants) for centuries, due to their relative ease of access to coastal farmers and their diverse chemical composition with bioactive properties (Craigie 2011; Hafting et al. 2015). Studies using extracts from brown algae, e.g. *Ascophyllum nodosum* (AMPEP) and *Ecklonia maxima* (Kelpak®; Robertson-Andersson et al. 2006, Hafting et al. 2015) on the growth and cultivation of selected macroalgae showed promising results by either ameliorating growth rates and final extract quality and quantity, by means of mitigating the damaging effects from either abiotic, physiological stress or biotic, epiphytic infestations.

The beneficial applications of AMPEP is supported by a number of recent publications on its effects on *Kappaphycus alvarezii* and its cultivar 'seedlings' (see: Hurtado et al. 2009, 2012; Loureiro et al. 2010, 2012, 2014a, b; Yunque et al. 2011; Borlongan et al. 2011; Marroig et al. 2016). AMPEP (see Table 6.1) is the most prominently studied of seaweed extracts with applications for the cultivation of another seaweed genus, clearly demonstrating effectiveness to improve *Kappaphycus alvarezii* daily growth rates (both *in situ* and *in vitro*) and carrageenan gel quality and quantity while also mitigating the presence of *Neosiphonia* epiphytes, thus proving to be a cost-effective solution for poor coastal communities (i.e. the majority of *K. alvarezii* farmers), and reducing the complications encountered while managing a carrageenophyte farm.

Table 6.1 Composition of Acadian Marine Plant Extract Powder 1-1-17 from *Ascophyllum nodosum* (Hurtado et al. 2009)

Physical data	
Appearance	Brownish-black crystals
Odor	Marine color
Solubility in water	100%
pH	10.0–10.5
Typical analysis	
Maximum moisture	6.50%
Organic matter	45–55%
Ash (minerals)	45–55%
Total nitrogen (N)	0.8–1.5%
Available phosphoric acid (P_2O_5)	1–2%
Soluble potash (K_2O)	17–22%
Sulfur (S)	1–2%
Magnesium (Mg)	0.2–0.5%
Calcium (Ca)	0.3–0.6%
Sodium (Na)	3–5%
Boron (B)	75–150 ppm
Iron (Fe)	75–250 ppm
Manganese (Mn)	5–20 ppm
Copper (Cu)	1-5 ppm
Zinc (Zn)	25–50 ppm
Carbohydrates	Alginic acid, mannitol, laminarin
Amino acids (total 4.4%)	
Alanine	0.32%
Arginine	0.04%
Aspartic acid	0.62%
Cystine	0.01%
Glutamic acid	0.93%
Glycine	0.29%
Histidine	0.08%
Isoleucine	0.26%
Leucine	0.41%
Lysine	0.16%
Methionine	0.11%
Phenylalanine	0.25%
Proline	0.28%
Serine	0.08%
Threonine	0.04%
Tyrosine	0.17%
Valine	0.28%
Tryptophan	0.07%

The first evidence of AMPEP's effectiveness against epiphytes was reported by Borlongan et al. (2011), although the extract's ability to increase *K. alvarezii*'s daily growth rate

Table 6.2 Average (±SD) daily growth rate and perceantge occurrence of *Neosiphonia* sp. on *K. alvarezii* var. giant tambalang and var. tungawan with and without AMPEP dipping grown at four depths from April to November (n = 5) (Borlongan et al. 2011)

		K. alvarezii var. giant tambalang		*K. alvarezii* var. tungawan	
	Depth (cm)	With AMPEP dipping	Without AMPEP dipping (control)	With AMPEP dipping	Without AMPEP dipping (control)
Growth rate (% day^{-1})	0	3.1 ± 0.8^{a}	2.0 ± 1.2^{c}	4.1 ± 1.0^{b}	3.0 ± 0.9^{a}
	50	2.4 ± 0.7^{e}	1.4 ± 0.7^{g}	3.0 ± 1.0^{e}	2.2 ± 1.5^{e}
	100	1.8 ± 0.9^{i}	1.0 ± 0.7^{g}	2.5 ± 0.8^{i}	1.4 ± 1.1^{ig}
	150	1.3 ± 0.8^{j}	0.9 ± 0.8^{gj}	1.4 ± 1.0^{j}	1.2 ± 0.9^{gj}
% occurrence of *Neosiphonia* sp.	0	36 ± 27^{a}	69 ± 16^{b}	30 ± 21^{a}	54 ± 36^{b}
	50	34 ± 11^{c}	63 ± 18^{b}	21 ± 15^{c}	52 ± 32^{b}
	100	12 ± 6.5^{d}	20 ± 17^{de}	9 ± 11^{d}	11 ± 13^{de}
	150	6 ± 6.5^{d}	13 ± 11^{de}	4 ± 5.8^{df}	7 ± 8.7^{df}

P <0.05) - level of significance

had been already described by Hurtado et al. (2009) and Loureiro et al. (2010). Borlongan et al. (2011) reported that the use of AMPEP significantly ($p < 0.05$) increased the growth rate (Table 6.2) of both of the *Kappaphycus* varieties tested, while also decreasing the percentage occurrence of a *Neosiphonia* sp. epiphyte. The percentage occurrence of *Neosiphonia* sp. infection (i.e. 6–50% at all depths) of both *Kappaphycus* varieties tested with AMPEP treatment was significantly lower than the controls (i.e., 10–75% at all depths). Both, the growth rate of the cultivated seaweed and the percent occurrence of the epiphytes decreased as the cultivation depth increased. Plants dipped in AMPEP and suspended at the surface had the highest growth rates (i.e. 4.1%, Tungawan; 3.1%, Tambalang) after 45 days. Those control 'seedlings' which had not been dipped in AMPEP had the highest percentage occurrence of *Neosiphonia* sp. infection (viz. 70–75%; Figs. 6.3a and b). The occurrence of the *Neosiphonia* infestation was found to be correlated with changes in irradiance and salinity at the depths observed (Table 6.3).

Borlongan et al. (2011) claimed that both varieties of *K. alvarezii* used in this study had their fastest growth rates when suspended immediately at the water surface. However, it was recommended that in order to minimize damage caused by the occurrence of epiphytic *Neosiphonia, K. alvarezii* would be best grown within a depth range of 50–100 cm. These were important first observations for the improved management of the commercial farming of *Kappaphycus* and more importantly as a mitigating strategy which enhanced growth and improved resistance to endophytic filamentous algae (EFA) such as the deleterious *Neosiphonia* sp. infestations.

Driven by the amount of data generated confirming the beneficial uses of AMPEP, Marroig et al. (2016) tested the same to assess impacts on epibiosis of *K. alvarezii* on commercial, floating rafts in Brazil. The total epibiont mass attached to AMPEP-treated seedlings was less than the untreated (control treatment) samples; seedlings from the control had a greater concentration of epibionts [e.g. 0.76 ± 0.52 g dry mass tubular net^{-1}] when compared to the AMPEP-treated samples [e.g. 0.26 ± 0.27 g dry mass tubular net^{-1}; Fig. 6.4). The authors confirmed the earlier observations of Loureiro et al. (2012), that during the first and second sampling periods there was a 'vaccine-like' effect on AMPEP- treated *K. alvarezii* thereby reducing the early settlement and establishment of epibionts.

The 'vaccine-like' effect described by Loureiro et al. (2012) was forwarded under the hypothesis that AMPEP treatments of seedlings was responsible for the elicitation of *K. alvarezii*'s natural defense mechanisms, inducing an oxidative burst response thus shielding the samples exposed to the brown seaweed extract from the harmful effects of constant H_2O_2 exudation and exposure. The effect initiated by induced stress, activated a biochemical chain reaction that allowed the interaction between the β-1,3 glucan present in the extract and a β-1,3 glucanase enzyme known to be present in *K. alvarezii*'s biochemical composition. Such interactions were postulated to stimulate the production of volatile halogenated compounds which were elicited, through NADPH oxidase in the plasma membrane, as pulse-like, oxidative bursts (Fig. 6.5). The protection from the harmful effects of prolonged exposure to the H_2O_2 by the seedlings was attributed to the presence of betaines. The significant improvement of the 'seedlings' daily growth rates was also observed during this study and other publications, when *K. alvarezii* was treated with AMPEP. The presence of cytokinin-elicitors known to be present in most brown algal extracts are likely as the agents behind these results, and it is clear that there are action overlaps as to benefits of applications of brown seaweed extract to either seaweeds and/or land plants (Khan et al. 2009, 2011a, b, 2013; Hurtado et al. 2012; Loureiro et al. 2012; Wally et al. 2013).

Hurtado et al. (2012) conducted an in-depth investigation of the biochemical impacts of AMPEP on the growth of *K. alvarezii* (Tambalang purple) and *K. striatus* (Sacol green). While there were no significant effects of AMPEP-dipping in regard to total phenolics in either *K. alvarezii* or *K. striatus,* the sacol green with AMPEP treatment showed a significant

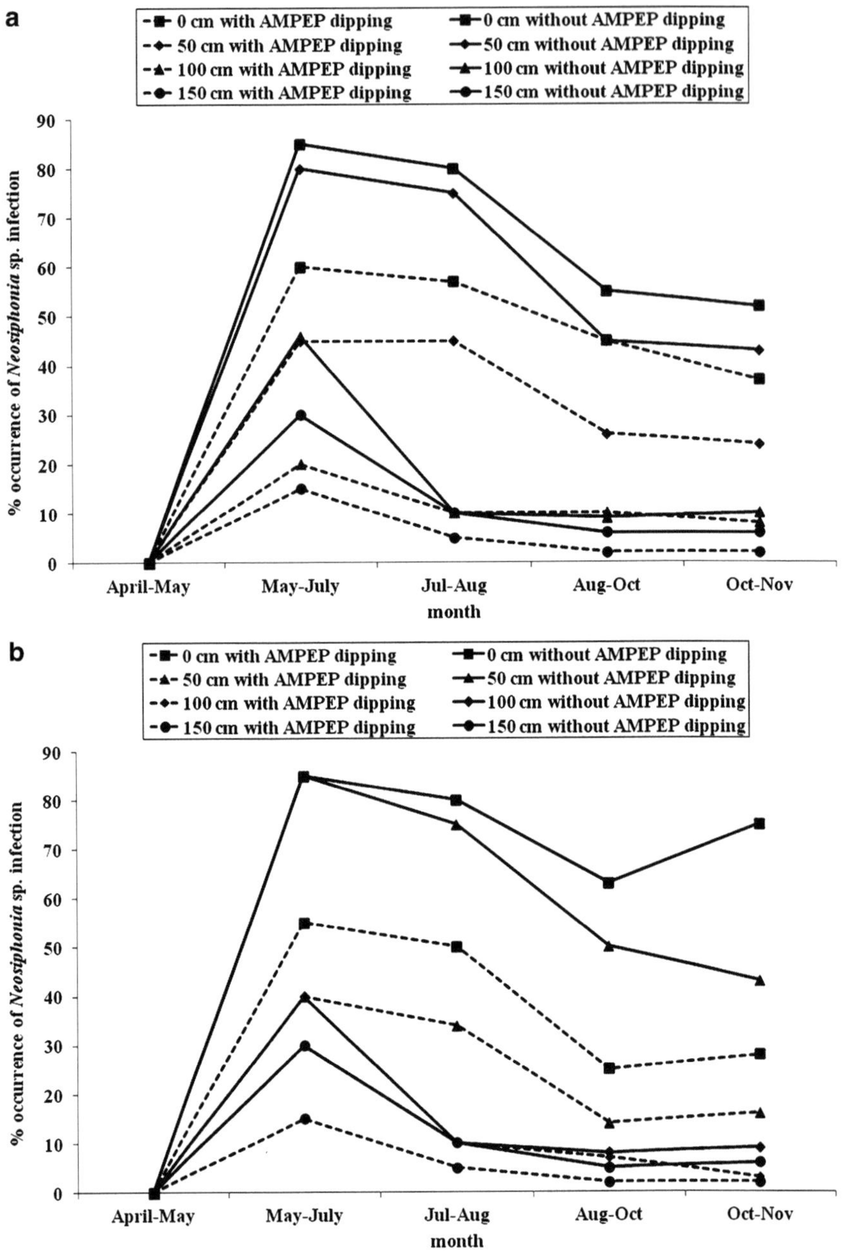

Fig. 6.3 (**a** and **b**) Percentage (%) occurrence of *Neosiphonia* sp. on *Kappaphycus alvarezii* (Tambalang strain; Borlongan et al. 2011)

Table 6.3 Correlation coefficient between percentage occurrence of *Neosiphonia* sp. and some environmental parameters in two varieties of *K. alvarezii* (Borlongan et al. 2011)

Varieties of *K. alvarez*ii	Irradiance μmols photons m^{-2} s^{-1}	Salinity (ppt)	Temperature (°C)	Turbidity (NTU)
Tungawan with AMPEP dipping	1.00	−0.99	0.21	−0.03
Tunagwan without AMPEP dipping (control)	0.94	−0.95	0.46	−0.21
Giant tambalang with AMPEP dipping	0.97	−0.98	0.36	−0.14
Giant tambalang without AMPEP dipping (control)	0.94	−0.96	0.47	−0.25

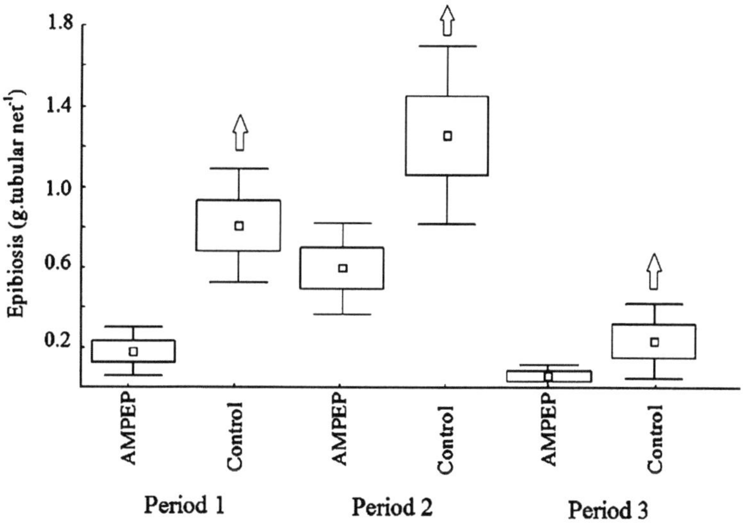

Fig. 6.4 Weight (g) of epi-bionts present on *K. alvarezii* grown in tubular nets (Marroig et al. 2016)

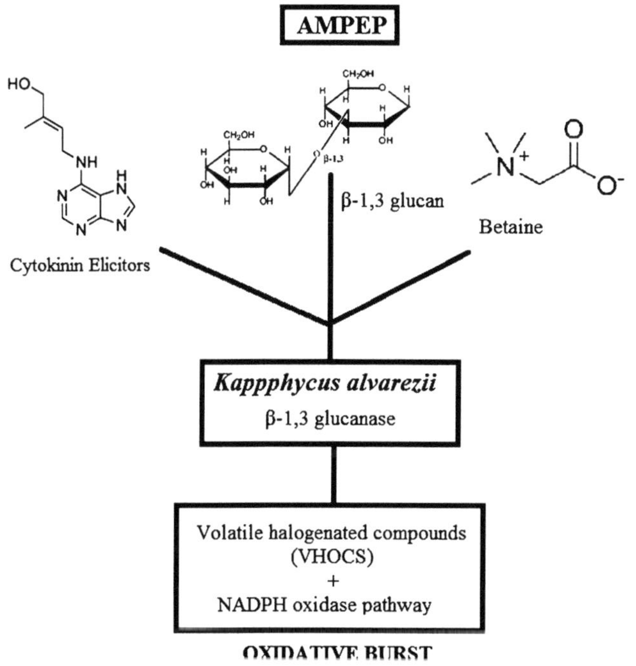

Fig. 6.5 Suggested biochemical pathway for the beneficial effects of AMPEP on *Kappaphycus alvarezii*

increase ($p < 0.05$) in total antioxidant capacity (in the October–November period). The Tambalang purple strain with AMPEP treatment, produced a sustained high Fe^{2+} -chelating ability from September–October and January–February, which coincided with the positive growth of *K. alvarezii* during that time. Increased free-radical scavenging and transition-metal chelating abilities might be of interest from the point of view of resistance of the seaweed strains to epiphytes/pathogens and abiotic stresses. Likewise, dipping *K. alvarezii* (Tambalang purple) in AMPEP showed encouraging results which are important for imparted resistance to pathogens and abiotic stressors resistance.

Moreover, Loureiro et al. (2014b) suggested that the gradual (but constant) accumulation of H_2O_2 may be responsible for the delay in responses to CAT (catalase) and APX (ascorbate peroxidase gene), as also first suggested by Barros et al. (2006). This process would guarantee a long-term acclimatization process, permitting *K. alvarezii* to increase its growth rate (i.e., biomass increments) under adverse environmental conditions; this is something that could benefit seedlings either affected by 'ice-ice' or to be farmed in areas that are known to have variable (and at times stressful) environmental conditions.

6.5 Conclusions

AMPEP has so far shown to be very promising in mitigating epiphytes and countering the effects of diseases, however, it is yet to be introduced into industrial-sized farms, nor given as an alternative to small family farms cooperatives around the world. All the knowledge that was generated with the use of AMPEP on *K. alvarezii* and other carrageenophytes should be made available to those communities that rely on macroalgal farming as their only source of income. However, the benefits of using a commercial seaweed extract is likely not to be exclusive to those few extracts which have been tested so far. Further research (i.e. rates of concentration and durations of dipping times) for other commercial seaweed extract products should be conducted by cultivation extension officers in affected areas of the world, in order to verify local efficacy and application "protocols". There is no one panacea and it is likely that additional challenges to cultivation will arise and farmers should have a toolbox of cost-effective, treatments at their disposal.

It is unfortunate but most outbreaks (epiphytes and disease) remain uncontrolled, either due to the lack of training and/or proper management of farming sites; the out-come is the loss of entire production cycles, not least the effect on the farmers, some of whom lose all income potential and/or have to relocate considerable distances to new "unaffected" sites. Obviously these are only short term solutions. Initiatives between academia and local farming cooperatives, within both the large- and small-scale producing countries, should start with proper and inclusive farmer education – this would be a starting point to popularize the use of AMPEP (or its alternatives) as cost-effective, viable solutions to crop-loss and disease/epiphyte prevention/control, thereby eventually increasing the value of a better quality cultivated raw material from which a greater yield and quality of carrageenan or bioactive extract maybe obtained.

References

Achmad M, Alimuddin A, Widyastuti U, Sukenda S, Harris E, Suryanti E (2016) Molecular identification of new bacterium causing 'ice-ice' diseases on seaweed *Kappaphycus alvarezii*. Paper presented at the Asian-Pacific Aquaculture, Surabaya Indonesia, 26–29 April 2016 (Abstract only)

Barros MP, Necchi O Jr, Colepicolo P, Pedersén M (2006) Kinetic study of the plastoquinone pool availability correlated with H_2O_2 release in seawater and antioxidant responses in the red alga *Kappaphycus alvarezii* exposed to single or combined high light, chilling and chemical stresses. Biochim Biophys Acta 1757:1520–1528

Borlongan IA, Tibubos KR, Yunque DAT, Hurtado AQ, Critchley AT (2011) Impact of AMPEP on the growth and occurrence of epiphytic *Neosiphonia* infestations on two varieties of commercially cultivated *Kappaphycus alvarezii*, grown at different depths in the Philippines. J Appl Phycol 23:615–621

Buchholz CM, Krause G, Buck BH (2012) Chapter 22: Seaweed and man. In: Wiencke C, Bischof K (eds) Seaweed biology. Springer, Berlin/Heidelberg, pp 471–493

Butardo VM, Ganzon-Fortes ET, Silvestre V, Bacano-Maningas MBB, MNE M, Lluisma AO (2003) Isolation and classification of culturable bacteria associated with 'ice-ice' disease in *Kappaphycus* and *Eucheuma* (Rhodophyta, Solieriaceae). Philipp Scientist 40:223–2379

Craigie JS (2011) Seaweed extract stimuli in plant science and agriculture. J Appl Phycol 23:371–393

Critchley AT, Largo DB, Wee W, Bleicher-L'Honneur G, Hurtado A, Shubert J (2004) A preliminary summary of *Kappaphycus* farming and the impacts of epiphytes. Jpn J Phycol 52:231–232

Food and Agriculture Organization of the United Nations (2014) Fisheries and Aquaculture Information and Statistics Services [WWW document] URL: http://www.fao.org/figis/ . Accessed 2 July 2016

Gachon CMM, Sime-Ngando T, Strittmatter M, Chambouvet A, Kim GH (2010) Algal diseases: spotlight on a black box. Trends Plant Sci 15(11):633–640

Hafting JT, Craigie JS, Stengel DB, Loureiro RR, Bushmann AH, Yarish C, Edwards MD, Critchley AT (2015) Prospects and challenges for industrial production of seaweed bioactives. J Phycol 51:821–837

Hurtado AQ (2005) Effect of epiphytes on the productivity of *Kappaphycus*: impact of diseases on the economics of producing and processing *Kappaphycus*. Terminal report submitted to Degussa Texturant system (now Cargill Texturizing Solutions), France

Hurtado AQ, Yunque DA, Tibudos K, Critchley AT (2009) Use of Acadian marine plant extract powder from *Ascophyllum nodosum* in tissue culture of *Kappaphycus* varieties. J Appl Phycol 21:633–639

Hurtado AQ, Joe M, Sanares RC, Fan PB, Critchley AT (2012) Investigation of the application of Acadian Marine Plant Extract Powder (AMPEP) to enhance the growth, phenolic content, free radical scavenging, and iron chelating activities of *Kappaphycus* Doty (Solieriaceae, Gigartinales, Rhodophyta). J Appl Phycol 24:601–611

Hurtado AQ, Reis RP, Loureiro RR, Critchley AT (2014) *Kappaphycus* (Rhodophyta) cultivation: problems and impacts of Acadian Marine Plant Extract Powder. In: Pereira L, Neto JM (eds) Marine algae biodiversity, taxonomy, environmental assessment, and biotechnology. CRC Press, Boca Raton, pp 251–299

Khan W, Rayirath UP, Subramanian S et al (2009) Seaweed extracts as biostimulants of plant growth and development. J Plant Growth Regul 28:386–399

Khan W, Hiltz D, Critchley AT, Prithiviraj B (2011a) Bioassay to detect *Ascophyllum nodosum* extract-induced cytokinin-like activity in *Arabidopsis thaliana*. J Appl Phycol 23:409–414

Khan ZH, Kahn MA, Aftab T, Idrees M, Naeem M (2011b) Influence of alginate oligosaccharides on growth, yield and alkaloid production of opium poppy (*Papaver somniferum* L.) Front Agric China 5:122–127

Khan W, Palanisamy R, Critchley AT, Smith DL, Papadopoulos Y, Prithiviraj R (2013) *Ascophyllum nodosum* extract and its organic fractions stimulate *Rhizobium* root nodulation and growth of *Medicago sativa* (Alfalfa). Commun Soil Sci Plant Anal 44:900–908

Largo DB, Fukami F, Nishijima T (1995a) Occasional bacteria promoting 'ice-ice' disease in the carrageenan-producing red algae *Kappaphycus alvarezii* and *Eucheuma denticulatum* (Solieriaceae, Gigartinales, Rhodophyta). J Appl Phycol 7:545–554

Largo DB, Fukami F, Nishijima T, Olino M (1995b) Laboratory-induced development of the 'ice-ice' disease of the farmed red algae *Kappaphycus alvarezii* and *Eucheuma denticulatum* (Solieriaceae, Gigartinales, Rhodophyta). J Appl Phycol 7:539–543

Loureiro RR, Reis RP, Critchley AT (2010) *In vitro* cultivation of three *Kappaphycus alvarezii* (Rhodophyta, Areschougiaceae) variants (green, red and brown) exposed to a commercial extract of the brown alga *Ascophyllum nodosum* (Fucaceae, Ochrophyta). J Appl Phycol 22:101–104

Loureiro RR, Reis RP, Berrogain FD, Critchley AT (2012) Extract powder from the brown alga *Ascophyllum nodosum* (Linnaeus) Le Jolis (AMPEP): a vaccine-like effect on *Kappaphycus alvarezii* (Doty) Doty ex P.C. Silva. J Appl Phycol 24:427–432

Loureiro RR, Reis RP, Berrogain FD, Critchley AT (2014a) Effects of a commercial extract of the brown alga *Ascophyllum nodosum* on the biomass production of *Kappaphycus alvarezii* (Doty) Doty ex P.C. Silva and its carrageenan yield and gel quality cultivated in Brazil. J Appl Phycol 24:427–432

Loureiro RR, Reis R, Marroig R (2014b) Effect of the commercial extract of the brown alga *Ascophyllum nodosum* Mont. on *Kappaphycus alvarezii* (Doty) Doty ex P.C. Silva in situ submitted to lethal temperatures. J Appl Phycol 26:629–634

Marroig RG, Loureiro RR, Reis RP (2016) The effect of *Ascophyllum nodossum* (Ochrophyta) extract powder on the epibiosis of *Kappaphycus alvarezii* (Rhodophyta) commercially cultivated on floating rafts. J Appl Phycol 28:2471–2477

Pedersén M, Collén J, Abrahamsson K, Ekdahl A (1996a) Production of halocarbons from seaweeds: an oxidative stress reaction? Sci Mar 60:257–263

Pedersén M, Collén J, Abrahamsson K, Mtolera M, Semesi A, Garcia Reina G (1996b) The 'ice–ice' disease and oxidative stress of marine algae. In: Björk M, Semesi AK, Pedersén M, Bergman B (eds) Current trends in marine botanical research in the east African region. Ord & Vetande AB, Uppsala, pp 11–24

Rebours C, Marinho-Soriano E, Zertuche-Gonzalez JA, Hayashi L, Vasquez JA, Kradolfer P, Soriano G, Ugarte R, Abreu MH, Bay-Larsen I (2014) Seaweeds: an opportunity for wealth and sustainable livelihood for coastal communities. J Appl Phycol 26:1939–1951

Robertson-Andersson DV, Leitao D, Bolton JJ, Robertson RJ, Njobeni A, Ruck K (2006) Can kelp extract (KELPAK) be useful in seaweed mariculture? J Appl Phycol 18:315–329

Solis MJL, Draeger S, de la Cruz TEE (2010) Marine-derived fungi from *Kappaphycus alvarezii* and *K. striatum* as potential causative agents of 'ice-ice' disease in farmed seaweeds. Bot Mar 53:587–594

Sutherland JE, Lindstrom SC, Nelson WA, Brodie J, Lynch MD, Hwang MS, Choi HG (2011) A new look at an ancient order: generic revision of the Bangiales (Rhodophyta). J Phycol 47:1131–1151

Vairappan CS (2006) Seasonal occurrences of epiphytic algae on the commercially cultivated red alga *Kappaphycus alvarezii* (Solieriaceae, Gigartinales, Rhodophyta). J Appl Phycol 18:611–617

Vairappan CS, Chung CS, Hurtado AQ, Soya FE, Bleicher-L'honneur G, Critchley A (2008) Distribution and symptoms of epiphyte infection in major carrageenophyte producing farms. J Appl Phycol 20:477–483

Valderama D, Cai J, Hishamunda N, Ridler R (2013) Social and economic dimensions of carrageenan seaweed farming. Fisheries and Aquaculture Technical Paper No. 580, Rome. FAO:204 pp

Wally OSD, Critchley AT, Hiltz D et al (2013) Regulations of phytohormone biosynthesis and accumulation in Arabidopsis following treatment with commercial extract from the marine macroalga *Ascophyllum nodosum*. J Plant Regul 32:324–339

Yunque DAT, Tibubos KR, Hurtado AQ, Critchley AT (2011) Optimization of culture conditions for tissue culture production of young plantlets of the carrageenophyte *Kappaphycus*. J Appl Phycol 23:433–438

Impacts of Climate change on *Eucheuma-Kappaphycus* Farming

7

Danilo B. Largo, Ik Kyo Chung, Siew-Moi Phang, Grevo S. Gerung, and Calvyn F.A. Sondak

Abstract

Climate change impacts all forms of life – including seaweeds! For farmed *Eucheuma/Kappaphycus* the direct and indirect impacts have caused not only physical damage to the crops but has also affected the eco-physiological, reproductive and metabolic processes of the seaweed. As a result of their importance as sources of different types of carrageenan, *Eucheuma* (iota)/*Kappaphycus* (kappa) have spread, through facilitated transfer, from its original tropical farming sites in the Philippines, across several localities in the Pacific, Atlantic and Indian Oceans. Ocean warming has caused the 'tropicalization' of sub-tropical waters of northeast Asia and South America, and more areas are becoming increasingly conducive to *Eucheuma/Kappaphycus* farming. Increases in surface seawater temperatures could have deleterious effects on reproductive capacity, including spore production, germination, recruitment and growth which, coupled with the long-practiced clonal propagation, have resulted in the declining productivity and quality of extracted colloids. Global climate change and anomalous climatic events such as El Niño and La Niña have increased the susceptibility of the selected cultivars of farmed eucheumatoids to bacterial pathogens as a result of stressful abiotic conditions which are conducive to 'ice-ice' disease and damaging impacts of epiphytism which have combined and literally wiped out activity in some farms. Typhoons and storms, as well as heavy rainfall during the wet season, increased in intensity by climate change, have also destroyed farms resulting in the loss of investments and income for those workers dependent on seaweed cultivation as a cash generating activity.

In addition, increased CO_2 dissolution in the oceans has resulted in its acidification, coupled with exposure of the surface-cultivated seaweeds to increased UV radiation (due to ozone depletion), which could decrease the concentration of the seaweeds' own protective pigments, thereby resulting in photo-inhibition and abiotic stresses. All of the foregoing,

D.B. Largo (✉)
Office of Research/Department of Biology, University of San Carlos, Talamban Campus, Talamban, Cebu City 6000, Philippines
e-mail: largodb@yahoo.com

I.K. Chung
Department of Oceanography, Pusan National University, Busan, Metro City 46241, Republic of Korea

S.-M. Phang
Institute of Ocean and Earth Sciences (IOES), University of Malaya, 50603 Kuala Lumpur, Malaysia

G.S. Gerung • C.F.A. Sondak
Faculty of Fisheries and Marine Science, Sam Ratulangi University, Manado 95114, Indonesia

A.Q. Hurtado et al. (eds.), *Tropical Seaweed Farming Trends, Problems and Opportunities*, Developments in Applied Phycology 9, DOI 10.1007/978-3-319-63498-2_7

individually and collectively, integrate to reduce daily growth rates. Reduced pH could also lessen the spore germination capacity of cultivated eucheumatoids.

The combination of the above factors has produced an overall decline in cultivated *Eucheuma/Kappaphycus* production. However, the efforts of many countries to increase their total area of farming and improvement in farming techniques, ensures the world's overall production trend for these seaweeds of industrial value to be on the rise, in general.

7.1 Introduction

The Earth's climate is changing and becoming increasingly warmer (IPCC 1990). Climates have indeed changed over millennia, however, the current changes are amongst the toughest ever faced with a resident human population. Global warming, sea-level rise and ocean acidification are serious threats to the structure and function of this planet. Evidence of the observed impacts of climate change is most obvious in natural systems including oceans and especially for those organisms which are benthic (and/or have been adapted for cultivation – often at the sea/air interface).

The risk of harmful impacts on ecosystems and human communities increased with the rates and magnitudes of warming oceans, ocean acidification, sea-level rise (and accompanying storm surge), as well as other dimensions of climate change along the coasts. The population and coastal assets projected to be exposed to risks, as well as human pressures on coastal ecosystems, will increase significantly in the coming decades due to population growth, economic development and coastal urbanization. Many (mobile) marine species have already shifted their geographic ranges, modified seasonal activities, migration patterns, abundances and species interactions and a large segment of biological diversity already faces increased extinction risks during and beyond the twenty-first century, especially as climate change interacts with other stressors (IPCC 2014).

Seaweeds are ecologically important primary producers, competitors and ecosystem architects and engineers that play a central role in coastal habitats. Human-driven production of selected algal species has a long history and large-scale, seaweed cultivation has been an attractive activity because of decades-proven, low-tech, low-cost technologies and the multiple uses that can be made of the resulting farmed biomass. Recent advances in mariculture techniques led to greater supplies of seaweeds as a true "marine crop" (Critchley et al. 1998; White and Wilson 2015; Bjerregaard et al. 2016). However, the potential of seaweed production is far from being fully exploited and or realized for the benefits of coastal communities. Macroalgal-vegetated coastal, marine habitats account for less than 2% of the global sea surface, yet contribute 210–244 Tg C year^{-1}, or approximately 50% of all carbon sequestered in the global coastal oceans (Duarte et al. 2005). The importance of coastal, benthic marine vegetation, a.k.a. "Blue Carbon", receives ever-increasing interest for their mitigation and adaptation potential (Laffoley and Grimsditch 2009; Nellemann et al. 2009). However, seaweed beds and kelp forests were not considered as blue carbon when UNEP (1994) proposed the concept, although they have gradually increased in importance due to their now elevated profile. Highly productive seaweed species can contribute significantly to the annual biological drawdown of CO_2 and the global carbon cycle (Turan and Neori 2010) – and importantly seaweeds selected for cultivation are amongst the most productive marine species.

It was only recently that the influence of climate change on fisheries and aquaculture sectors began to be addressed. Harley et al. (2012) noted that the impacts of climate change in seaweed-dominated ecosystems remained poorly understood. They reviewed how global climate change influenced marine macroalgae and their associated ecosystems and services they provide. They also pointed out certain gaps in the eco-physiological framework (e.g. thermal, ocean acidification, abiotic stress and algal life cycle at an individual-level), the importance of diversity rates of change, environmental history and multiple stressors and non-additive effects of other stressful agents. Harley et al. (2012) went on to discuss the wide array of responses of competitive, trophic and symbiotic relationships, plusshifts in the structure and function of ecosystems at the population- and community-levels as well as ecosystem goods and services.

Although a great deal of progress has been made in recent years, there are still significant gaps in our understanding which hampers our ability to predict the outcomes of global climate changes on seaweed-dominated systems. Some of the most important areas in marine ecosystems where climate change could have a significant negative impacts are the farmed seaweeds.

While current fisheries and some aquaculture industries with high-technology and/or large investments indeed have high capacities for adaptation (due to the greater development/application/adoption of environmental monitoring, modelling and resource assessments; IPCC, 2014), here we consider the impacts of climate change in the current *Eucheuma-Kappaphycus* farming industry. In assessing the vulnerability of seaweed aquaculture to climate change (FAO 2016) some measures of adaptation and mitigation are being proposed (N'Yeurt et al. 2012; Chung et al. 2013). It was perhaps an unrealistic expectation when seaweed culture was introduced as part of geo-engineering to enhance natural sinks of carbon against global warming in the middle of 1980s, as proposed by Matthews (1996). More recently, this interest has been revived by a network of the Asian-Pacific Phycological Association (APPA) and Korean Project. Since then, there have been interesting outcomes from several other projects (Chung et al. 2013). The concept of seaweed as a carbon sink has been a contentious issue largely due to the fact that the life span of seaweeds is relatively short and does not build extensively long-lasting body tissue, unlike trees on land which can lock up carbon in their body mass and remain for long periods, moreso if they are utilized into long-lasting products such as materials used in construction.

There are very few direct studies dealing with the impacts of climate change on seaweeds, but many studies have already demonstrated the biological and eco-physiological response characteristics in terms of temperature, pH and other abiotic, environmental factors on marine organisms. In particular, many cultivated species (and their strains and variants) have been translocated from their original habitat to many countries for commercial cultivation; a number of studies were conducted for these successful transplantations (see elsewhere in this book). The impacts of climate change, in terms of surface seawater temperature increases for geographic spread, reproductive cycles, general health of farming activities, pH, climate change-induced storms and typhoons and ultimately photosynthetic efficiency of eucheumatoid algae will be considered in this chapter, for the purposes of which, the genus names *Eucheuma* and *Kappaphycus* will be treated as synonymous (see elsewhere discussion on the characteristics of each genus).

7.2 *Eucheuma/Kappaphycus* Production and Its Vulnerability to Climate Change

Generally, biodiversity is affected by climate change, for indeed, change in environmental conditions affects all forms of life; this has occurred over the millennia of the Earth's existence. Although no specific literature on *Kappaphycus* and *Eucheuma* (or simply referred to as 'eucheumatoids') has dealt with the impacts of change in climate on these farmed seaweeds, studies from other marine species provide a clue to how changing climate may affect seaweeds, either as natural populations, or as farmed marine crops. Just as terrestrial production of crops adapts, so too must marine agronomy. Climate change impacts on seaweed communities can be summarised into two broad contexts which Harley et al. (2012) identified as mainly warming – with consequent sea level rise and ocean acidification.

7.2.1 Warming Oceans: Effects of Surface Seawater Temperature Rise on the Geographic Spread of *Eucheuma/Kappaphycus*

The effects of seawater temperature rise can be either subtle or abrupt. Subtle effects of global warming can be seen in an organism's slow, physiological adaptation to a gradual rise in temperature over time. The geographic shift in distribution of seaweeds, in general, is the best example of this adaptation. Although the success of the deliberate spread of *Eucheuma* and *Kappaphycus*, to other places in the world, from its tropical origin to sub-tropical destinations has been a 'facilitated transfer' (with farming operations being the main "vector" and motivation for the spread); the seeming adaptation of the seaweeds (and their cultivars) to these new (sub-tropical) habitats could be that the warming water temperature became more conducive to the successful farming of the tropical eucheumatoids. In other words, it could be postulated that the natural adjustment of seaweeds to a subtle rise in surface seawater temperature enabled (and encouraged) this facilitated transfer as a means of introducing seaweeds of economic value to many places around the world, in a circumferential belt of coastal locations, where it was predicted they would become rapidly adapted to local conditions. For instance, several countries in Latin America, e.g. Brazil, Mexico, Venezuela and Panama are now cultivating eucheumatoids that originated from The Philippines. Initially in Brazil in the 1990s, de Paula et al. (1998) brought live *K. alvarezii* from Japan (which originated from the Philippines), after which its cultivation began to spread in a matter of a few years to other parts of the South America's (some of this was by unofficial sources – which did not originate from Japan, see Chaps. 1 and 4). It is likely that when a sub-tropical region such as southern Japan becomes "tropicalized" by the warm Western Pacific current, *K. alvarezii* will gradually have an increasingly suitable habitat where its farming activities can be practiced throughout the year. The current production is seasonal, commencing after the onset of spring, seedstock being maintained indoors over winter (Mairh et al. 1986; Ohno et al. 1994; Gerung and Ohno 1997).

Tropicalization of the southern temperate boundary of Japan has long been observed notably when naturally occurring species of kelp, e.g. *Ecklonia cava* and *Sargassum* in Tosa Bay, southern Japan, began to disappear due to grazing by warm-water herbivorous fish (e.g. siganids and parrot fish) which encroached and foraged upon once lush, wild kelp beds (Verges et al. 2014). The habitats so vacated possibly could be taken over by possibly more resilient and herbivore-resistant tropical, seaweeds. Surface seawater temperatures of between 22–33 °C are reportedly required for *Kappaphycus* to succeed in commercial cultivation, although Lideman et al. (2013) suggested a slight difference in optimal temperature of between 23–32 °C for *E. denticulatum* and 22–33 °C for *Kappaphycus* spp. The temperature ranges for *Eucheuma* and *Kappaphycus* are almost identical in having similar adaptations to temperature and light conditions (Harley et al. 2012).

7.2.2 Warming Oceans: Effects of Temperature Rise on the Reproductive Cycle of *Eucheuma/Kappaphycus*

An important facet to consider is that temperature plays a central role in the reproductive cycle of seaweeds. Based on studies of the kelp *Macrocystis pyrifera* – one of the most studied seaweeds, increased seawater temperature has negative effects on spore production, germination, recruitment and sporophytegrowth of this large brown seaweed. This could be one of the reasons why reproductive thalli of eucheumatoids are rarely found in the natural grounds of the Philippines where coastal areas are more susceptible to higher temperature. While attempting to culture *Kappaphycus* in open water, using tetraspores, in the sub-tropical part of Brazil in the 1990s, de Paula et al. (1999) reported limited success. Thus, warming waters, due to climate change, would likely increase the likelihood of these test cultivation grounds approaching the optimal temperature for spore production and germination in Brazil.

It appears that the vulnerability of farmed seaweeds to surface seawater temperature rise is mainly due to effects on reproductive processes. *Kappaphycus* is capable of producing tetraspores in Brazil, but the reproductive structures fall short of completing their germination to successful thallus development (Bulboa et al. 2008), obviously an important factor in the field cultivation of the seaweed. In contrast, vegetative, monoclonal propagation (the main cultivation method in the Philippines and in most other countries where the seaweeds are transplanted), has been recognized as not the ideal mode of sustainable seaweed production, primarily due to the absence of genetic exchange. This has contributed to the declining productivity and quality of *Eucheuma/Kappaphycus* farmed in the Philippines (Hayashi et al. 2010). Resilience to changing environmental conditions is built on a strong genetic foundation which accrues to species which possesses both sexual and asexual methods of reproduction. This is also more important for species that have triphasic alternation of diploid, sporophytic (tetrasporophyte and carposporophyte) and haploid, gametophytic generations, making them more resistant to changes in climatic conditions, hence their evolutionary longevity and long-term survival. Repeated monoclonal propagation results in a seaweed or strain developing weakened resistance (strain fatigue) to changes in water temperature, with consequent impacts on metabolic activity (i.e. photosynthetic and respiratory) and reproductive processes, thereby to loss of seedling vigor. It would therefore be ideal if a genus, such as *Eucheuma,* could be freshly propagated from naturally occurring stocks, possessing genes from both male and female lineages, rather than repeated production by asexual means. In the example of *Gracilaria* farming, reduced genetic diversity was also attributed to continuous clonal propagation of the alga, and that the predominance of diploid individuals in the farming areas suggested that the agronomic practices had significantly modified the life-history traits of this agarophyte (when compared with wild populations, Mantri et al. 2009).

7.2.3 Warming Oceans: Effects of Temperature Rise on the Health of Farmed *Eucheuma/Kappaphycus*

Declining productivity is one impact of subtle surface seawater temperature rise on seaweeds due to the effects on the reproductive cycle which occurs over time. In contrast, the "abrupt effects" of temperature is when a rise takes place over an extended, yet relatively short, time-scale. Anomalous oceanic events such as El Niño have already exposed seaweeds to extreme conditions of temperature and increased their susceptibility to pathogens resulting in the disease commonly known as 'ice-ice' and infestations of algal epiphytes (e.g. *Neosiphonia*). Using temperature alone as a controlling factor, in the growth of *Kappaphycus alvarezii*, Largo et al. (1995a) demonstrated in tank (Aquatron) experiments, that exposure to water temperatures beyond 30°C could trigger the 'ice-ice' disease within just one - two weeks. Aside from observations in the Philippines (Largo et al. 1995b; Largo 2002; Butardo et al. 2003) the same condition was also observed in eucheumatoids farmed in different places, e.g. Indonesia (Aris 2011), Kenya (Wakibia et al. 2006), Madagascar (Ateweberhan et al. 2015), Zanzibar, Tanzania (personal observations of first author in 2017), China (Pang et al. 2015) and Vietnam (Terada et al. 2015). A warming sea will likely have a significantly negative impact on farmed carrageenophytes, primarily inducing the 'ice-ice' disease in *Kappaphycus* and *Eucheuma.* Based on the report of Mendoza et al. (2002), this could result in a de-polymerized carrageenan with lowered levels of iota and methyl constituents, as well as decreased yield, gel strength and viscosity. Therefore the rise of seawater temperature will have repercussions on seaweed health which will further exacerbate the persistent issue amongst farmers in the Philippines and elsewhere. With the impact of global climate change, scientists have warned of an increase in the rate and severity of bacterial disease in marine systems, including macroalgae (Harvell et al. 1999). The rationale is that environmental stress, as a result of climate change, can render macroalgae susceptible to opportunistic pathogens. Furthermore, Egan et al. (2013) in their mini-review entitled: "Bacterial Pathogens, Virulence Mechanisms and Host Defense in Marine Macroalgae" predicted that current and future anthropogenic changes in the marine environment, especially global climate change, would lead to an increase in the occurrence of macroalgal diseases – as indicated by a number of studies providing evidence of the involvement of bacterial pathogens that were otherwise previously under-appreciated.

Abrupt temperature change may induce stress in an organism due to the deleterious effects on enzymatic processes and metabolic functions. *Eucheuma denticulatum* has been shown to produce volatile, halogenated, organic compounds when the seaweed is under stress such as elevated pH and irradiance (Mtolera et al. 1996). Rising temperatures would exacerbate this effect.

7.2.4 Warming Oceans: Effects of Salinity and Sea Level Rise on Farmed *Eucheuma/Kappaphycus*

The optimal salinity for growth of *Eucheuma denticulatum* and *Kappaphycus alvarezii* was determined to range between 30–35 ppt (Largo et al. 1995a). A change in salinity below or above may have negative consequences to growth of the two genera. Low salinity stresses the seaweed to the extent that it becomes vulnerable to 'ice-ice'-causing bacteria (Largo et al. 1995b). Salinity change could be a result of freshwater intrusion in coastal areas, something that could happen during heavy rains causing salinity to drop rapidly and considerably during low tide. This is the main reason why *Eucheuma/Kappaphycus* farms do not exist near river mouths, or at least where coastal waters are diluted to below 30 ppt. To what extent seawater is affected by global warming and consequent sea surface temperature anomalies in terms of salinity is difficult to measure in the light of precipitation records, but there is an indication that increased intensity of rainfall events during a La Niña season could lead to freshening of the coastal waters – an event observed in Panama in 2010 where salinity was recorded between 31–34‰ during the dry season; in estuaries and coastal areas salinity was reduced to an unprecedented low level of 20‰ during the wet season (Valiela et al. 2012). This events were preceded by a drop in global mean sea level by 5 mm due to a strong El Niño during 2009/2010 (Boening et al. 2012). In order to avoid the negative effects of low salinity, the pattern of the planting season for *K. alvarezii* in the Teluk Gerupuk, West Nusa Tenggara, Indonesia, needed to be modified since farming generally occurs in the month where rainfall is low (dry season) and with concomitant lower water temperatures (i.e. 24–27 °C, Radiarta et al. 2013).

On the other extreme, low salinity as a result of gradual melting of ice in the polar regions, especially the in Greenland and the West Antarctic ice sheets, due to warming oceans and its effects on farmed seaweeds is hard to measure. This effect could be too subtle to detect in the short term, but could be discernible in the longer term. This does not disregarding the fact that global warming is causing coastal surges and sea level to rise, which was predicted by the Intergovernmental Panel on Climate Change in its Fourth Assessment Report (AR4) which indicated that the rise in level could reach as much as 60–100 cm (if not more) by 2100 (Nicholls and Cazenave 2010). It is therefore imperative that seaweed farms, on a global basis, will have to gradually adjust to such sea level rise and additional water forces on farming structures in the future.

7.3 Climate Change-Induced Storms and the Impact of Typhoons on *Eucheuma/Kappaphycus* Farming

Global warming is known to result in the increase of sea surface temperature (SST) that will most likely increase the strength and frequency of occurrences of tropical cyclones (typhoons), although analyses based on computer models seem to be inconclusive (Chan and Liu 2004; Stowasser et al. 2007). Fixed off-bottom mono-line, which is the most popular cultivation method applied in shallow water *Eucheuma* farming, is most vulnerable to even "ordinary" storms. Entire farms are commonly lost during the typhoon seasons in the Philippines, forcing farmers to harvest seaweeds early, or before they could reach their maximum biomass (normally achieved after approximately 45 farming days). Exercises in improvements to, and perhaps complete re-engineering of, farming techniques for *Eucheuma* and *Kappaphycus* are therefore necessary, if only to minimize the impact of typhoons and increasing storm surges. Typhoons are unheard of in such country as Indonesia, the reason why it has overtaken the Philippines and now leads in the world production of eucheumatoids. Although the firmly-anchored, off-shore farming of seaweeds has been explored for several species of brown seaweeds such as *Macrocystis* and *Laminaria* for biofuel biomass production (Roesijadi et al. 2008), very little improvement in farming technology has been observed for eucheumatoid seaweeds (Ask and Azanza 2002; Ask et al. 2003; Hurtado et al. 2013, 2015). However, it is patently obvious that many (most) rural, coastal farmers cannot yet afford shifting methods from simple mono-line farming (comprising a few stakes and ropes) to more sophisticated floating technologies often requiring boats and cranes and large, engineered, underwater mooring structures, such as those described by Bjerregaard et al. (2016) and Hayashi et al. (2016). Unless governments subsidizes or provide innovative ways of funding appropriate loans as part of a development package there will be major issues in providing the capital investments required by these modifications. This appears as a great opportunity for the upliftment of farming techniques in general, a common approach to farming required to address climate change that could have many beneficial, knock-on effects. Also, a paradigm shift may be necessary to make seaweed farming in the tropics more sustainable such as incorporating it with other high-value species following multi-tropic and extractive aquaculture practices (Bindu and Levine 2011; Largo et al. 2015).

7.4 pH Decreases Resulting from Dissolved CO_2

While seaweeds are considered as short term carbon sinks, an increase in CO_2 concentration could have a subtle effect on photosynthetic rates and growth performance, notwithstanding the fact that CO_2 rise could result in ocean acidification that affects calcification rates in calcareous algae and shelled molluscs. A pH which is lower than the normal value of around 8.2, would have stressful effects on most seaweeds including *Kappaphycus* and *Eucheuma*. There are indications from the recent study of Borlongan et al. (2016) and an older one by Collen et al. (1992) that the lowering of pH below 8 might result in increased photosynthetic rates in *Kappaphycus alvarezii* and *Eucheuma denticulatum.* Tee et al. (2015), however, showed that the stress induced in *K. alvarezii* due to pH 6 resulted in decreased daily growth rate and protein profiles (proteomes). Ribulose bisphosphate carboxylase (Rubisco) has been identified as a biomarker for pH-induced, abiotic stress. Arnold et al. (2012) showed evidence that low pH could lead to decreased, rather than increased, concentrations of phenolic, protective metabolites in marine plants (as far as seagrasses were concerned). Will this be the case too for marine algae, including eucheumatoids? In a study by Roleda et al. (2012) on the kelp *Macrocystis pyrifera,* a lower pH of 7.59–7.60 caused a significant reduction in the germination of spores, which might also be true in the case of the eucheumoid seaweeds. On the other hand, Gao et al. (1992, 1993) showed that increases in carbon dioxide levels, by up to five times that of normal atmosphere, in fact increased the growth of red seaweeds such as *Porphyra yezoensis* and *Gracilaria* spp. Algal photosynthesis is not saturated at ambient atmospheric CO_2 levels and an increase in its levels may be rather beneficial. Also the utilisation of bicarbonate with its increasing pH would offset the pH decrease due to dissolution of CO_2 in the ocean (Gao and McKinley 1994). While the lowering of pH may have negative implications for biomass production in *Kappaphycus* and *Eucheuma,* we do not know exactly to what extent such factors may have on other parameters, e.g. the molecular weight of carrageenan and or any of the bioactive metabolites.

7.5 Photo-Inhibition of Photosynthesis in *Eucheuma/Kappaphycus* Due to UV Irradiation

Analyzing the present and potential impacts of climate change on benthic, marine organisms, including seaweeds, is complicated by yet another factor of global scale, i.e. ozone depletion. This phenomenon occurs due to depleting compounds which have been shown to contribute in increasing levels of ultra-violet radiation reaching the earth's surface. Ozone depletion has been predicted to increase the "greenhouse effect" via reductions in the phytoplankton-driven, oceanic CO_2-pump (Franklin and Forster 1997, citing reports of United Nations Environmental Program 1994 and a study of Worrest and Hader 1997). The nature of eucheumatoid cultivation is that the seedlings are held at, or close to, the water surface (normally these cultivated plants of sub-tidal origins, would be attached in water depths exceeding 1–2 m, even at low tide). The impact of UVB radiation on *K. alvarezii* has been studied by Eswaran et al. (2001) and Schmidt et al. (2010). The former authors found that a short exposure of *K. alvarezii* to UVB from 60 - 180 min in a culture room strongly inhibited pigment accumulation (36–90% for phycocyanin and 12–69% for phycoerythrin, as compared to control), carrageenan yield (54–39%), gel strength (166–41%) and protein (11–58%, as computed from Table 1 of Eswaran et al.'s study) were also negatively affected. Pigment inhibition in this case decreased with increasing exposure for phycocyanin and phycoerthrin, while chlorophyll a increased, indicating that damage to the main photosynthetic pigment was more pronounced than for the accessory pigments. Schmidt et al. (2010) confirmed this effect in a more detailed study conducted in Brazil where *K. alvarezii* was exposed to UVB radiation which resulted in a significant decrease in both chlorophyll-a and phycobiliprotein (i.e. allophycocyanin, phycocyanin and phycoerythrin) contents, as compared to the control. Aside from decreased growth rates in the first few days, UVB exposure, as observed in Schmidt et al. (2010), also resulted in changes in ultra-structure, such as increased thickness of the cell wall and a concomitant increase in microfibrils (i.e. a more robust cellular structure which could be viewed as a defense mechanism). Also, cell walls became irregular and lost their normal "cup-shaped" appearance. Decreased growth in algae exposed to UV radiation, according to van de Poll et al. (2001, as cited by Schmidt et al. 2010), is a combined effect of damage to several cellular components. However, *K. alvarezii* has developed two morphotypes which could be a strategic adaptation to photoinhibition resulting from photosynthetically active radiation (PAR) to some extent and UV radiation to a lesser extent (Aguirre-von-Wobeser et al. 2001) where absorption spectra of its biliproteins, namely phycoerythrin (PE), phycocyanin (PC) and allophycocyanin (APC) vary. Green morphotypes lack expression for chlorophyll and PE, or an over-expression of PC and APC, hence its lack in photosynthetic efficiency (alpha) which Aguirre-von-Wobeser et al. (2001) suggested as may be due to the much lower energy absorbance of PC and APC than that of PE and chlorophyll. However, whether the presence of these two morphotypes in nature vary in terms of depth as an adaptation to irradiance, is not clear and needs further assessment. Increased levels of photosynthesis in *K. alvarezii* in both red and green morphotypes with irradiance reaching a limit of about 100 μmol quanta.$m^{-2}.s^{-1}$ indicated that this species could tolerate a higher irradiance curve when seawater temperature was increased. The findings of Aguirre-von-Wobeser et al. (2001) wherein P_{max} reached its maximum when water temperature was maximal, at about 30 °C. This clearly indicated that *K. alvarezii* was a highly flexible species thereby allowing it to be relocated from tropical to sub-tropical regions. Hence, the phenomenon of 'globalization' of *K. alvarezii* culture from the Philippines to the many different countries of the world as mentioned above; as long as a seawater temperature range of approximately 15–32 °C prevaild. Global climate change is therefore making it possible for the transfer of this species from a purely tropical origin to non-traditional, sub-tropical regions where water temperature is shifting to increasingly warmer levels.

With hind-sight it is possible to speculate that the massive epiphyte infestations in Camarines Norte, Philippines in early 2000s could have been attributed to a prolonged hot season caused by El Niño (Largo 2002; Critchley et al. 2004; Hurtado et al. 2006). This event was also repeated in Malaysian *Eucheuma* (Vairappan 2006) and Tanzanian farms (Vairappan et al. 2008).

7.6 Conclusion

Based on a limited number of studies involving eucheumatoids (and those of other seaweeds), climate change clearly has an existing impact on *Eucheuma/Kappaphycus* at the population level (as farmed seaweeds). At the species level, the exposure of the seaweeds to warming oceans, along with increased surface seawater temperatures (although effects may be presently only "subtle"), could have direct and indirect effects on the eco-physiology as well as reproductive functions resulting in compromised productivity and quality in some farming areas. Warming oceans may have also facilitated the successful transfer of these seaweeds - highly valued for their carrageenan, to sub-tropical regions (e.g. Brazil and Japan). Although a general atmospheric CO_2 increase seems to be favorable for the growth of eucheumatoids, there is no direct evidence of the impact of the associated lower pH in the field. However, controlled experiments showed that lower pH could add stress to the eucheumatoids resulting in decreased growth and a change in protein profile.

At the population level, climate change is already impacting farmed seaweeds by increasing the intensity of typhoons and storm surges causing wide-scale damage to the infrastructure. Likewise, the intensity of El Niño and La Niña phenomena, as well as UV radiation from ozone depletion have compounded the impacts. Exposure of the farmed seaweeds to increased seawater temperature during El Niño and lowered salinity from unusually heavy rainfall during La Niña, could all interact to increase the stressful conditions experienced by seaweeds cultured at the surface. Added to these abiotic impacts are the fact that the farmed "seedstock" is mainly produced using monoclonal propagation, leading to "strain fatigue" resulting in reduced vigour, compromised growth and susceptibility to 'ice-ice' disease and epiphyte/pathogen attacks.

Despite the above factors, the production trendline for *Eucheuma/Kappaphycus* continues to increase on a global scale, mainly due to expansion of farming areas (i.e. more area farmed, yet yield and productivity may be declining in certain regions), even reaching some sub-tropical waters where 'tropicalization' gradually allows for the establishment and successful farming of tropical species.

Acknowledgment The authors appreciate the invitation of the Editors to write this article. Also the Asian Network of Algae as Mitigation and Adaptation Measures (ANAMAM) (formerly the Asian Network for Using Algae as Carbon Dioxide Sink) is acknowledged for its role in advocating the promotion of seaweed aquaculture beds as a CO_2 mitigating measure.

References

(UNEP) United Nations Environmental Program (1994) Effects of increased solar ultraviolet radiation on aquatic ecosystems

Aguirre-von-Wobeser E, Figueroa FL, Cabello-Pasini A (2001) Photosynthesis and growth of green and red morphotypes of *Kappaphycus alvarezii* (Rhodophyta) from the Philippines. Mar Biol 138:679–686

Aris M (2011) Identification, pathogenicity of bacteria and the use of gene 16S rRNA for ice-ice detection on seaweed aquaculture, *Kappaphycus alvarezii* (Linn., 1758), M.S. thesis, Graduate School of Bogor Agricultural University (IPB), Bogor, 127 pp

Arnold T, Mealey C, Leahey H, Miller AW, Hall-Spencer JM, Milazzo M, Maers K (2012) Ocean acidification and the loss of phenolic substances in marine plants. PLoS One 7(4):e35107

Ask EI, Azanza RV (2002) Advances in cultivation technology of commercial eucheumatoid species: a review with suggestions for future research. Aquaculture 206:257–277

Ask EI, Azanza R, Simbik M, Cay-An R, Lagahid J (2003) Technological improvements in commercial *Eucheuma* cultivation (a short communication). Sci Diliman 15(2):47–51

Ateweberhan M, Rougier A, Rakotomahazo C (2015) Influence of environmental factors and farming technique on growth and health of farmed *Kappaphycus alvarezii* (cottonii) in south-west Madagascar. J Appl Phycol 27:293–934

Bindu MS, Levine IA (2011) The commercial red seaweed *Kappaphycus alvarezii* – an overview on farming and environment. J Appl Phycol 23:789–796

Bjerregaard R, Valderrama D, Radulovich R, James D, Capron M, Mckinnie CA, Cedric M, Hopkins K, Yarish C, Goudey, C, Forster J (2016) Seaweed aquaculture for food security, income generation and environmental health in Tropical Developing Countries. World Bank Group, Washington, DC. http://documents.worldbank.org/curated/en/947831469090666344/Seaweed-aquaculture-for-food-security-income-generation-and-environmental-health-in-Tropical-Developing-Countries

Boening C, Willis JK, Landerer FW, Nerem RS, Fasullo J (2012) The 2011 La Niña: so strong, the oceans fell. Geophys Res Lett 39:L19602

Borlongan IAG, Luhan MRJ, Padilla PIP, Hurtado AQ (2016) Photosynthetic responses of '*Neosiphonia* sp. epiphyte-infected' and healthy *Kappaphycus alvarezii* (Rhodophyta) to irradiance, salinity and pH variations. J Appl Phycol 28:2891–2902

Bulboa C, de Paula EJ, Chow F (2008) Germination and survival of tetraspores of *Kappaphycus alvarezii* var. *alvarezii* (Solieriaceae, Rhodophyta) introduced in subtropical waters of Brazil. Phycol Res 56:39–45

Butardo VM, Ganzon-Fortes ET, Silvestre VA, Bacano-Maningas MBB, Montaño MNE, Lluisma AO (2003) Isolation and classification of culturable bacteria associated with *ice-ice* disease in *Kappaphycus* and *Eucheuma* (Rhodophyta, Solieriaceae). Philipp Sci 40:223–237

Chan JCL, Liu KS (2004) Global warming and Western North Pacific typhoon activity from an observational perspective. J Clim 17:4590–4602

Chung IK, Oak JH, Lee JA, Shin JA, Kim JG, Park K-S (2013) Installing kelp forests/seaweed beds for mitigation and adaptation against global warming: Korean project overview. ICES J Mar Sci 70:1038–1044

Collen J, Pedersen P, Ramazanov Z, Mtolera M, Ngoile M, Semesi A (1992) Carbon assimilation of *Eucheuma denticulatum*. In: Mshigeni K, Bolton J, Critchley AT, Kiangi G (eds) Proceedings of the first international workshop on sustainable seaweed resource development in sub-Saharan Africa. University of Namibia, Windhoek, pp 265–273

Critchley AT, Ohno M, Largo D, Gillespie R (eds) (1998) The seaweed resources of the world. JICA, Kanagawa

Critchley AT, Largo D, Wee W, Bleicher L'honneur G, Hurtado AQ, Schubert J (2004) A preliminary summary on *Kappaphycus* farming and the impact of epiphytes. Jpn J Phycol (Suppl) 52:231–232

de Paula EJ, Pereira RTL, Ostini S (1998) Introdução de espécies exóticas de *Eucheuma* e *Kappaphycus* (Gigartinales, Rhodophyta) para fins de maricultura no litoral brasileiro: abordagem teórica e experimental. In: de Paula EJ, Cordeiro-Marino M, Pupo Santos D, Fujii M, Plastino EM, Yokoya N (eds) IV Congresso latino Americano de Ficologia. II Reuniao Ibero-Americana de Ficologia e VII Reuniao Brasileira de Ficologia: 340–357

de Paula EJ, Pereira RTL, Ohno M (1999) Strain selection in *Kappaphycus alvarezii* var. *alvarezii* (Doty) Doty ex P. Silva (Rhodophyta, Solieriaceae) using tetraspore progeny. J Appl Phycol 11:111–112

Duarte CM, Middelburg JJ, Caraco N (2005) Major role of marine vegetation on the oceanic carbon cycle. Biogeosciences 2:1–8

Egan S, Fernandes ND, Kumar V, Gardiner M, Thomas T (2013) Bacterial pathogens, virulence mechanism and host defence in marine macroalgae. Environ Microbiol 16:925–938

Eswaran K, Subba Rao PV, Mairh OP (2001) Impact of ultraviolet-B radiation on a marine red alga *Kappaphycus alvarezii* (Solieriaceae, Rhodophyta). Indian J Mar Sci 30:105–107

FAO (2016) The state of fisheries and aquaculture 2016. Contributing to food security and nutrition for all, Rome, 200 pp

Franklin LA, Forster RM (1997) The changing irradiance environment: consequences for marine macrophyte physiology, productivity and ecology. Eur J Phycol 32:207–232

Gao K, McKinley KR (1994) Use of macroalgae for marine biomass production and CO_2 remediation: a review. J Appl Phycol 6:45–60

Gao K, Aruga Y, Asada K, Ishihara T, Akano T, Kiyohara M (1992) Photorespiration and CO_2 fixation in the red alga *Porphyra yezoensis* Ueda. Jpn J Phycol (Sorui) 40:373–377

Gao K, Aruga Y, Asada K, Kiyohara M (1993) Influence of enhanced CO_2 on growth and photosynthesis of the red algae *Gracilaria* sp. and *G. chilensis*. J Appl Phycol 5:563–571

Gerung GS, Ohno M (1997) Growth rates of *Eucheuma denticulatum* (Burman) Collins et Harvey and *Kappaphycus striatum* (Schmitz) Doty under diferent conditions in warm waters of Southern Japan. J Appl Phycol 9:413–415

Harley CDG, Anderson KM, Demes KW, Jorve JP, Kordas RL, Coyle TA, Graham MH (2012) Effects of climate change on global seaweed communities. J Phycol 48:1064–1078

Harvell CD, Kim K, Burkholder JM, Colwell RR, Epstein PR, Grimes DJ et al (1999) Review: marine ecology – emerging marine diseases – climate links and anthropogenic factors. Science 285:1505–1510

Hayashi L, Hurtado AQ, Msuya FE, Bleicher-Lhonneur G, Critchley AT (2010) A review of *Kappaphycus* farming: prospects and concerns. In: Israel A, Einav R, Seckbach J (eds) Seaweeds and their role in globally changing environments. Cellular origin, life in extreme habitats and astrobiology 15, 251–283. doi:10.1007/978-90-481-8569-6_15, © Springer Science+Business Media B.V. 2010

Hayashi L, Novaes A, Santos A (2016) Design and prototype of *Kappaphycus alvarezii* mechanical harvester. In: Abstract, 22nd international seaweed symposium 2016, Copenhagen, Denmark, p 111

Hurtado AQ, Critchley AT, Trespoey A, Bleicher-Lhonneur G (2006) Occurrence of *Polysiphonia* epiphytes in *Kappaphycus* farms at Calaguas Is. Camarines Norte. Philipp J Appl Phycol 18(3–5):301–306

Hurtado AQ, Montaño NME, Martinez-Goss MR (2013) Commercial production of carrageenophytes in the Philippines: ensuring long-term sustainability for the industry. J Appl Phycol 25:733–742

Hurtado AQ, Neish IC, Critchley AT (2015) Developments in production technology of *Kappaphycus* in the Philippines: more than four decades of farming. J Appl Phycol 27:1945–1961

IPCC (1990) Climate change – the IPCC scientific assessment [Houghton JT, Jenkins GI, Ephraums JJ (eds)]. Cambridge University Press, Cambridge (UK)

IPCC (2014) Climate change 2014: impacts, adaptation, and vulnerability. Part A: Global and sectoral aspects. Contribution of working group II to the fifth assessment report of the Intergovernmental Panel on Climate Change [Field CB, Barros VR, Dokken DJ et al (eds)]. Cambridge University Press, Cambridge/New York, 1132 pp

Laffoley D, Grimsditch GD (eds) (2009) The management of natural coastal carbon sinks. IUCN, Gland, 53 pp

Largo DB (2002) Recent developments in seaweed diseases. In: Hurtado AQ, Guanzon NG Jr, de Castro-Mallare TR, Luhan MRJ (eds) Proceedings of the National Seaweed Planning Workshop held on August 2–3, 2001. SEAFDEC Aquaculture Department, Tigbauan, Iloilo, pp 35–42

Largo DB, Fukami K, Nishijima T (1995a) Occasional pathogenic bacteria promoting ice-ice disease in the carrageenan-producing red algae *Kappaphycus alvarezii* and *Eucheuma denticulatum* (Solieriaceae. Gigartinales, Rhodophyta). J Appl Phycol 7:545–554

Largo DB, Fukami K, Nishijima T, Ohno M (1995b) Laboratory-induced development of the ice-ice disease of the farmed red algae *Kappaphycus alvarezii* and *Eucheuma denticulatum* (Solieriaceae, Gigartinales, Rhodophyta). J Appl Phycol 7:539–543

Largo DB, Diola AG, Marababol MS (2015) Development of an integrated multi-trophic aquaculture (IMTA) system for tropical marine species in southern Cebu, Central Philippines. Aquac Rep 3:67–76

Lideman NGN, Noro T, Terada R (2013) Effect of temperature and light on the photosynthesis as measured by chlorophyll fluorescence of cultured *Eucheuma denticulatum* and *Kappaphycus* sp. (Sumba strain) from Indonesia. J Appl Phycol 25:399–406

Mairh OP, Htun US, Ohno M (1986) Culture of *Eucheuma striatum* (Rhodophyta, Solieriaceae) in subtropical waters of Shikoku, Japan. Bot Mar 29:185–191

Mantri VA, Reddy CRK, Jha B (2009) Reproductive processes in red algal genus *Gracilaria* and impact of climate change. In: Israel et al (eds) Seaweeds and their role in globally changing environments. Cellular origin, life in extreme habitats and astrobiology 15, 319–338. doi:10.1007/978-90-481-8569-6_15, © Springer Science+Business Media B.V. 2010

Matthews B (1996) Climate engineering: a critical review of proposals, their scientific and political context, and possible impacts. http://www.chooseclimate.org/

Mendoza WG, Montaño NE, Ganzon-Fortes ET, Villanueva RD (2002) Chemical and gelling profile of *ice-ice* infected carrageenan from *Kappaphycus striatum* (Schmitz) Doty "sacol" strain (Solieriaceae, Gigartinales, Rhodophyta). J Appl Phycol 14:409–418

Mtolera MSP, Collén J, Pedersén M, Ekdahl A, Abrahamsson K, Semesi AK (1996) Stress-induced production of volatile halogenated organic compounds in *Eucheuma denticulatum* (Rhodophyta) caused by elevated pH and high light intensities. Eur J Phycol 31(1):89–95

N'Yeurt ADR, Chynoweth DP, Capron ME, Stewart JR, Hasan MA (2012) Negative carbon via ocean afforestation. Process Saf Environ Prot 90:467–474

Nellemann C, Corcoran E, Duarte CM et al (eds) (2009) Blue carbon. A rapid response assessment. GRID-Arendal: United Nations Environment Programme. ISBN: 978-82-7701-060-1

Nicholls RJ, Cazenave A (2010) Sea-level rise and its impact on coastal zones. Science 328:1517–1520

Ohno M, Largo DB, Ikumoto T (1994) Growth rate, carrageenan yield and gel properties of cultured kappa-carrageenan producing red alga *Kappaphycus alvarezii* (Doty) Doty in the subtropical waters of Shikoku, Japan. J Appl Phycol 6:1–5

Pang T, Liu J, Liu Q, Li H, Li J (2015) Observations on pests and diseases affecting a eucheumatoid farm in China. J Appl Phycol 27:1975–1984

Radiarta IN, Erlania E, Rusman R (2013) Pengaruh iklim terhadap musim tanam rumput laut, Kappaphycus alvarezii di Teluk Gerupuk, Kabupaten Lombok Tengah, Nusa Tenggara Barat. Jurnal Riset Akuakultur 8(3):453–464

Roesijadi AE, Huesemann MH, Forster J, Benemann JR (2008) Techno-economic feasibility analysis of offshore seaweed farming for bioenergy and biobased products. Independent Research and Development Report IR Number: PNWD-3931 Battelle Pacific Northwest Division. 115 pp

Roleda MY, Morris JN, McGraw CM, Hurd CL (2012) Ocean acidification and seaweed reproduction: increased CO_2 ameliorates the negative effect of lowered pH on meiospore germination in the giant kelp *Macrocystis pyrifera* (Laminariales, Phaeophyceae). Glob Chang Biol 18:854–864

Schmidt EC, Nunes BG, Maraschin M, Bouzon ZL (2010) Effect of ultraviolet-B radiation on growth, photosynthetic pigments, and cell biology of *Kappaphycus alvarezii* (Rhodophyta, Gigartinales) macroalgae brown strain. Photosynthetica 48(2):161–172

Stowasser M, Wang Y, Hamilton K (2007) Tropical cyclone changes in the Western North Pacific in a global warming scenario. J Clim 20:2378–2396

Tee MZ, Yong YS, Rodrigues KF, Yong WTL (2015) Growth rate analysis and protein identification of *Kappaphycus alvarezii* (Rhodophyta, Gigartinales) under pH induced stress culture. Aquac Rep 2:112–116

Terada R, Duy Vo T, Nishihara GN, Shioya K, Shimada S, Kawaguchi S (2015) The effect of irradiance and temperature on the photosynthesis and growth of a cultivated red alga *Kappaphycus alvarezii* (Solieriaceae) from Vietnam, based on in situ and in vitro measurements. J Appl Phycol 28:457

Turan G, Neori A (2010) Intensive Sea weed aquaculture: a potent solution against global warming. In: Israel A et al (eds) Seaweeds and their role in globally changing environments. Cellular origin, life in extreme habitats and astrobiology 15:357–372

Vairappan CS (2006) Seasonal occurrences of epiphytic algae on the commercially cultivated red alga *Kappaphycus alvarezii* (Solieriaceae, Gigartinales, Rhodophyta). J Appl Phycol 18:611–617

Vairappan CS, Chung CS, Hurtado AQ, Soya FE, Bleicher-Lhonneur G, Critchley A (2008) Distribution and symptoms of epiphyte infection in major carrageenophyte-producing farms. J Appl Phycol 20:477–483

Valiela I, Camilli L, Stone T, Giblin A, Crusius J, Fox S, Barth-Jensen C, Monteiro RO, Tucker J, Martinetto P, Harris C (2012) Increased rainfall remarkably freshens estuarine and coastal waters on the Pacific coast of Panama: magnitude and likely effects on upwelling and nutrient supply. Glob Planet Chang 92–93:130–137

van De Poll WH, Eggert A, Buma, AGJ, Breeman AM (2001) Effects of UV-B induced DNA damage and photoinhibition on growth of temperate marine red macrophytes: habitat-related differences in UV-B tolerance. J Appl Phycol 37:30.37

Verges A, Steinberg PD, Hay ME, Poore AGB, Campbell AH, Ballesteros E, Heck KL, Booth DJ, Coleman MA, Feary DA, Figueira W, Langlois T, Marzinelli EM, Mizerek T, Mumby PJ, Nakamura Y, Roughan M, van Sebille E, Gupta AS, Smale DA, Tomas F, Wernberg T, Wilspon SK (2014) The tropicalization of temperate marine ecosystems: climate-mediated changes in herbivory and community phase shifts. Proc R Soc B 281:20140846. http://dx.doi.org/10.1098/rspb.2014.0846

Wakibia JG, Bolton JJ, Keats DW, Raitt LM (2006) Factors influencing the growth rates of three commercial eucheumoids at coastal sites in southern Kenya. J Appl Phycol 18:565–573

White WL, Wilson P (2015) World seaweed utilization. In: Tiwari BK, Troy DJ (eds) Seaweed sustainability. Food and man-food applications. Academic Press, Elsevier, London, pp 7–25

Post-Harvest Handling of Eucheumatoid Seaweeds

8

Majid Khan Majahar Ali, Ahmad Fudholi, Jumat Sulaiman, Mohana Sundaram Muthuvalu, Mohd Hafidz Ruslan, Suhaimi Md. Yasir, and Anicia Q. Hurtado

Abstract

The post-harvest handling of cultivated seaweed biomass is a crucial stage in the whole value-chain of the carrageenophyte industry. The quantity and quality of carrageenan derived from any given biomass depends largely on the post-handling treatment and management of the supply-chain for the harvested seaweeds. Since the successful farming of *Kappaphycus* and *Eucehuma* began in the early 1970s in the Philippines, the generally low technologies and cheaper conventional methods of drying the harvested seaweeds (i.e. hanging, platform, shade drying, 'sauna-like method,' etc.) are still prefered by the majority of farmers. However, these methods pose serious problems for carrageenophyte quality. A new day is coming when the traditional methods will not be sufficient in order to add value to the biomass. Recently, new methods for the drying of various eucheumatoid seaweeds were introduced, e.g. passive drying (chimney and greenhouse-based solar houses), active drying (de-humidifier drier) and hybrid (wind turbine, double-pass solar collectors with fins and v-groove, solar driers). However, these are still in their early stages of development and technology transfer to the industry.

The main types of drier technology, as applied to harvested biomass of *Kappaphycus* and *Eucheuma,* are discussed in this chapter.

M.K.M. Ali (✉)
School of Mathematical Sciences, Universiti Sains Malaysia, 11800 Gelugor, Penang, Malaysia

Seaweed Research Unit (UPRL), Faculty of Science and Natural Resources, Universiti Malaysia Sabah, 88400 Kota Kinabalu, Sabah, Malaysia
e-mail: majidkhanmajaharali@usm.my

A. Fudholi • M.H. Ruslan
Solar Energy Research Institute (SERI), Universiti Kebangsaan Malaysia, 43600 UKM, Bangi, Selangor, Malaysia

J. Sulaiman
Mathematics with Economics Programme, Faculty of Science and Natural Resources, Universiti Malaysia Sabah, 88400 Kota Kinabalu, Sabah, Malaysia

M.S. Muthuvalu
Department of Fundamental and Applied Sciences, Faculty of Science and Information Technology, Universiti Teknologi PETRONAS, Bandar Seri Iskandar, 32610 Tronoh Perak, Malaysia

S.M. Yasir
Seaweed Research Unit (UPRL), Faculty of Science and Natural Resources, Universiti Malaysia Sabah, 88400 Kota Kinabalu, Sabah, Malaysia

A.Q. Hurtado
Integrated Services for the Development of Aquaculture and Fisheries (ISDA) Inc., McArthur Highway, Tabuc Suba Jaro, 5000 Iloilo City, Philippines
e-mail: anicia.hurtado@gmail.com

A.Q. Hurtado et al. (eds.), *Tropical Seaweed Farming Trends, Problems and Opportunities*, Developments in Applied Phycology 9, DOI 10.1007/978-3-319-63498-2_8

8.1 Introduction

The global seaweed industry provides a variety of products directly or indirectly for human consumption, with a total value of approximately US$ 10 billion a year (Bixler and Porse 2011; FAO 2013). A comparison of estimated volumes for farmed seaweed production (2005–2014) is shown in Fig. 8.1 (FAO 2016). It can be seen that *Kappaphycus* and *Eucheuma* have estimated, higher production values than even the brown seaweeds such as *Saccharina* and *Undaria*, combined together, these kelps were once the top of the rankings for cultivated seaweed biomass. It was in 2012 when the scale of cultivated production of these two brown seaweeds was overtaken by the red seaweeds *Kappaphycus* and *Eucheuma* (FAO 2013, 2014). The latest production figures (2015) for the major carrageenan-bearing seaweeds are as follows: *Kappaphycus* (170,000 MT, dwt), followed by *Eucheuma* (4,500 MT, dwt), *Gigartina radula* and *G. skottsbergii* (14,000 MT, dwt) and finally, *Chondrus crispus* and others (3,200 MT, dwt; Fig. 8.2), with corresponding equivalent volumes of carrageenan, i.e.35,700 MT; 9,900 MT; 6,300 MT and 450 MT, respectively (Fig. 8.3; Porse and Rudolph 2017).

In 2016, 96% of the total estimated production of wet seaweed was dominated by Asian countries (FAO 2016) due to favorable enviromental and geographical factors such as the availability of seawater, surface area designated and licenced for cultivation, adequate surface seawater temperatures, salinity, turbidity, nutrient availability and pH of the ocean (Ali et al. 2014c, d). Amongst the Asian countries, at the time of writing (April, 2017) the most active in seaweed farming, particularly of the red seaweeds *Kappaphycus* and *Eucheuma*, were: Indonesia, the Philippines and Malaysia (Hurtado et al. 2008; Hurtado et al. 2014). Both genera are the main

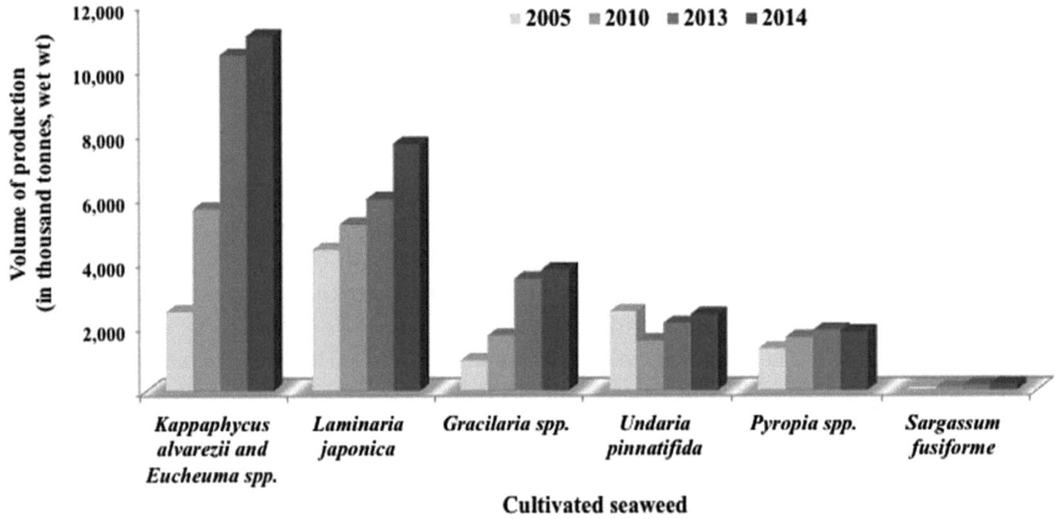

Fig. 8.1 A comparison of estimated volumes of farmed seaweed production (2005–2014)

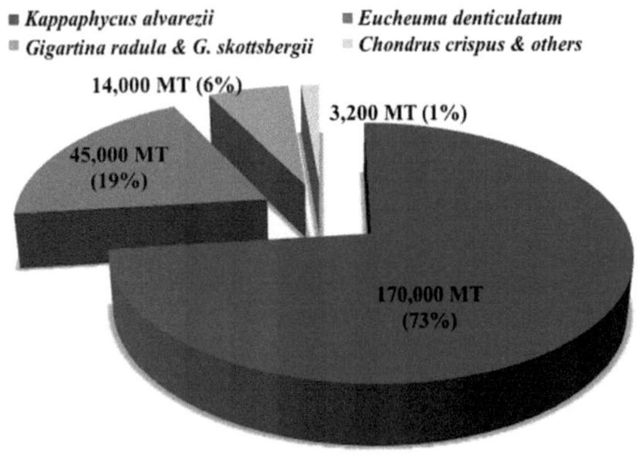

Fig. 8.2 Major sources of carrageenophytes and volumes of production (MT, dwt)

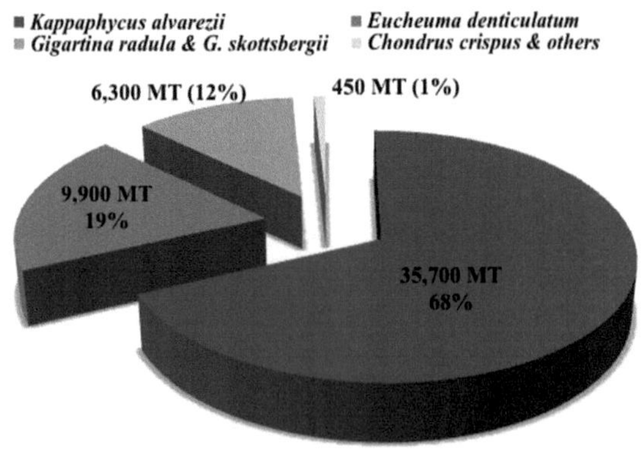

Fig. 8.3 Carrageenan volume of production (MT)

tropical carrageenonphytes used in the processing of semi-refined and refined carrageenan (McHugh 2003).

The carrageenan-bearing seaweed industry involves several processes such as: nursery, cultivation, harvesting, drying, processing and marketing. Post-handling (drying) management includes sorting and segregating the kappa and iota-bearing seaweeds. This stage is crucial because it greatly affects the quality of the seaweed and consequently the quantity and quality of the carrageenan yield. During the drying stage, a decrease in water activity retards microbial growth, this helps conserve the desirable qualities of the carrageenan and reduces the volume required for storage (Gupta et al. 2011). There are important factors attributed to seaweed biomass drying such as initial and desired final moisture content (MC), average, relative humidity (%) and water activity. Furthermore, the drying process not only reduces the amount of internal water in the biomass, but also reduces the costs related to time and labor during transportation and consequently adds value to the biomass through quality (Ali et al. 2013). According to the FAO (2013), seaweeds that are dried properly, with a final MC of less than 38% can be stored for a number of years, without appreciable loss to their gel properties. The quantity and quality of the carrageenan yield from the processed biomass depends on the duration and care taken during the post-harvest handling of the raw seaweed.

8.2 Types of Drying Methods

Only a few studies have been published on the post-harvest handling of *Kappaphycus* and *Eucheuma* (Vairappan et al. 2014; Djaeni and Sari 2015). Seaweed drying can be divided into four basic types, namely: (1) conventional (e.g. platform, hanging, indoor (shade) and 'sauna'); (2) passive (e.g. greenhouse concept, chimney concept drier); (3) active (e.g. freeze drier, de-humidifier and oven); and (4) hybrid (e.g. wind turbine drier, double-pass solar collector with fins).

In this chapter, passive, active and hybrid drying will be described in the detail available.

8.2.1 Conventional Drying

Conventional drying of carrageenophyte seaweed biomass is practiced in an open area, generally without a roof. This method depends largely on sunshine, prevailing weather conditions, wind speed and acceess to, and availability of, a large surface area. The MC of the seaweed drops dramatically when dried at an ambient temperature of 35 °C, but at lower temperatures (i.e. 28–29 °C), drying requires a longer time to reach the desired moisture content. According to the Philippines National Grade standard (PNGS 2007), the standard MC for *Kappaphycus* and *Euchema* is 35–40% and 33–38%, respectively. Moreover, *Eucheuma* seaweeds are relativery stable for periods of up to 12 months at moisture contents between 25–35% (Blakemore 1990) which also means the thalli remain flexible for efficient baling. Between 15–25% MC, bales are extremely stable, but the thalli are quite brittle and resist pressure. *Eucheuma* seaweeds below 15% MC remain stable, but can cause processing problems during carrageenan extraction as they require some rehydration before processing (Faria et al. 2014). At present, seaweed farmers prefer to use the lower technology, conventional drying methods largely due to the lower costs of investment required thereby providing them with a higher, apparent income (Ahemad 2006).

8.2.1.1 Hanging Drying

Immediately the cultivation lines (with the attached seaweed, at the farm site), are pulled out, they are hung vertically from posts (Fig. 8.4a–d), without removing any unwanted drift or attached seaweeds (epiphytes), floats, barnacles, shells or attached foreign material. It is much easier to remove these unwanted contaminants when dried, by shaking the whole line. However, this type of drying method needs a lot of clean space and only few lines can be hung on any given day (depending on the size of the area available). Sometimes, farmers cover their entire structures with polyethylene in order to prevent rain from re-wetting the drying seaweed. Sometimes the biomass is moved even to the inside the farmer's house. It can take 4–9 days of drying (depending on the relative humidity, temperature and rate of air flow) for the seaweed to reach 43–45% MC. Once the seaweeds are dried to this level, they are stripped off the cultivation line and any unwanted contaminants including attached seaweeds (epiphytes), sand, shells or other debris are removed.

8.2.1.2 Platform

There is a positive correlation between the effort required to dry the seaweed correctly and price received. Experienced and skilled traders can estimate the MC by merely looking and feeling the dried seaweed. Generally, drier seaweed has a higher value. Drying seaweed using an elevated platform made of bamboo slats is popular in the Philippines, Indonesia and Malaysia (Fig. 8.5a–d). A fish net is normally laid on a platform before the seaweeds are scattered for drying. The fish net is turned over at the edges in order to prevent losses especially when the seaweed is already dried and can be blown by high winds. Likewise, use of the fish net makes the work more convenient at the end of the day in order to aggregate the harvested biomass. Normally, it takes 3–4 sunny days, or 7–10 not so sunny days, of drying in order to obtain an MC of 43–45%.Turning the seaweed every 2–3 h obtains uniformity of dryness. Similar to the hanging method, the seaweed is dried directly by the sun, without any pre-treatment method.

Fig. 8.4 (**a–d**) Hanging-type of drying (Photos courtesy of: (**a–c**) MKM Ali, (**c**) I Indrayani

8.2.1.3 Shade-Drying

Vairappan et al. (2014) conducted a comparative study of three types of drying (i.e. freeze-drying, shade-drying and direct sun-drying) of *Kappaphycus*. The authors claimed that freeze- and shade-drying demonstrated insignificant differences in terms of the extracted carrageenan qualities, as compared to the direct, sun-drying technique (Table 8.1).

8.2.1.4 'Sauna-Like' Conditions as a Pre-treatment Method

Ali et al. (2014a) reported the use of 'sauna-like' techniques to be used as pre-treatments for drying carrageenan-bearing seaweed biomass. It was then a new, alternative method to decrease the period of drying, without compromising the quality of the carrageenan held within the biomass. The seaweed was first completely covered with polyethylene for two to three days until it became whitish in color, after which, the plastic sheet was removed and sun drying was used directly for a further one to two days (Fig. 8.6a–c).

In the Philippines, local processors employed the practice of paying the traders or farmers 75% of the total value of the seaweed upon collection, with the remaining 25% being paid after a determination of the carrageenan yield was completed (Balicuatro, personal communications). It is now a common practice in the Philippines to include not only the MC but also the percentage yield of carrageenan from the harvested biomass when determining a price to be paid to the farmers. Seaweed dried to the correct specifications of the processor is normally given a premium to compensate for the additional care and efforts required. Ali et al. (2014e) made a comparative study between direct, sun-drying of fresh - and sun-drying – of 'sauna' pre-treated seaweed. The latter technique required only two days in order to reduce the initial MC of the thalli from 96.3 to 35%, equivalent to 2,500 kg w/b (wet basis; fresh) to 281.15 kg (w/b), as compared to

Fig. 8.5 (**a–d**) Platform form of drying (Photos courtesy of: (**a**) AQ Hurtado, (**c–d**) MKM Ali

Table 8.1 Average (±SD, n = 3) physico-chemical properties of carrageenan extracted from *Kappaphycus* using three post-harvest, drying techniques

	Post-harvest treatments		
Parameters tested	Freeze-drying	Shade-drying	Direct, sun-drying
Carrageenan yield (%)	56.5 ± 2.5[a]	58.3 ± 2.1[a]	42.6 ± 3.8[b]
Melting point (°C)	58.6 ± 0.4[a]	57.5 ± 0.8[a]	49.2 ± 0.6[b]
Gelling temperature (°C)	35.3 ± 0.5[a]	35.6 ± 0.4[a]	31.5 ± 0.8[b]
Viscosity (cPs)	58.4 ± 4.2[a]	57.8 ± 2.6[a]	42.4 ± 2.2[b]
Gel strength (g cm^{-2})	1454.4 ± 12.0[a]	1424.6 ± 14.2[a]	898.2 ± 8.2[b]
Syneresis index (%)	15.4 ± 2.3[a]	16.6 ± 2.1[a]	24.1 ± 3.6[b]
Molecular weight (kDa)	720 ± 38 (80%)[a]	710 ± 42 (81%)[a]	460 ± 18 (55%)[b]
	542 ± 24 (16%)	520 ± 32 (9%)	210 ± 28 (25%)
	230 ± 15 (4%)	280 ± 26 (10%)	60 ± 8 (15%)
			<30 ± 6 (5%)

Means within a row with different superscript letters were significantly different at $P < 0.05$ (Vairappan et al. 2014)

direct, sun-drying of the same initial biomass, which took four days to produce dried seaweed from 5,000 kg w/b, fresh to 563.1 kg w/b dried. The efficiency of the time saved was approximately 58% – equivalent to almost three days; adoption of the 'sauna' technique was therefore a considerable improvement in time required. The carrageenan qualities of conventionally dried seaweed, as compared to 'sauna'-pretreatment followed by sun drying are shown in Table 8.2.

8.2.2 Passive Drying

Due to some disadvantages of the platform and hanging drying methods, further studies were conducted to improve the quality of dried seaweed for the carrageenan industry. However, these techniques are still at the developmental stage and 'fine tuning' is in progress in order to test their economic efficiencies for the industry to adopt, such that

Fig. 8.6 (**a–c**) 'Sauna'-like pre-treatment of drying (Photos courtesy of: (**a**) AQ Hurtado, (**b–c**) MKM Ali

Table 8.2 Average (±SD, n = 3) for the carrageenan properties of *Kappaphycus* conventionally sun-dried and 'sauna'-pre-treated, then sun-dried

	Types of drying	
Properties	Platform	'sauna'
Protein (%)	3.76 ± 1.33^a	2.46 ± 1.33^a
Ash content (%)	39.19 ± 0.42^a	36.19 ± 0.42^a
Sulfate content (%)	21.75 ± 0.04^a	14.59 ± 0.45^a
Gel strength (g cm^{-2})	492.6 ± 0.25^a	530.9 ± 0.52^b
Viscosity (cp)	56.7 ± 0.42^a	50.7 ± 0.23^a
Carrageenan yield (%)	30.4 ± 0.21^a	33.5 ± 0.32^a

Means within a row with different superscript letters were singificantly different at $p < 0.05$ (Ali et al. 2015a)

higher carrageenan yields with improved quality characteristics will be obtained, which should then consequently translate to a higher value of the dried biomass for processing and hence prices paid to the farmers. Due to the generally lower quality of carrageenan obtained from biomass produced using the traditional drying methods, the industry is seeking the assistance of academia to help farmers increase the yield and quality of carrageenan obtained from the dried, harvested seaweed biomass.

Below are some newly developed techniques for drying seaweeds.

8.2.2.1 'Greenhouse' Concept

Passive drying infers direct, sun-drying, without any additional facilities such as a fan. This system was designed based on the priniciple of a greenhouse (40 m × 30 m; Yasir and Ramlan 2012, Fig. 8.7a–b).The conical roof was constructed from transparent polyethylene (0.25 mm thickness) and all sides were constructed from marine plywood (9.0 mm thick). Three stacks of removable trays, made from plastic with a mesh size of mesh 0.01 mm, were arranged in an inclined manner in order to drain excess water during the process (Yasir 2012, 2013). The trays were rotated daily from top to bottom in order to uniformly dry the seaweed. It took six to 12 days to dry the biomass to a 35% MC. The'greenhouse' drier, as sized above, could accommodate 2 MT of seaweed per drying cycle.

The study of Phang et al. (2015) used a chimney solar dryer for 'cottonii' seaweeds in Malaysia. The natural draft, solar dryer consisted of an opening for air at the bottom, a drying chamber with trays and a draft-enhancing chimney, fitted with wire mesh at the top. The moist air was discharged through air vents, or another chimney at the top of the chamber. This system must be properly insulated so as to minimize heat loss and also be constructed from durable materials (within economically justifiable limits). Construction from metal sheets, or water-resistant cladding, e.g. paint or resin, was recommended. Heated air flowed through the stack of trays until the entire product was dried. As the hot air entered through the bottom tray, this tray dried first. The last tray to dry was the one at the top of the chamber. The chimney, with wire mesh, prevented the inflow of cooler air and had, as high as 90% improvement over the efficiency of air velocity, as compared to a conventional chimney (Chu et al. 2012), as shown in Fig. 8.8. The capacity of the dryer, as tested, was a maximum of 100 kg fresh biomass, with a size of 10 m × 30 m. This drying device was designed to enhance drying rates through the enhancement of air flow by natural convection, without the additional use of fossil fuel or electrical devices.

The chimney-concept dryer was designed specificly to study the characteristics of drying seaweeds under natural convection and was also compared to a shade drying process. A de-watering stage, as a pre-treatment process, was initially applied to enhance the drying process for both methods. The initial weight of the seaweed before and after pre-treatment were recorded and the biomass was then introduced into the solar and shade drying systems. The air temperature and relative humidity inside the solar dryer and surroundings were recorded. Based on the results obtained, a representative sample from each tray was taken for the final MC deter-

Fig. 8.7 (**a** and **b**) 'Greenhouse'-type of seaweed drying (Yasir and Ramlan 2012)

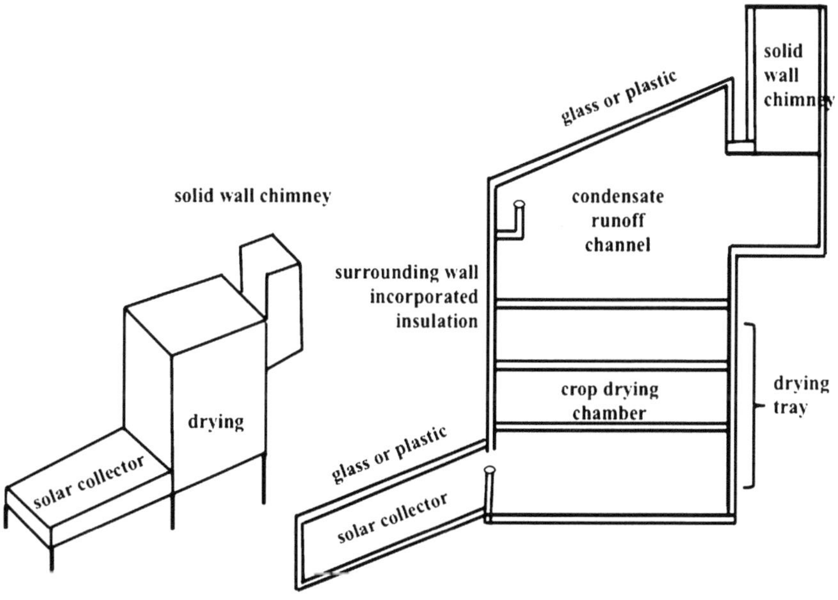

Fig. 8.8 Chimney-based drier (Phang et al. 2015)

mination. A difference of weight of less than 5% was measured, as compared to shade drying. The average loss of moisture during the pre-treatment was about 54%. The final MC of the seaweed from the chimney, concept drier was in the range of 24–61% (wb = wet basis) and shade drying resulted in seaweed samples which were in the range of 40–48% (wb), with the standard deviation of the final MC samples being 20.5% for solar drying and 3.8% for shade drying (Phang et al. 2015). The total time required for solar drying, inclusive of pre-treatment, was 6 days and shade drying was 9 days. The initial MC of seaweed on each tray was in the range of 86–89% (wb). The initial MC of samples from the solar dryer was slightly higher on the bottom tray, as compared to the upper tray, due to drips of moisture and also because of the draft created by the chimney effect (Kumaresan et al. 2013). The effect of solar irradiation on the top tray was found to be key and significant, as noted in the solar dryer. The initial MC under shade drying was almost the same at the initial stage, since the flow of hot air in to the drying chamber flowed at a constant rate, and the effects of solar irradiation were not obvious, as compared to solar drying. This study concluded that in the chimney-based solar drier, solar radiation was not the most important factor, as compared to solar drying. This maybe due to a self-shading effect between the drying trays. Hence, there is need for the trays to be rotated on a daily cycle. The MC of the seaweed samples in the middle tray, under shade drying was the highest, as compared to the other trays, but not as high as the MC of material on the middle tray of the solar dryer. The airflow under shade conditions may be steady at different stages throughout the drying process and was not significantly affected by solar radiation. Consequently, the MC of the middle tray biomass could be used as an operational indicator guiding assessment of the final MC.

8.2.3 Active Drying

8.2.3.1 De-humidified Air Drying

The study of Djaeni and Sari (2015) used air which had been de-humidified using zeolites in order to dry 'cottonii' seaweeds in Indonesia. The average MC for the fresh seaweed biomass, as harvested, was 78.9%. In order to maintain the quality of the carrageenan, it was suggested that the relative humidity of the air in contact with the drying seaweed should be less than 50% (Ali et al. 2015). Based on the results of Djaeni and Sari (2015), air drying using zeolites could provide positive effects on water removal from fresh seaweed biomass (as a non-mechanical method of drying). Results showed that for all cases, drying at 70°C or below (i.e. the average temperature recorded during the study) resulted in shorter drying times, as compared to conventional methods. The authors further reported that the drying time was shorter at the higher temperatures or lower relative humidity of the air. Higher temperatures increased moisture diffusion, as well as a constant of drying rate, whilst low relative humidity enhanced the drying process.

The higher the air temperature and airflow rate, the faster was the drying time. At higher temperatures, water evaporation increased (due to the moisture carrying capacity of the air) which simply indicated that more water was moved from the internal tissues to the surface of the seaweed.

The study of Djaeni and Sari (2015) further evaluated the rehydration of the seaweed. Their results demonstrated that seaweed drying was not just for biomass to supply carrageenan manufacturers, but could also treat the seaweed biomass as a source of human food, with a higher price being provided for a sea vegetable salad, for yogurt or energy/booster drink preparations. At present, Southeast Asian countries internally use fresh *Kappaphycus* or *Eucheuma* as a source of sea vegetable salad (Pedrosa, pers. commn.). However, the Philippines also exported young thalli of *K. striatus* and *E. denticulatum* to China and Taiwan and sometimes to USA (Abubakar, personal communications) for seaweed salad ingredients (Solante, personal communications)

8.2.3.2 Forced Convection Pump v-Groove Collector Using Generator

The Universiti Kebangsaan Malaysia (UKM) through the Solar Energy Research Institute (SERI) also helped in providing tools for the improvement of methods for drying seaweeds at an industrial scale. Cooperation between small and medium sized enterprises (SMEs) and the Scientific and Industrial Research Institute of Malaysia (SIRIM) successfully installed heat-pump driers, which required the use of an electric generator. This system was installed at Sebangkat Island, Semporna, with a capacity of 100 kg fresh seaweed for each experiment. A heat-pump powered drier (30 m × 20 m), with fresh seaweed arranged on trays; this took 8 days to attain a stable MC. Although there was an effort from the government to assist the seaweed farmers by installing the turbine drier, there were still challenges to resolve including: a backup-power plan, the overall efficiency of the system and the on-going costs of operation and

Fig. 8.9 Wind turbine solar drier (Ruslan et al. 2012)

maintenance, all of which posed issues in terms of operational sustainability. In short, further studies were needed in order to address these.

8.2.4 Hybrid Drying

8.2.4.1 Wind Turbine and Solar Drier

In 2009, the Scientific and Industrial Research Institute of Malaysia (SIRIM) installed another seaweed drying project using a wind turbine combined with solar driers on Karindingan Island, Semporna (Fig. 8.9). These driers had an area of 20.45 m^2. The wind turbine generated electricity which powered a fan, an auxilary heater with the solar system functioning to dry the seaweed. The chamber was able to accommodate a total of 300 kg fresh biomass, for each trial, with a duration of 50–78 h per cycle (Othman et al. 2012).

8.2.4.2 Double-Pass Solar Collector with Fins

In 2012, following the success and demonstration of the high efficiency of the solar-assisted drying system, the Small Medium Enterprise (SME) program helped by providing a grant to SERI in order to build a small solar dryer, with solar collectors and heat absorption fins (Fudholi et al. 2011). The dryer was designed and installed at the Solar Energy Technology Park, National Universiti of Malaysia. This laboratory-scale, pilot dryer (11.28 m^2) accommodated a maximum capacity of 40 kg (fresh biomass of seaweed, with a high moisture content). Currently, the drier is actively used to dry seaweeds, herbs and medicinal plants and provides an air flow rate of 0.005–0007 ms^{-1}. This drier reduced the fresh seaweed to 40–43% MC in 12–36 h. A schematic diagram of this kind of dryer is shown below (Fig. 8.10).

8.2.4.3 v-Groove Hybrid Solar Drier (v-GHSD)

A v-Groove Hybrid Solar Drier (v-GHSD) was installed at Selakan Island, Semporna, Sabah. This drier was a collobarative project between the Universiti Kebangsaan Malaysia (UKM) and the Deparment of Fisheries (DOF), Malaysia. This drier was partially installed in 2012 and completely finished in 2013 (Ali et al. 2014e). It was designed to meet all the criteria of a premium, dried seaweed such as: defined MC, high carrageenan yield recovery and ultimately reduced number of drying days. It was expected therefore that the seaweed dried through this system would command a higher price in the market. However, there were some issues which required specific considerations, i.e.: (1) calculation of the system efficiencies by developing a mathematical model which would explain the drying process and provide some projections for long-term operations; (2) economic efficiency; (3) presence of management tools for maintenance purposes; and (4) commercial capacity of the drying system, scalable to meet the demands of the monthly harvested biomass (~ 20,000 MT fwt). An estimated 20 drying units would be required to provide a maximum drying capacity of 10 MT fwt. This type of drier was categorised as a forced-convection, indirect type. A photograph is shown in Fig. 8.11. The solar drier consisted of an axial fan, drying chamber, v-shaped aluminium roof, solar collector and trays with Teflon screen tape. Based on the report of Ali et al. (2015b), the lowest recorded relative humidity (%) of the drier was 30%, the maximum internal temperature was 60 °C and the lowest seaweed MC obtained was 20%.

The flow of air in the system was controlled, begining when solar radiation fell on to the aluminum plate at the v-grooved, absorption collector. The radiation was convected to the back of the collector using an axial fan within the

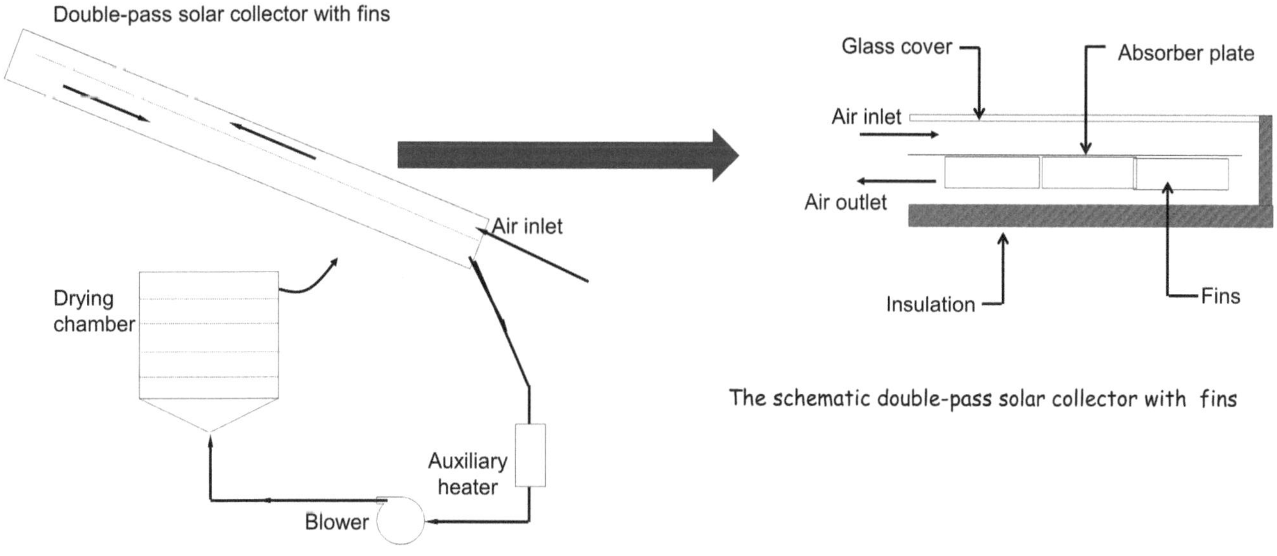

Fig. 8.10 Schematic diagram of a double-pass solar collector, with fins (Fudholi et al. 2011)

Fig. 8.11 Photograph of a v-GHSD dryer (Photo courtesy of MKM Ali)

chamber. During convection airflow into the chamber, the air passed through a v-groove causing the air to heat up over the collector which had a range of 50–69 °C. In this system, the axial fans would only start to rotate once the temperature of the collectors reached 55 °C, stopping when it reached 59 °C. The hot air flowed from the back of the chamber through the tray racks. The fan pushed hot air from the collector to the front of the chamber. A concentric fan on the wall recycled hot air to the seaweed trays. The hot moist air, so produced, was pushed out through a small window at the bottom of the drying chamber. The drying process continued until the relative humidity in the chamber was reduced and an equilibrium MC (35%) was reached. In this study, the lowest relative humidity recorded was 79% and the maximum temperature was 72°C.

Physical Performance of v-GHSD

The seaweed thalli were spread uniformly on each tray in order to achieve uniform drying. Freshly harvested seaweed (10,000 kg), at approximately 93% MC underwent a 'sauna-like' pre-treatment method, as explained by Ali et al. (2015). The same authors further reported that the results showed that 'sauna' drying, without auxiliary heating of 2,500 kg of dry red seaweed, reached about 35% MC within 16 h (two days of drying) and yielded 281.5 kg of dried biomass. The MC was decreased by approximately 50% in two days of 'sauna'-like drying. The time saved using this technique was calculated as 57.9%, at an average solar radiation of approximately 500 Wm^{-2}, with an air flow of 0.056 kg s^{-1}.

Water loss was calculated based on the formula:

$$water\ loss\ (\text{kg}) = \frac{(X_{i-}X_F)}{(100 - X_F)} \times m_0 \quad (8.1)$$

where,

X_i = *Initial moisture conten* (wb)
X_F = *Final moisture content* (wb)
m_0 = *Total weight* (kg)

the value of X_i was determined using the AOAC (2001) method which dried 25 g fresh seaweed sample at 105°C for 24 h, until it reached an equilibrium MC. The gathered data were inserted into the equation as indicated below:

$$5000.00 = \frac{(92.68 - X_F)}{(100 - X_F)} \times 10{,}000. \quad (8.2)$$

$$X_F = 85.36\% \quad (8.3)$$

Then, the same amount of 'sauna'-pre-treated seaweed was divided equally between each of the v-GHSD and conventional drying platforms. Approximately 114 h (equivalent to 15 days drying) were required in order to dry seaweed using the conventional method. The dried seaweed was then relatively stable for a period of 12 months if maintained between 25–38% MC. This compared to a range of 15–25%, where the thalli were too brittle and resistant to compression in order to bale conveniently.

Eucheumatoids having a MC below 15% remained stable, but are known to cause processing problems during carrageenan extraction (Blakemore 1990), hence, it is necessary to set the MC in the region of 38%. The values for the mass flow rate, ambient temperature (relating to the immediate surroundings), the rate of evaporation, average solar radiation and relative humidity of the ambient air through the systems were: 0.08 ms^{-1}, 30.1 °C, 21.93 kg h^{-1}, 610 Wm^{-2} and 78.1%, respectively.

Table 8.3 Average comparative parameters (±SD, n = 3) of a drying system using conventional and v-GHSD systems

	Factor	
System	Conventional	v-GHSD
Drying time (h)	114.00 ± 0.22	38.00 ± 0.57
Moisture evaporative capacity (kg)	21.93 ± 1.32	64.79 ± 0.12
Average chamber relative humidity (%)	78.40 ± 2.12	45.00 ± 3.12
Average chamber temperature (°C)	30.20 ± 0.19	48.00 ± 0.57
Average solar radiation (w m^2)	615.50 ± 10.3	650.10 ± 15.22
Mass flow rate (kg s^{-1})	0.08 ± 0.001	0.272 ± 0.02
Initial weight (kg)	2,506.80 ± 0.12	2,506.80 ± 0.11
Final weight (kg)	593.54 ± 0.01	593.54 ± 0.02

The final weight of seaweed, with 38% MC was 593.5 kg. In this drying process, the amount of surface area used was the same as the total surface area of the v-GHSD drying chamber (32.25 m^2). The performance of the v-GHSD dryer over 5 days of drying operation (using three replicates; 38 h were used, not including the night periods). The recorded average temperature (ambient and chamber), relative humidity (ambient and drying chamber) and solar radiation were: 43.2 °C, 30.1 °C, 49.5%,71.7% and 630 Wm^2, respectively. The results presented in Table 8.3 showed that the v-GHSD drying system was more efficient than a conventional drying platform. About 78 h of saving times was achieved using the v-GH, as compared to the conventional system. The evaporative water capacity of the seaweed drier was higher with an average of 64.79 kg h^{-1} drying, as compared to 21.93 kg h^{-1} on a conventional platform. The average humidity in the chamber was reduced by 33.4%, as compared to ambient air (on the days tested).

The average temperature in the chamber was 8.8 °C above ambient air temperature. Overall, drying in the sun drying chamber was faster than the conventional drying systems. Overall, the v-GHSD was an efficient drying system, taking five days to dry 2,500 kg of 'sauna'-treated seaweed (i.e. 5,000 kg of initial fresh mass). The results obtained were promising and could be provided as a recommendation to the industry. However, economic efficiency has to be further studied in order to make its capital costs more attractive, not only to small farmers, but also large processors who would like to contract seaweed farmers to use this equipment for drying purposes, so as to obtain premium dried seaweed and consequently, obtain premium grade carrageenan. The v-groove, hybrid solar drier remains perhaps the most advanced solar drier developed for seaweed drying in

Table 8.4 Standards of semi-refined carrageenan as compared to extracts from seaweed dried in the v-GHSD

	Yield (%)	Gel strength (g cm$^{-2)}$	Viscosity (cPs)
Philippines[a]	48–53	<1000	<5
FDA[b]	<40	200–700	<10
FAO[c]	<60	>250	<10
FAO[d]	38-55	>50	>5
v-GHSD	34	531	50.7

[a]**Philippines** – Philippine National Grade Standard for E407a and E407 (DPNS 601, 2008)
[b]**FDA** – Carrageenan Standard in Food-based Industry (2006)
[c]**FAO** – Monographs Specifications: Carrageenan (2007)
[d]**FAO** – Carrageenan and processed *Eucheuma* seaweed (FAO 2001)

Malaysia. This equipment could be employed in other seaweed producing countries; the invention is patented by the Universiti Kebangsaan Malaysia (Patent No. PR324185) and they are currently licensing the technology to interested parties.

The Effects of Drying on the Quality Characteristics and Performance of Extracted Carrageenan

Seaweeed biomass dried using the 'sauna' technique, which were then divided into the v-GHSD drying chamber and a conventional platform system, were taken to the laboratory to test the quality characteristics of the carrageenan obtained from the biomass. Carrageenan yield, gel strength, viscosity and molecular weight were determined in order to characterize the carrageenan, analyses were carried out in triplicate. The results were compared with conventionally dried samples with a MC of 38%.

Semi-refined carrageenan (SRC) samples were extracted from *Kappaphycusalvarezii* grown at the UMS mini-estate system Semporna, Sabah. The absolute moisture content of the material was determined by drying the seaweed in an oven at 60 °C for 5–6 h to obtain 38% MC (Ali et al. 2014b). Before extraction, the seaweed was cleaned from foreign material such as salt and epiphytes, prior to the alkaline treatment process. The seaweed was cut in to 1 cm pieces to aid extraction and blending. A 20 g sample was used, to which 6% KOH was added and heated at temperature 70 °C for 1 h. The seaweed was washed with clean water several times. The material was then placed into a ventilated oven at 60 °C for 12 h. After drying, the sample was ground to a powder, i.e. Semi-Refined Carrageenan. Table 8.4 below shows the quality of SRC extracted from seaweed dried in a v-GHSD, as compared to some commercial standards.

From the data presented above, it is shown that the implementation of the 'sauna' technique, as a pre-treatment and further drying using the v-GHSD, improved the quality of the seaweed, with a yield of higher molecular weight carrageenan.

Results from the implementation of the latest technique is within the acceptable range from an industry perspective.

8.3 Advantages and Disadvantages of the Different Drying System

Cash-strapped seaweed farmers will generally settle for the cheapest technology of drying seaweeds, especially those located in remote islands or those living in the 'pondohans', i.e. a cluster of families related by blood and marriage, living in shallow reaf areas, in the middle of the sea, far from the mainland, where access to modern living facilties are greatly limited. Their dwellings are small, e.g. a room (4 × 5 m) for sleeping, cooking and eating purposes and an extended platform for the drying of their seaweed harvest (10 × 5 m). This has been the common practice for the past 45 years.

Innovations are badly needed to meet the changing world conditions and enhance the income potential and living conditions of these farming communities. The introduction of drying systems to a local community would take a lot boldness on the part of the seaweed farmers. Hands-on training for the seaweed farmers as to how to operate the new systems of drying would be of paramount importance. It would also require full cooperation amongst the major stakeholders along the value-chain in order to accept such innovations and to improve the quality and quantity of their cultivated seaweeds. At the end of the day, all of the stakeholders would be in a win-win situation. Table 8.5 lists the adavantages and disadvantages of the different drying system.

Table 8.5 Summary of advantages and disadvantages of the varied drying systems as applied to eucheumatoid seaweeds

Types of drying system	Advantages	Disadvantages
Conventional		
Hanging	Preserved color	Large surface area (footprint) required; not hygienic (possible avian contamination); low quality RDS
Platform	Cheap	Large surface area required
Shade-drying	Maintained quality	Longer time needed
'Sauna-like'	Shortened pre-treatment	Maintained MC
Passive		
Greenhouse-concept	Maintained quality	Daylight operation only
Chimney	Hygienic product	Longer time required
Active		
De-humidified air drying	Reduce mold growth	High cost; small capacity
Forced-convection dryer with a v-groove collector	Higher temperature needed	High cost
Hybrid		
Wind-turbine & solar dryer	Renewable energy	High cost maintenance
Double-ass finned collector	Preserved quality; high temperature needed; high efficiency; economical	High cost; small capacity
v-groove hybrid solar drier (v-GHSD)	High efficient management tools needed	Medium cost

8.4 Economics of a Double-Pass, Finned Collector

Solar drying systems represent amongst the most attractive and promising applications for this energy technology in tropical and sub-tropical countries. There are several earlier reports on the economic analysis of drying seaweed biomass using conventional types of drying (e.g. platform and hanging, see the chapter of Samonte, in this book), however, the report of Fudholi et al. (2011) on the techno-economic analysis of using double-pass finned collector for the solar drying of seaweed biomass was only made amongst the active, passive and hybrid drying devices. The design and application of a suitable air collector is one of the most important factors controlling the operational economics of a solar drying system (Fudholi et al. 2013).

This section will focus on the economic analysis of a Solar Drying Systems (SDS), which utilized a double-pass, solar finned collector. The design was based off the solar-assisted, forced convection drying system installed at the Green Energy Technology Innovation Park, Universiti Kebangsaan Malaysia (Fudholi et al. 2011). The key parameters (Table 8.6) of the re-designed solar drying systems were as follows: (1) collector area; (2) drying chamber; (3) capacity of drier; (4) mass flow rate; and (5) average temperature of the drying chamber. The main components of the system, with their corresponding values in Malaysian Ringgit (RM), were: (1) double-pass solar collector with finned absorber, (2) ducting system, (3) the blower, (4) the auxiliary heater, (5) distribution system and installation, and (4) flooring and drying chamber (values provided in Table 8.7).

Table 8.6 Key parameters of the solar drying system

Parameters	Unit	Value (RM)
Collector surface area	m^2	11.52
Drying chamber area	m^2	4.8
Capacity of dryer	kg	250–300
Mass flow rate	kg s^{-1}	0.05–0.12
Average temperature of drying chamber	°C	40–65

Table 8.7 Estimated component costs of the solar drying system

Components/parameters	Cost (RM)
Double-pass solar collector with finned absorber	15,000
Ducting system	250
Blower	1,600
Auxiliary heater	1,200
Distribution system and installation	10,000
Flooring and drying chamber	4,000
Total	32,050

Table 8.8 Cost-benefit analysis of a solar drying system

Parameters	Cost (RM)
Capital cost	32,050
Operating cost	15,421
Benefit	12,000

Table 8.9 Net Present Value (NPV) of the solar drying system, without the operational costs of a worker, NPV = RM 34,144.1

	FC (RM)	NCF (RM)	i	Present value	
Year			(10%)	FC (RM)	Pn
0	32,050	5,959	1	32,050	
1		5,959	0.9091		5,417.3
2		5,959	0.8264		4,924.8
3		5,959	0.7513		4,477.1
4		5,959	0.6830		4,070.1
5		5,959	0.6209		3,700.1
6		5,959	0.5645		3,363.7
7		5,959	0.5132		3,057.9
8		5,959	0.4665		2,779.9
9		5,959	0.4241		2,527.2
10		5,959	0.3855		2,297.5
	6,410			2,471.3	36,615.5

Table 8.10 Simple Payback (SPB) of the solar drying, without operational cost of a worker, SPB = 5.4 years

Year	FC (RM)	Annual Benefit (RM)	Benefit Cumulative (RM)
0	32,050	0	0
1		5,259	5,959
2		5,259	11,918
3		5,259	17,877
4		5,259	23,836
5		5,259	29,795
6		5,259	35,754

Based on the results of Fudholi et al. (2011), the following economic analysis was performed in order to test the efficiency: (1) Cost-benefit analysis which included capital cost, operating costs and benefits (Table 8.8); (2) Net-Present-Value (NPV) of the SDS and a Simple Payback (SPB) calculation, without the operational cost of a worker (Tables 8.9 and 8.10, respectively); and NPV and SPB for the opera-

Table 8.11 Net Present Value (NPV) for operational costs, including the salary of a worker, NPV = RM 82,071.8

			i	Present value	
Year	FC (RM)	NCF (RM)	(10%)	FC (RM)	Pn
0	32,050	13,759	1	32050	
1		13,759	0.9091		12,508.2
2		13,759	0.8264		11,371.1
3		13,759	0.7513		10,337.3
4		13,759	0.6830		9,397.6
5		13,759	0.6209		8,543.3
6		13,759	0.5645		7,766.6
7		13,759	0.5132		7,060.5
8		13,759	0.4665		6,418.7
9		13,759	0.4241		5,835.2
10		13,759	0.3855		5,304.7
	6,410			2,471.3	84,543.1

Table 8.12 Simple Payback (SPB) for operational costs including the salary of a worker, SPB = 2.3 years

Year	FC (RM)	Annual Benefit (RM)	Benefit Cumulative (RM)
0	32,050	0	0
1		13,759	13,759
2		13,759	27,518
3		13,759	41,277

Table 8.13 Net Present Value (NPV) of the solar drying system with assumption of the potential market prices, without the salary of a worker, NPV = RM 137,372.9

	FC (RM)	NCF (RM)	i	Present value	
Year			(10%)	FC (RM)	Pn
0	32,050	22,759	1	32,050	
1		22,759	0.9091		20,690.0
2		22,759	0.8264		18,809.1
3		22,759	0.7513		17,099.2
4		22,759	0.6830		15,544.7
5		22,759	0.6209		14,131.5
6		22,759	0.5645		12,846.9
7		22,759	0.5132		11,679.0
8		22,759	0.4665		10,617.2
9		22,759	0.4241		8,652.0
10		22,759	0.3855		8,774.6
	6,410			2,471.3	

Table 8.14 Simple Payback (SPB) of the solar drying system with assumptions made for potential prices, without the salary of a worker, SPB = 1.4 years

Year	FC (RM)	Annual Benefit (RM)	Benefit Cumulative (RM)
0	32,050	0	0
1		22,759	22,759
2		22,759	45,518

tional costs including salary for a worker (Tables 8.11 and 8.12, respectively); and NPV and SPB of the SDS assuming potential market prices, without a salaried worker (Tables 8.13 and 8.14, respectively).

It can be concluded that the use of a double-pass, solar finned collector for seaweed drying was economically feasible, with a payback period of 2.33 years.

8.5 Conclusions

Post-harvest handling of carrageenophyte biomass is a crucial phase in maintaining the quality of the seaweed which quality dictates the buying price. This is a major issue amongst farmers, traders and processors. Since the desired moisture content and quality criteria, based on the standards required by the processing industry, are often not met. The conventional methods of platform and hanging drying are too labor intensive, taking too much time to reach the desired MC, especially at the volumes required for the intensive through-put of an industrial facility. Though the hybrid types of drying systems are still all at various developmental stages, it is expected that after a thorough R&D phase, they could be considered for commercial implementation at the farm level. Improved yields and quality of the extracted carrageenan are of paramount interest amongst the seaweed farmers in order to achieve the highest value for their cultivated biomass. The double-pass, finned collector, as a hybrid drying system and its economic efficiencies, are highly recommended for the industry to review and consider for widespread adoption.

References

Ahemad AR (2006) Agricultural cooperatives in Malaysia: innovations and opportunities in the process of transition towards the 21st century model. Paper presented at the 2006 FFTC-NACF international seminar on agricultural cooperatives in Asia: innovations and opportunities in the 21st century, Seoul, Korea, 11-15 September

Ali MKM, Rafiq S, Sakari M, Sulaiman J, SMd Y (2013) Seaweed modelling for drying and the efficiency as heavy metal removal in *Kappaphycus striatum* var. sacol using solar dryer. J Environ Bioenergy 8:41–55

Ali MKM, Wong J, Muthuvalu MS, Ruslan MH, Sulaiman J, Yasir SMd (2014a) Mathematical modelling for the drying method and smoothing drying rate using cubic spline for seaweed *Kappaphycus striatum* var. durian in a solar dryer. In: AIP conference proceedings: 113–120

Ali MKM, Wong J, Muthuvalu MS, Ruslan MH, Sulaiman J, Yasir SMd (2014b) Improvement of growth and mass of *Kappaphycus striatum* by using plant density study at Selakan Island in Semporna Malaysia. In: The 3rd international conference of biological, chemical and environmental sciences, Penang, Malaysia: 14–15 June 2014. doi: http://dx.doi.org/10.15242/IICBE.C614020

Ali MKM, Ruslan MH, Sulaiman J, Yasir SMd (2014c) Optimization of process condition of refined carrageenan (RC) produced from seaweed *Kappaphycus striatum* using response surface methodology (RSM) in Malaysia. In: The 3rd international symposium on processing of foods, vegetables and fruits (ISPFVF 2014), Penang, Malaysia: 11–13 August 2014

Ali MKM, Rafiq S, Sakari M, Sulaiman J, SMd Y (2014d) Effectiveness the drying time and kinetic of seaweed *Kappaphycus alvarezii* var. tambalang in green V-roof hybrid solar drier. J Mod Math Sci 10:125–136

Ali MKM, Fudholi A, Sulaiman J, Ruslan MH, SMd Y (2014e) Sauna technique, drying kinetic modelling and effectiveness compared with direct drying in drying process of *Kappaphycus striatum* in Selakan Island Malaysia. J Energy Power Eng 36:300–315

Ali MKM, Sulaiman J, Yasir SMd RM (2015a) The effectiveness of sauna technique on the drying period and kinetics of seaweed *Kappaphycus alvarezii* using solar drier. Adv Envitl Agric Sci 1:86–95

Ali MKM, Fudholi A, Muthuvalu MS, Ruslan MH, Sulaiman J, Yasir S Md (2015b) The effectiveness of sauna technique on the drying period and quality of seaweed *Kappaphycus alvarezii* using v-Groove Hybrid Solar Drier. In: The 8th Asia-Pacific Drying Conference (ADC 2015) in Kuala Lumpur, Malaysia: 10–12 August 201e

AOAC (2001) Official methods of analysis of AOAC international (17th ed) Gaithersburg, 23(1):23–56

Bixler HJ, Porse H (2011) A decade of change in the seaweed hydrocolloids industry. J Appl Phycol 23:321–335

Blakemore WR (1990) Post harvest treatment and quality control of *Eucheuma* seaweeds. In: Adams T, Foscarini R (eds) Proceedings of the regional workshop on seaweed culture and marketing. South Pacific aquaculture development project, Food and Agriculture Organization of the United Nations, Suva, Fiji, pp 48–52

Chu CM, Rahman MM, Kumaresan S (2012) Effect of cold inflow on chimney height of natural draft cooling towers. Nucl Eng Des 249:125–131

Djaeni M, Sari DA (2015) Low temperature seaweed drying using dehumidified air. Procedia Envitl Sci 23:2–10

FAO (2007) FAO JECFA monographs 4. Specifications: Carrageenan. Available at: http://www.fao.org/ag/agn/jecfa-additives/details.html?id=830

FAO (2013) Food and agricultural commodities production:2011

FAO (2014) Food and agricultural commodities production:2012

FAO (2016) The state of world fisheries and aquaculture – contributing to Food Security and Nutrition for All Rome:200pp

FAO/WHO Expert Committee on Food Additives (2001) Compendium of food additives specifications, Addendum 9: FAO food and nutrition paper No. 52. FAO, Rome, pp 192–194

Faria GSMD, Hayashi L, Monteiro AR (2014) Effect of drying temperature on carrageenan yield and quality of *Kappaphycus alvarezii* (Rhodophyta, Solieriaceae) cultivated in Brazil. J Appl Phycol 26:917–922

Food and Drug Administration (2006) Part 172. Food Additives, permitted for direct addition to food for human consumption, code of federal regulations, USA

Fudholi A, Otman MY, Ruslan MH, Yahya M, Zaharim A, Sopian K (2011) Techno-economic analysis of solar drying system for seaweed in Malaysia. Rec Res Energy Environ Landsc Archi 89–95

Fudholi A, Sopian K, Ruslan MH, Othman MY, Bakhtyar B (2013) Energy analysis and improvement potential of finned double-pass solar collector. Energy Convers Manag 75:234–240

Gupta S, Cox S, Abu-Ghannam N (2011) Effect of different drying temperatures on the moisture and phytochemical constituents of edible Irish brown seaweed. LWT Food Sci Technol 1–7

Hurtado AQ, Critchley AT, Bleicher-L'honneur G (2008) *Kappaphycus* 'cottonii' farming, rev edn. Degussa Texturant Systems, 26 p

Hurtado AQ, Gerung SG, SMd Y, Critchley AT (2014) Cultivation of tropical red seaweeds in the BIMP-EAGA region. J Appl Phycol 26:707–718

Kumaresan S, Rahman MM, Chu CM, Phang HK (2013) A chimney of low height to diameter ratio for solar crops dryer. Developments in Sustainable Chemical and BioprocessTechnology, Springer International Publishing

McHugh DJ (2003) A guide to the seaweed industry. FAO Fisheries Technical Paper, FAO, Rome

Othman MY, Fudholi A, Sopian K, Ruslan MH, Yahya M (2012) Analisis kinetik pengeringan rumpai laut *Gracilaria changii* menggunakan sistem pengering suria (drying kinetics analysis of seaweed *Gracilaria changii* using solar drying system). Sains Malaysiana 41:245–252

Phang HK, Kumaresan S, Rahman MM, Chu CM, Suhaimi MY (2015) Preliminary study of seaweed drying under a shade and in a natural draft solar dryer. Int J Sci Eng 8:10–14

DPNS 601:2008. Carrageenan-Food Grade-Specification. https://tsapps.nist.gov/notifyus/docs/wto_countr/PHL/full... PHL100(english).pdf

Porse H, Rudolph B (2017) The seaweed hydrocolloid industry: 2016 updates, requirements, and outlook. J Appl Phycol. doi:10.1007/s10811-017-1144-0

Ruslan MH, Fudholi A, Othman MY (2012) The double pass solar dryer for drying palm oil fronds. In: Proceedings of 10th WSEAS international conference on system Sscience and simulation engineering (ICOSSSE '11), Penang, Malaysia, pp 143–149

Vairappan CS, Razalie R, Elias UM, Ramachandram T (2014) Effects of improved post-harvest handling on the chemical constituents and quality of carrageenan in red alga, *Kappaphycus alvarezii* Doty. J Appl Phycol 26:909–916

Yasir SM (2012) Kertas kerja Algae Farming via Mini Estate System in Sabah. Bio-Borneo. Kuching, Sarawak, 24–26 Februari

Yasir SM (2013) Penyelidikan dan Pengkomersilan. Jabatan Perikanan Malaysia, Putrajaya

Yasir SM, Ramlan A (2012) Sistem Mini Estet Industri Rumpai Laut. Jabatan Perikanan, Putrajaya

Economics of *Kappaphycus* spp. Seaweed Farming with Special Reference to the Central Philippines

9

Giselle P.B. Samonte

Abstract

The farming of *Kappaphycus* spp. in coastal marine waters has brought significant socio-economic benefits to coastal communities and fishermen. Seaweed farming, along with tourism and fishing, is considered an important direct-use value for coastal and marine ecosystems generating positive, net present values. In addition, there is increasing economic dependence of local coastal families on seaweed farming making farmers important stakeholders in the judicious management of coastal and marine resources. The investment required for 2,500 m^2 (0.25 ha) fixed off-bottom and raft long-line methods were valued at USD561 and USD1,051, respectively. Net farm income was estimated to be USD 565 per crop for the fixed off-bottom method and USD 739 per crop for the raft long-line method. Higher net farm income to gross returns was obtained from the fixed off-bottom method (56%), as compared to the raft long-line method (44%). It was estimated that the economic rate of return on investment in seaweed farming could be relatively high (14–20 months), the pay-back period for this activity on most farms surveyed was less than a year (6–9 months). Information on the relative values and investments in, and contributions of seaweed farming to, coastal and marine resources are essential in policy-making as this information can be used in the cost analysis of management interventions, programs and investments required.

9.1 Introduction

Culturing seaweeds in intertidal and sub-tidal zones is economically or monetarily attractive to coastal households. These households engage in this form of coastal aquaculture because of the opportunities for relatively higher cash incomes that can be derived, as compared to small-scale fishing. Seaweed farming provides additional income to fishermen as it is a high yielding investment (Padilla and Lampe 1989). Seaweed farming is a viable venture for coastal communities in Asia; however, there are few studies which provide evidence for the economic benefits of seaweed aquaculture in tropical regions (Padilla and Lampe 1989; Firdausy and Tisdell 1991; Valderrama et al. 2013; Alin et al. 2015).

G.P.B. Samonte (✉)
ERT, Inc., Silver Spring, MD, USA
e-mail: giselle.tan@ertcorp.com; samonte_g@yahoo.com

A survey was conducted amongst 72 seaweed (*Kappaphycus* spp.) farmers in the Western Visayas region, Philippines, from March – July, 1990 in order to assess their culture practices, in terms of production and economic efficiency (Samonte et al. 1993). Yields of 5.8 MT ha^{-1} crop^{-1} (dry weight) and 7.6 MT ha^{-1} crop^{-1} (dry weight) were obtained from bottom long-line and raft long-line methods, respectively. The investment required was USD 1,090 ha^{-1} for bottom long-line culture and USD 2,270 ha^{-1} for raft long-line culture. The fixed off bottom-line method was more profitable with a net farm income of USD 1,330 ha^{-1} crop^{-1}, as compared to USD 1,050/ha per crop for the floating raft long-line method. The fixed, off bottom long-line method of culturing seaweed was the most cost-effective, as compared to the raft long-line method. Production costs averaged USD 0.13 kg^{-1} for the former and USD 0.22 kg^{-1} for the latter method (Samonte et al. 1993).

A.Q. Hurtado et al. (eds.), *Tropical Seaweed Farming Trends, Problems and Opportunities*, Developments in Applied Phycology 9, DOI 10.1007/978-3-319-63498-2_9

In 1996, the farming of *K. alvarezii* brought tremendous economic benefits to marginal fishermen in the Western Visayas region (Hurtado-Ponce et al. 1996). A socio-economic survey was conducted amongst the *K. alvarezii* planters of Panagatan Cay, Caluya, Antique, Philippines in order to determine some social information, farming practices and the costs and return on effort of farming carrageenan-bearing seaweeds. Cultivation at that time was dominated by brown and green morphotypes, using the fixed bottom and hanging long-line methods. Approximately 9.3 MT d. wt ha^{-1} and 7.2 MT d. wt ha^{-1} were produced respectively, after 60–90 days of culture. The former method required a working capital and total investment of USD 288 and USD 72, respectively, compared to the hanging long-line which required USD 325 and USD 980, respectively. A higher total revenue (i.e. USD 5,300), net income (USD 7,300) was obtained from the bottom long-line rather than the hanging long-line system. Lower total expenses were incurred using the bottom long-line method (USD 821) rather than the hanging long-line (USD 944).

A large number of seaweed farms are located in the Bohol, Philippines region, where coastal residents consider seaweed farming as their primary source of income and main source of livelihood (Samonte et al. 2016). As compared to seven years ago (2010), there was an increase in the number of seaweed farmers from 100 to 300. The stated reason for this was that seaweed farming ensured a greater level of income, as compared to fishing, and was considered a viable livelihood and also a potential attraction for tourists. Much of the seaweed farming activity is in the coral reef area. Most farms were of a 1 ha size and usually applied the fixed off-bottom method, with approximately 2,000 single lines ha^{-1}. The biomass produced was mostly sold after drying (7:1 fresh-dry ratio). This study determined that the peak months for seaweed farming were August–December. On average, the seaweed farmers invested USD 67.65 mo^{-1} to purchase *Kappaphycus* spp. seedlings, bamboo and labor. During the lean months, net revenues ranged from USD 19–70 mo^{-1}. During peak months, net revenues ranged from USD 58–140 mo^{-1} (Samonte et al. 2016).

In this chapter, the economics of *Kappaphycus* spp. farming is shown via a cost and returns structure. This is followed by comparative budgets of various farming systems (after Valderrama et al. 2015). A case study in the central Philippines, as provided here outlines who were the users of the coastal and marine resources and how seaweed farming revenue contributed to their overall income. Finally, the potential contribution of seaweed farming to the economic value of marine ecosystems as a whole is highlighted.

9.2 Costs and Returns

A direct relationship can be expected from seaweed culture between the farm size, initial capital costs and farm revenue. The investment required for 2,500 m^2 (0.25 ha) fixed off-bottom and raft long-line methods was calculated as: USD 561 and USD1,051, respectively (Table 9.1).

Table 9.2 presents the cost of establishing a farm and the returns from *Kappaphycus* spp. production. Estimates were derived from background information (Hurtado 2013) and 2016 prices. For the raft long-line method, the cost for a

Table 9.1 Investment required for a 2,500 m^2*Kappaphycus* spp. farm in the central Philippines (USD 2016)

	Fixed-off bottom long-line			Raft long-line		
Item	Value (USD)	Economic life (year)	Depreciation (USD)	Value (USD)	Economic life (year)	Depreciation (USD)
Capital outlay						
Non-motorized boat	133	5	27	133	5	27
long-line	139	2	70	309	2	154
Bamboo	41	2	20	114	2	57
Tools	7	2	3	7	2	8
Polyethylene rope	–	–	–	10	2	5
Total capital outlay	320		120	573		246
Working capital (1 crop)						
Seedlings	103			277		
Plastic straw	68			105		
Hired labor	49			53		
Other expenses	22			43		
Total working capital	241			478		
Total investment	561			1,051		

Data source: Estimated values were derived from various socio-economic baseline surveys of seaweed farmers and focus group discussions conducted by the author in the Visayas region, Philippines, 2000–2010

Table 9.2 Costs and returns of 2,500 m^2 *Kappaphycus* spp. farming in the central Philippines (USD 2016)

	Fixed-off bottom long-line				Raft long-line		
Item	Unit cost (USD)	Quantity	Unit	Total	Quantity	Unit	Total
Returns	0.7	1500	kg, dwt	1005	2500	kg, dwt	1675
Cash costs							
Seedlings	0.5	230	kg	115	622	kg	311
Plastic straw	3	23	kg	75	35	kg	117
Hired labor	3	19	man day	63	20	man day	67
Others	10% of seed, straw, labor			25			
Total cash costs				277			544
Non-cash costs							
Opportunity cost of labor	3	13	man day	43	44	man day	147
Depreciation				120			246
Total non-cash costs				163			393
Total production cost				440			936
Operating income				728			1,131
Net farm incomecrop^{-1}				565			739
Annual net farm income (2 crops)				1,130			1,478

Data source: Estimated values were derived from various socio-economic baseline surveys of seaweed farmers and focus group discussions conducted by the author in the Visayas region, Philippines, 2000–2011

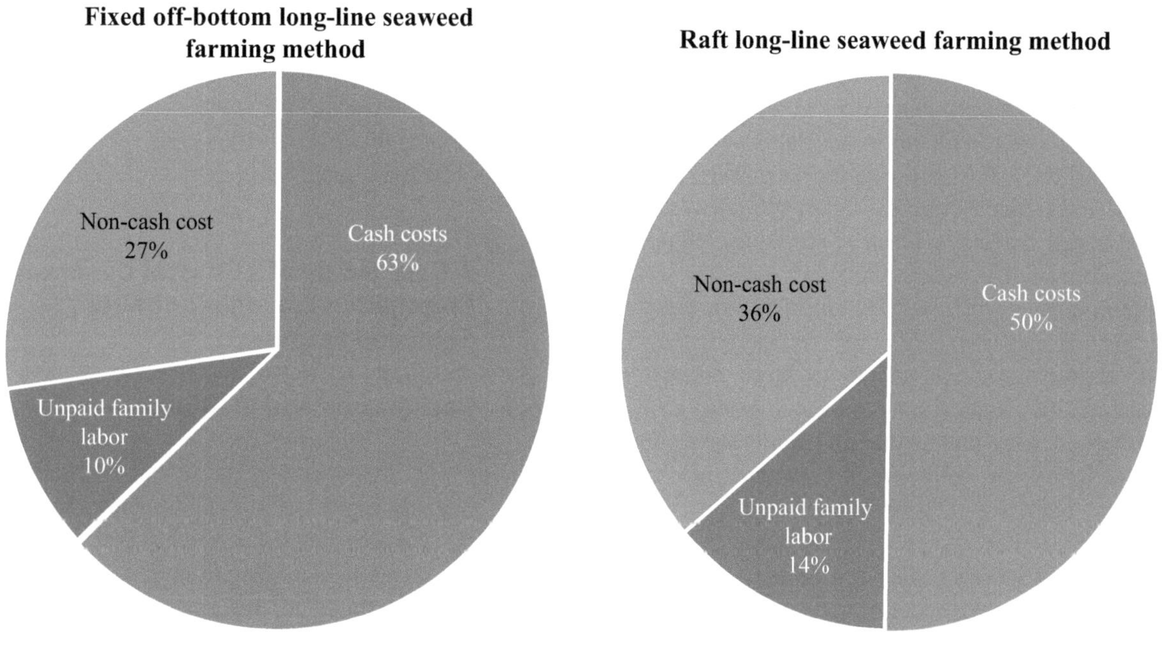

Fig. 9.1 Cost structure for *Kappaphycus* spp. farming (Data source: Estimated values were derived from various socio-economic baseline surveys of seaweed farmers and focus group discussions conducted by the author in the Visayas region, Philippines, 2000–2011)

2,500 m^2 farm was estimated at USD 936 crop^{-1}. This was more than double the cost incurred for the fixed off-bottom method (USD 440 crop^{-1}). Higher seeding rates and materials such as nylon lines and labor inputs are required for the raft long-line method.

Labor was one of the major operating costs involved in the seaweed farms studied, depending on size, it accounted for more than 20% of the total annual expenses. This covered labor costs for seeding, weeding, harvesting, drying and related maintenance activities. In practice, labor expenses were found to be lower as farmers may employ their own family members (Firdausy and Tisdell 1991).

The non-cash costs were found to contribute at least 25% of the total costs incurred (Fig. 9.1). Non-cash costs, specifically the owner or family labor and depreciation, comprised approximately 30% of the total production costs. Unpaid labor was primarily the imputed cost of family labor. The actual family labor input was valued at the hired labor

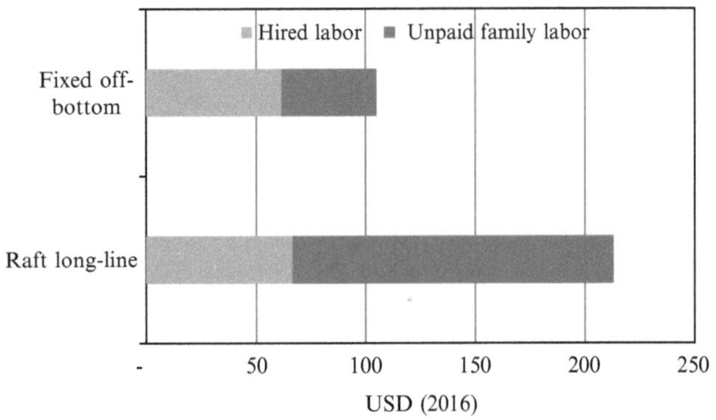

Fig. 9.2 Estimated labor costs for 2,500 m^2 *Kappaphycus* spp. farms in the central Philippines (Data source: Estimated values were derived from various socio-economic baseline surveys of seaweed farmers and focus group discussions conducted by the author in the Visayas region, Philippines, 2000–2011)

wage rate. These seaweed farmers, were usually fishermen and did not assign costs for their labor contribution to the farm, or for their management skills. For such fishermen, only the cash capital outlay had a measureable opportunity cost.

Many small-scale fishers and aquaculture producers are poor and often depend on unpaid family labor, including that of women and children (FAO 2016). Fisheries and aquaculture are a significant source of livelihoods and contribute to the food security of hundreds of millions of people. Notwithstanding, small-scale fishers, fish famers, boat crew members, fish-processors, fish traders and fish processing workers often face serious labor-related challenges which in turn limits the poverty reduction potential of the sector (Fig. 9.2).

Net farm income was USD 565 crop^{-1} for the fixed off-bottom method and USD 739 crop^{-1} for the raft long-line method. Higher net farm income to gross returns was obtained from the fixed off-bottom method (56%), as compared to the raft long-line method (44%). It was estimated that the economic rate of return on investment in seaweed farming was high (e.g. 141–201%) and that the pay-back period for this activity, on most farms surveyed, was less than a year (6–9 months). Seaweed farmers were able to depend on carrageenophyte cultivation as their main source of household income, in those areas which were ecologically suitable for sustainable seaweed culture, with the proviso that access to markets was not too difficult (Firdausy and Tisdell 1991).

9.3 Comparative Analysis of Farming Systems

The farming of the red seaweed *Kappaphycus alvarezii* and related species, as raw material for the hydrocolloid carrageenan, rapidly spread from the Philippines in the late 1960s to Indonesia, Tanzania, and other tropical countries around the world. Using standard budgeting techniques, Valderrama et al. (2015) adapted production and market data from a FAO-led, global review of seaweed farming in order to develop comparative enterprise budgets for eight farming systems, in six countries (Table 9.3). The systems analysis illustrated a variety of production scenarios such as small-scale operations in Tanzania; large-scale farms in Indonesia; and intermediate systems in the Philippines, India, Solomon Islands and Mexico.

Farm prices were significantly higher (e.g. >USD 0.80 kg^{-1} for dry seaweed) in Indonesia, the Philippines and Mexico, due to their proximity to markets. At USD 0.06 kg^{-1}, Tanzania achieved the lowest cost of production, while the highest cost per kg was for the Philippines and Mexico at USD 0.70 and USD 0.65 kg^{-1}, respectively. Production costs for the other systems were around USD 0.27 kg^{-1}.

Assuming a farm price of at least USD 0.80 kg^{-1}, production costs of around USD 0.25 kg^{-1} and a minimum of 2000 m of production long-lines, seaweed farms were calculated to generate enough revenue to bring a five-family household over the national poverty line. This economic potential acquires even greater importance in places with few additional employment opportunities for coastal inhabitants. Even in under-performing regions such as Tanzania, seaweed farming has been instrumental in raising the socio-economic status of female villagers, who were traditionally able to access fewer employment opportunities than males.

9.4 A Case Study of a Coastal Community Engaged in Carrageenophyte Seaweed Farming[1]

9.4.1 Background and Methods

A survey was undertaken in the Bohol Marine Triangle, Central Visayas region of the Philippines (Samonte-Tan et al. 2007). The area covered over 1,120 km^2, 92% of which was deep-sea. This area, ranging from 32 to 357 m deep provided a migratory route for whales and dolphins, as well as a habitat for these species and other marine life. The major ecosystem types in the area included coral reefs, seagrass and *Sargassum* beds and mangroves. Users of the marine resources within the Bohol Marine Triangle were interviewed using a pre-tested questionnaire (including questions on: production, revenue, cash and non-cash expenses and socio-demographic information). Almost 8% of the coastal

[1] This section is drawn from a component of a broader economic analysis performed by the author for the Foundation for the Philippine Environment Bohol Marine Triangle (BMT) Project of the United Nations Development Project (UNDP)/Global Environment Facility (GEF).

Table 9.3 Comparative enterprise budgets for *Kappaphycus* spp. farming systems in six developing countries, 2009

Item	Unit	Indonesia	Philippines	Tanzania		India	Solomon Islands	Mexico	
		Floating	Floating	Off-bottom	Floating	Floating	Floating	Off-bottom	Floating
Production parameters									
Total length of lines	m	30,000	2,000	270	288	2,565	4,000	10,000	10,000
Number of cycles per year	Cycles	8	5	7	8	6		4	4
Length of a cycle	Days	45	63	45	45	45		60	60
Annual yield of dry seaweed	kg	33,000	2,850	662	806	5,400	21,700	53,778	53,778
Annual productivity	kg/m/year	1.10	1.43	2.45	2.80	2.11	5.43	5.38	5.38
Cycle productivity	kg/m/cycle	0.14	0.29	0.35	0.35	0.35		1.34	1.34
Farm-gate price	USD/kg	0.85	1.09	0.27	0.27	0.33	0.38	1.00	1.00
Gross receipts	USD	28,050	3,107	179	218	1,785	8,246	53,778	53,778
Variable costs									
Propagules	USD							13,264	13,264
Labor	USD	4,320	759	26	28	1,041	3,556	8,853	8,853
Fuel	USD	29	332				1,117		
Maintenance and repairs	USD	420							
Sales and marketing	USD	600						7,115	7,115
Total variable costs	USD	5,369	1,091	26	28	1,041	4,672	29,232	29,232
Fixed costs									
Depreciation	USD	2,501	906	26	24	432	1,157	2,274	2,934
Administrative costs	USD	900							
Utilities	USD	120							
Fees for coastal land usage	USD							3,109	3,109
Total fixed costs	USD	3,521	906	26	24	432	1,157	5,383	6,043
Total costs	USD	8,890	1,997	52	52	1,473	5,829	34,615	35,275
Net returns	USD	19,160	1,109	127	166	312	2,417	19,163	18,503
Production cost	USD/kg	0.27	0.70	0.08	0.06	0.27	0.27	0.64	0.66

Source: Valderrama et al. (2015)

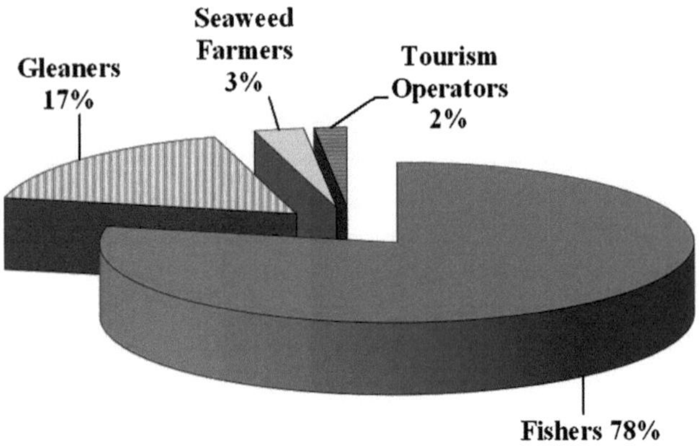

Fig. 9.3 Proportion of seaweed farmers amongst resource users in Bohol, Philippines (Samonte-Tan et al. 2004)

Table 9.4 Socio-demographic characteristics of seaweed farmers in Bohol, Philippines

Socio-economic profile		Seaweed farmers n = 42
1. Gender	Male	92.9%
	Female	7.1%
2. Age	Minimum	29
	Maximum	71
	Average	52
3. Civil status	Single	7.1%
	Married	92.9%
4. Number of years living in village	Minimum	4
	Maximum	64
	Average	42
5. No. of people living in household	Minimum	1
	Maximum	14
	Average	6
6. Monthly income from fishing[a]	US$	140
7. Monthly income from other occupation[b]	US$	70
8. Member of organization	Yes	82.10%
	No	17.90%
9. Education	No education	0.00%
	Elementary level	3.60%
	Elementary grad.	35.70%
	High Sch. Level	14.30%
	High Sch. Grad.	25.00%
	Vocational	7.10%
	College level	7.10%
	College grad.	7.10%

[a]USD 2016
[b]Land-based occupation such as hired farm labor, small retail store; tricycle driver

population comprised resource users with marine-based occupations, including marine fishers, seaweed farmers, gleaners and tourism-related business owners and boat operators. Marine fishers comprised 78% of the total resource users, while the remaining 22% were gleaners, seaweed farmers and owner/operators of tourism-related activities (Fig. 9.3).

Producer surplus was calculated for coastal and marine-based economic sectors (i.e., fishery, seaweed farming, tourism). Producer surplus was the excess revenue minus costs, received by the resource users (i.e., fishers, gleaners, seaweed farmers, tourism business operators). Costs represented the minimum amount a resource user would be willing to accept in order to maintain his current effort/operation level. Gross revenue included the value of products (fish, shellfish, seaweed) and services (tourism operations) generated from the economic activity. Total cost consisted of the variable costs (e.g. fuel, supplies, repair, packing cost, labor shares) and fixed costs. The net revenue for the economic sector (R_i), gross revenue (GR) and total cost (TC) were calculated, respectively.

$$NR_i = GR - TC$$
$$GR_i = Q_i P_i$$
$$TC = VC + FC$$

Where, Q is quantity of produce, P is ex-vessel or farm-gate price, VC is total variable cost and FC is fixed cost.

As natural assets, coral reefs provided a stream of valuable ecosystem services to society over time. Economic benefits derived from this area's coastal and marine resources were calculated as the sum of the present value of the stream of revenues (NPV), over a 20-year period, as follows:

$$NPV = \left(\sum_{T}^{T1} Bi - \sum_{T}^{T1} Ci \right) / (1+r)^t$$

where, NPV = net present value; B = benefits; C = costs; i = coastal and marine-based economic activities; t = year; r = social discount rate.

9.4.2 Socio-demographic Characteristics

It was found that many seaweed farmers were in their late forties and had lived in the area for an average of 38 years (Table 9.4). The average household size was six. Income from seaweed farming ranged from US$66–122 mo^{-1} and supplemented land-based occupations such as the processing/drying of fish products, laboring, agricultural farming, raising livestock and other small business opportunities. The majority of the seaweed farmers were members of the local organizations such as Songculan Guso Farmers' Association (SOGUFA), Tabalong Seaweed Association and Panglao Seaweed Farmers' Association (PASFA).

9.4.3 Revenue from Seaweed Farming

Seaweed farming in Bohol was another marine-based activity, where local residents engaged in activities through their local seaweed organizations. Long-lines were the preferred culture method adopted by those farmers. Peak months were given by the seaweed farmers as April to June, whilst the remainder of the year was considered to be the lean months due to the monsoon winds. Farm size ranged from 1,034–3,650 m^2. The seaweed of choice was *Kappaphycus alvarezii*. The duration of the culture period was typically 45 days. With two crops per year, the annual yield was 40.6–67.6 kg m^{-2}. The farm-gate price was US$0.10–0.20 kg^{-1} for fresh seaweed and US$0.82–1.90 kg^{-1} for dried seaweed. The average net revenue, obtained from a survey of 42 seaweed farmers, was US$100–182 $crop^{-1}$. The annual net revenue was US$200–364. Seaweed farming therefore provided alternative and/or additional income opportunities to fishermen, hence, coastal aquaculture generated income for coastal populations without adding to the alleged over-exploitation of fishery resources in the country (Rebours et al. 2014; Islam and Haroon 2016; van den Burg et al. 2016).

9.5 Conclusion

The farming of *Kappaphycus* spp. in coastal marine waters has brought tremendous economic benefits to coastal communities and fishermen. Human activities along the coast were sustained by the services provided by the coastal ecosystems. Seaweed farming, along with tourism and fishing, were considered as an important, direct-use value of coastal and marine ecosystems, generating positive, net-present values. In addition, it was noted that there was increasing dependence of local coastal families on seaweed farming making the farmers key stakeholders in the management of coastal and marine resources. Information on and quantification of the value and contribution of seaweed farming to coastal and marine resources is essential in policy-making, as this information can be used in cost-benefit analyses to guide management interventions, programs and investments. For example, multi-sectoral, stakeholder consultation discussions identified concerns related to disease, siganid herbivores and the lack of market outlets for farmed seaweed, all of which had negative impacts on the overall economic returns. Increasing improvements in culture technology, local training on farming techniques and monitoring were considered as challenges requiring attention in order for sustainable seaweed farming to succeed and reachest its fullest potential.

The calculation of economic value helped to identify the beneficiaries of conservation and also the magnitude of those benefits which they received and thus could help to develop mechanisms to capture some of these financial and ecological benefits and make them available for conservation. The economic valuation provided an estimate of the stream of benefits and also opportunities over a given time scale. The net present value was an estimate of all benefits (including future benefits) that would be lost in the absence of conservation.

The expansion of seaweed farming in tropical developing countries could have a large, positive impact on local poverty, ecosystem management and climate change mitigation (Bjerregaard et al. 2016). In 2012, the global production of seaweeds was approximately 3 million MT dry weight; growing by 9% per annum. Increasing the expansion of *Kappaphycus* farming by up to 14% per year was calculated to generate 500 million MT dry weight by 2050, adding about 10% to the world's present supply of food, generating revenues and improving environmental quality. In the Philippines, the Comprehensive National Fisheries Industry Development Plan (Medium Term 2016–2020) targeted a 25% increase in production of seaweeds over a period of 5 years (DA-BFAR 2016).

The levels of education, training and expertise were not important criteria for the ability to work in seaweed culture (Firdausy and Tisdell 1991). This suggested that seaweed cultivation had the potential to contribute to employment opportunities in countries where unskilled labor is relatively abundant. The revenue generated from seaweed farming in coastal areas would not only be used to satisfy day-to-day household consumption, but many farmers could also save and 'invest' in durables from their income.

In conclusion, the farming of carrageenophyte seaweeds is a suitable, coastal enterprise for small-scale, family farms with the potential to enhance the socio-economic well-being of coastal communities which have been traditionally dependent on marine resources (Valderrama et al. 2015).

Acknowledgement The author is grateful to Prof. Michael Pido, Palawan State University, Philippines, for a technical review of this paper.

References

Alin JM, Eranza DRD, Bahron A, Mahmud R (2015) Profit and loss analysis of *Eucheuma* seaweed farming in Green Island, Palawan, Philippines. Mediterr J Soc Sci 6:125–128

Bjerregaard R, Valderrama D, Sims N, Radulovich R, James D, Carpon M, Forster J, Goudey C, Yarish C, Hopkins K, Rust M, McKinnie C (2016) Seaweed aquaculture for food security, income generation and environmental health in tropical developing countries. Washington, DC, World Bank Group

DA-BFAR (Department of Agriculture – Bureau of Fisheries and Aquatic Resources) (2016) Comprehensive National Fisheries Industry Development Plan (Medium Term 2016–2020). BFAR, Quezon City, Philippines. 33 p. + Annexes

FAO (2016) Scoping study on decent work and employment in fisheries and aquaculture: issues and actions for discussion and programming. Food and Agriculture Organization of the United Nations, Rome

Firdausy C Tisdell C (1991) Economic returns from farming different types of seaweed (*Eucheuma*) and for farms of different sizes in Nusa Penida, Bali, Indonesia. Working Paper No. 27. The University of Queensland, St. Lucia campus, Brisbane QLD 4072, Australia. 27 p

Hurtado AQ (2013) Social and economic dimensions of carrageenan seaweed farming in the Philippines. In: Valderrama D, Cai J, Hishamunda N, Ridler N (eds) Social and economic dimensions of carrageenan seaweed farming, pp 91–113. Fisheries and Aquaculture Technical Paper No. 580. Rome, FAO. 204 pp

Hurtado-Ponce AQ, Agbayani RF, Chavoso EAJ (1996) Economics of cultivating *Kappaphycus alvarezii* using the fixed-bottom line and hanging-long line methods in Panagatan Cays, Caluya, Antique, Philippines. J Appl Phycol 8:105–109

Islam S, Haroon (2016) Seaweed farming as an alternative income generation to improve the livelihood of the coastal community of Bangladesh. J Aquac Res Dev 7:6

Padilla J, Lampe H (1989) The economics of seaweed farming in the Philippines. *Naga* ICLARM Quarterly 1–5

Rebours C, Marinho-Soriano E, Zertuche-González JA, Hayashi L, Vásquez JA, Kradolfer P, Soriano G, Ugarte R, Abreu MH, Bay-Larsen I, Hovelsrud G, Rødven R, Robledo D (2014) Seaweeds: an opportunity for wealth and sustainable livelihood for coastal communities. J Appl Phycol:261939–261951

Samonte GPB, Hurtado-Ponce AQ, Caturao RD (1993) Economic analysis of bottom line and raft monoline culture of *Kappaphycus alvarezii* var. tambalang in Western Visayas, Philippines. Aquaculture 110:1–11

Samonte GPB, Eisma-Osorio L, Rizaller A, White A (2016) Economic value of a large marine ecosystem: Danajon double barrier reef, Philippines. Ocean Coast Manag 122:9–19

Samonte-Tan GPB, Tabara E, Caballes C (2004) Coastal and marine biodiversity resource economic valuation: Bohol Marine Triangle, The Coastal Conservation and Education Foundation, Inc. and Bohol Marine Triangle Project, Cebu City, Philippines, 114 p

Samonte-Tan GPB, White A, Tercero MA, Diviva J, Tabara E, Caballes C (2007) Economic valuation of coastal and marine resources: Bohol Marine Triangle, Philippines. Coast Manag 35:319–338

Valderrama D, Cai J, Hishamunda N, Ridler N (2013) Social and economic dimensions of carrageenan seaweed farming. Fisheries and Aquaculture Technical Paper No. 580, FAO, Rome. 204 pp.

Valderrama D, Cai J, Hishamunda N, Ridler N, Neish I, Hurtado A, Msuya F, Krishnan M, Narayanakumar R, Kronen M, Robledo D, Gasca-Leyva E, Fraga J (2015) The economics of *Kappaphycus* seaweed cultivation in developing countries: a comparative analysis of farming systems. Aquac Econ Manag 19:251–277

van den Burg SWK, van Duijn AP, Bartelings H, van Krimpen MM, Poelman M (2016) The economic feasibility of seaweed production in the North Sea. Aquac Econ Manag 20:235–252

Carrageenan and More: Biorefinery Approaches with Special Reference to the Processing of *Kappaphycus*

10

José G. Ortiz-Tena, Doris Schieder, and Volker Sieber

Abstract

In the age of increasing awareness of our planet's limited resources, utilization of plant biomass not only for food but also for fuel and chemicals becomes an essential part of a more sustainable economy. Marine plants such as seaweeds have an especially high potential compared to land plants as the earth's surface is mainly covered by water and they can be considered still under-utilized. This scenario is currently changing with the increasing areas of seaweed farms close to the shore in the Indo-Pacific area as well as the development of off-shore concepts on floats in the Atlantic Ocean. Of the different types of seaweeds brown and red seaweed have been utilized most so far. The major products obtained from red seaweeds today are carrageenans. *Kappaphycus* here plays a central role as it is the dominant source for kappa-carrageenan. While *Kappaphycus*, as any other seaweed, contains many more components, the production process is optimized for highest carrageenan yields. The modern zero-waste policy, however, requires production processes that will allow utilization of any by-product. Biorefinery approaches have been developed for many land plants over the last decade. Can these concepts be transferred to seaweeds as well? This chapter evaluates the challenges and opportunities for optimal utilization of the crude biomass from *Kappaphycus* seaweeds, with special focus on the latest applications and processing technologies for its main components, including carrageenan as a major valorization product.

10.1 Constituents of *Kappaphycus*

From a general valorization perspective, seaweed biomass can be conveniently divided into: (a) insoluble polymeric material that confers structural support or serves as energy storage in the cell (i.e. carbohydrates and proteins), and (b) low molecular weight compounds, which, for example, may act as defenses against pathogens, or as enzyme co-factors. Polymers can either be hydrolyzed to their monomeric components or directly extracted and utilized in different applications, depending on their physical and chemical properties. Furthermore, the small molecules are usually extracted using organic solvents and then concentrated by thermal or filtration processes. The biomass from marine origin differs considerably in composition from terrestrial feedstock e.g. wood, straw, grasses, etc., and also requires different processing techniques (Jung et al. 2013). It is known that the chemical composition of *Kappaphycus* seaweeds varies according to many factors, including the weed strain, growth conditions (duration) and especially harvesting season (Zuldin et al. 2016; Suresh Kumar et al. 2015; Yong et al. 2015) (Table 10.1). In general, carbohydrates are the most abundant class of polymer present in red algae, such as *K. alvarezii*, (at least 30% DW). The major monosaccharides

J.G. Ortiz-Tena • D. Schieder
Chemistry of Biogenic Resources, Technical University of Munich, 16 Schulgasse, 94315 Straubing, Germany

V. Sieber (✉)
Chemistry of Biogenic Resources, Technical University of Munich, 16 Schulgasse, 94315 Straubing, Germany

Fraunhofer IGB, Straubing Branch BioCat, Straubing, Germany
e-mail: Sieber@tum.de

A.Q. Hurtado et al. (eds.), *Tropical Seaweed Farming Trends, Problems and Opportunities*, Developments in Applied Phycology 9, DOI 10.1007/978-3-319-63498-2_10

Table 10.1 Reported composition of *Kappaphycus* in the recent literature (% DW)

Carbohydrates[a]	Proteins	Lipids + insoluble aromatics	Ash	Species	Comments	References
46–59	5	NR	20–23	*K. striatum*	Culture in customized tank systems	Zuldin et al. (2016)
				K. alvarezii		
52–56	2–4	2–4	15–17	*K. alvarezii*	Red, brown, green and G11 strains	Masarin et al. (2016)
30–55	12–22	0.4–0.9	21–34	*K. alvarezii*	Evaluation of seasonal variation	Suresh Kumar et al. (2015)
NR	9.9	2–3	38	*K. alvarezii*	Comparison of farm and micro-propagated cultures	Yong et al. (2015)
56.8	16.2	0.8	19.7	*K. alvarezii*	Samples collected from the west coast of India	Fayaz et al. (2005)

[a]Carbohydrates accounts as the sum of reported sugars and fiber content
NR not reported

present in acid hydrolysates are galactans from carrageenan (ca. 73% of total carbohydrates), followed by glucans from cellulose (ca. 23% total carbohydrates). Mannose and xylose are also present in lower concentrations (Masarin et al. 2016). The protein amounts reported in the literature vary, ranging from 2–6% DW (Masarin et al. 2016; Zuldin et al. 2016) up to 20% DW, depending on the specific strain and collecting season (Suresh Kumar et al. 2015). Compared to other types of biomass, the inorganic ash content is relatively high, from 20% DW up to 38% DW being reported (Fayaz et al. 2005; Yong et al. 2015). A substantial part of this is derived from sulphate, which is bound to the galactans forming the carrageenan and essentially increasing the carbohydrate content when included in the weight analysis. Lipids and other small molecules constitute barely 3% DW of the biomass (Masarin et al. 2016).

Kappaphycus species and variants also contain a number of low molecular weight compounds, which are bioactive and as such of interest for valorization. Extracts of seaweeds made with water, alcohols (methanol, ethanol), ether or acetone have been found to possess anti-oxidant, anti-microbial, anti-tumor or neurotrophic activity (Ling et al. 2013; Lau et al. 2014; Tirtawijaya et al. 2016). Phenols and flavonoids are responsible for parts of these activities but the general analysis of the chemical structures of isolated compounds of this genus is rare. Only one patent referring to a bromophenol describes a specific compound with biological activity (Rajasulochana et al. 2012; Rajasulochana et al. 2015). Valorization of simple extracts to be used as nutritional supplements should be possible without knowing the identity of each compound, however.

10.2 Carrageenan – Types and Uses

Over the last 50 years, carrageenan has been the main product extracted from *Kappaphycus* seaweeds, especially for their application in the food industry as thickener, gelling agent, emulsion stabilizer and recrystallization inhibitor (also elsewhere in this book) (De Ruiter and Rudolph 1997; Seta et al. 2013; Leiter et al. 2017). Carrageenan constitutes a very broad group of sulphated polysaccharides made up of galactans. The disaccharide repeating unit, carrabiose consists of alternating 3-linked β-D-galactopyranose (G-units, also referred to as "A-residue") bonded to 4-linked α-D-galactopyranose (D-units), or to 3,6-anhydro-α-D-galactopyranose (DA-units), either being also called "B-residue". Depending on the sulphation grade of the repeating disaccharide, the major carrageenans are classified as kappa (κ)-, iota (ι)- and theta (θ)- carrageenan (G plus DA-unit) and the corresponding, mu (μ)-, nu (ν)-, and lambda (λ)-carrageenan (G plus D unit) (Fig. 10.1) (Campo et al. 2009). This sulphation grade, as well as the presence of the DA-unit, are determining parameters for the physical and chemical properties and therefore for applications in different fields. *Kappaphycus* sp. and variant strains contain predominantly κ – (90–93%) with some minor amounts of ι – carrageenan (7–10%) including varying amounts of their corresponding precursors (μ and ν) depending on the plant condition at harvest (van de Velde et al. 2005). Kappa carrageenan forms strong gels in the presence of potassium ions, and rather brittle, stiff gels with calcium ions. Very low concentrations are enough for gel formation to occur (0.5% in water, 0.2% in milk) (Campo et al. 2009). Small amounts of kappa (0.01–0.05%) are commonly added to cheese, ice cream or cocoa beverages to prevent separation of whey proteins and to control stability. Other applications include thickener in low-calorie foods (e.g., jellies, salad dressings, sorbets), as fat replacer in sausages, ingredient in pet food

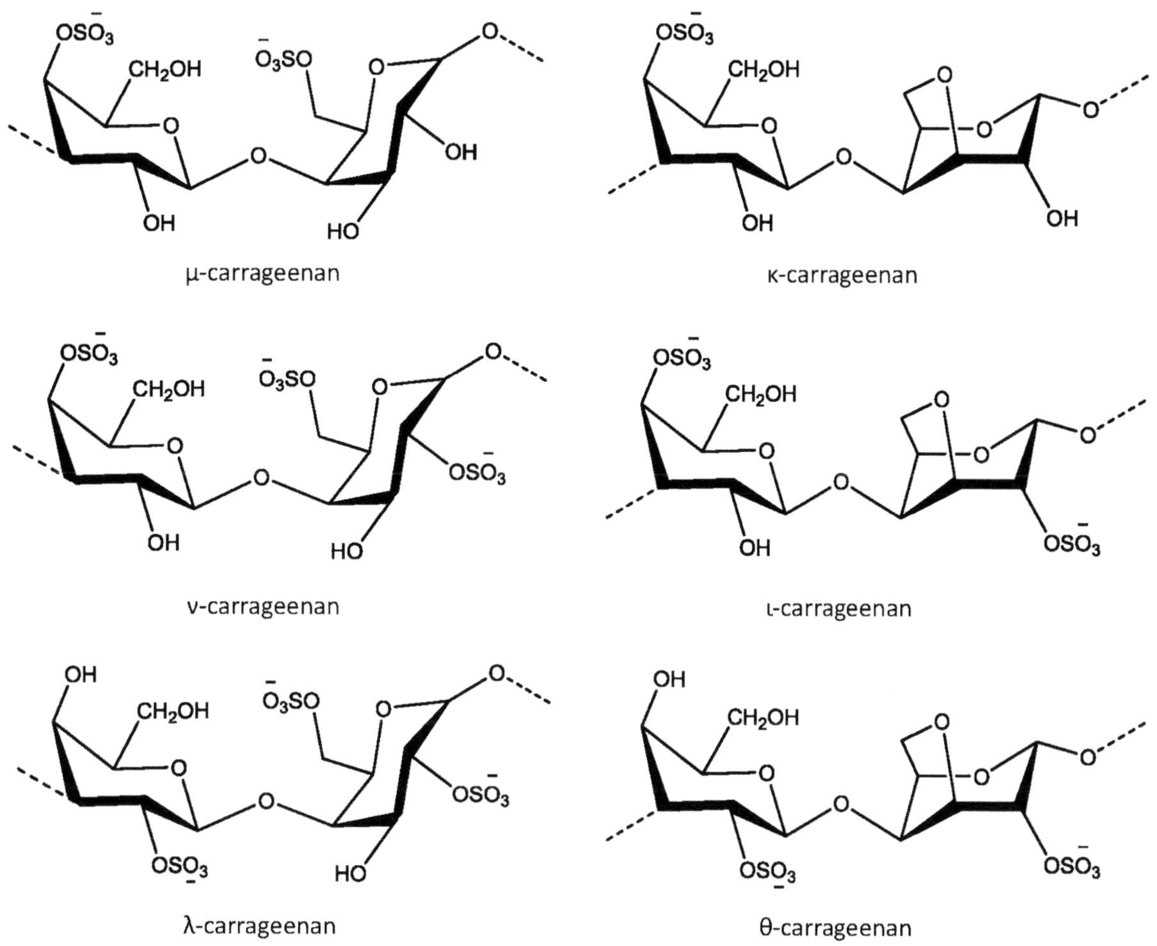

Fig. 10.1 Chemical structure of the major building blocks of different types of carrageenan

and air freshener gels (McHugh 2003). On the other hand, ι – carrageenan forms better gels with calcium than with potassium ions and are rather soft, resilient and free of bleeding. They have the advantage of being stable to freeze-thaw cycles and display thixotropic flow. This property is desirable in toothpaste, because the gel should stand but also flow when pressure is applied. For this reason iota is added in about 1% concentration. Further, ground beef can be treated with iota to provide meat with fat-like characteristics, such as moisture retention for better product acceptance (McHugh 2003). Finally, lambda is a good thickening agent and is used mainly in combination with the other types of carrageenan.

Prior to carrageenan extraction, seaweeds are usually dried using different methods (freeze-, shade- and sun-drying), which have an effect on the chemical composition and properties of the final product (Vairappan et al. 2014). The extraction process plays a significant role for the quality of the carrageenan extracted and its subsequent applications. If the extraction of the hydrophilic carrageenan is performed in hot water, the carrageenan removed displays poor gel strength. One reason for this is the presence of significant amounts of μ – carrageenan, the precursor of kappa. Therefore, the dried seaweeds are subjected to alkali at high temperatures with either KOH, NaOH or $Ca(OH)_2$ to improve the gel strength (Mishra et al. 2006). Under such alkaline conditions, an intra-ring cyclization of the D-units occurs to form the 3,6-anhydro bridge of the DA-units (van de Velde et al. 2002) and μ – and ν – are converted into κ – and ι – carrageenan, respectively. The desired properties (e.g. gelling or viscosity behaviour) can be tailored by choosing the key parameters, such as time, temperature and alkali concentration (Azevedo et al. 2013, 2015). Alternative extraction procedures have been proposed without alkali treatment, such as atomization drying and the use of deep eutectic solvents (Webber et al. 2012; Das et al. 2016). One of the most commercialized types of carrageenan is the so called hybrid carrageenan, or "Kappa-2". It was largely debated whether this type of carrageenan was made of mixtures of κ – and ι – homo-polymers or if both types were present in a single chain forming a copolymer (Hilliou 2014). It was only recently demonstrated that this product is actually a block copolymer comprising iota and kappa with different lengths and distributions within the chain (Guibet et al. 2008). The gel formation of this blend depends on the block pattern of the conforming units, showing an intermediate gelation strength between κ – and ι – carrageenan (van de Velde et al. 2005). Other properties, such as the cation type used for gel formation, gel setting or temperature do not differ from the homo-polymer (van de Velde et al. 2005; Hilliou et al. 2014). This "technical grade carrageenan" is employed in different applications within the food industry, such as milk beverages (Bixler et al. 2001) and as edible films with enhanced properties (Larotonda et al. 2016).

On the other hand, the anionic character of carrageenan allows its electrostatic interaction with various interesting molecules which can be targeted to useful applications, such as methylene blue for colorimetric oxygen indicators in food packaging (Vu and Won 2014). The more common polysaccharides, i.e. starch and cellulose are often chemically modified to improve their performance for specific purposes, such as flocculation or wet-end additives. This type of processing is also feasible for seaweed polysaccharides. Barahona et al. (2015) created an amphoteric carrageenan by adding 3-chloro-2-hydroxypropyltrimethylammonium chloride, a cation that substitutes the hydroxyl groups of the galactan units to different degrees. Another example includes the complexing of ι -carrageenan with ammonium bromide to improve its conductivity towards the production of biofilms for fuel cell applications (Karthikeyan et al. 2016). Furthermore, a graft carrageenan polymer with N-isopropylacrylamide which exhibited thermo-sensitive properties and improved interaction with cationic species was produced (Gaweł et al. 2013). This type of polymer finds applications in water treatment or metabolite encapsulation. Consequently, the interest for carrageenan in the pharmaceutical industry has increased over the last few years for two main reasons. Firstly, diverse biomaterials applications have been devised, such as vehicle for drug delivery in oral formulations and tissue engineering (Liu et al. 2015). This has been driven by novel developments of carrageenan-based, nano-structures, which can be produced either from partially degraded pure carrageenan (Sun et al. 2016) or by electrostatic interactions with different cationic species, such as chitosan or protamine (Dul et al. 2015; Long et al. 2015). Secondly, diverse biological activities of carrageenan in polymeric and oligomeric form have been published, including anti-tumor, anti-coagulant, anti-hyperlipidemic, anti-viral and immune-modulatory properties (Campo et al. 2009; Hu et al. 2006). The requirements for the use of carrageenan in medical applications are much stricter than in the food industry, since the chemical composition can have significant effects in the desired property. The sulphate regiochemistry, for instance, has a considerable influence in the anti-coagulant activity and cyto-toxicity of carrageenans (Liang et al. 2014; de Araújo et al. 2013). For this reason, the technical hybrid carrageenans must be refined to the pure homo-polymeric forms for pharmaceutical applications. In the recent past, the bio-catalytic potential of diverse marine micro-organisms has been explored to specifically modify the chemical structure of carrageenan (Helbert 2012). Diverse carrageenan sulphatases that modify the sulphation pattern of the carrabiose repeating units of κ -, ι – and hybrid carrageenans have been isolated from cell lysates of various species of *Pseudoalteromonas* and recombinants produced (Genicot et al. 2014; Préchoux et al. 2016, 2013). This opens the possibility of further genetic exploration of similar marine organisms for the biotechnological utilization of this valuable polysaccharide. In addition to the afore-mentioned applications of carrageenan as a polymer, monomeric sugars can be obtained by chemical or enzymatic hydrolysis, which increases its valorization potential as a fermentation source for the production of commodity chemicals and biofuels (Yun et al. 2016). This strategy is addressed in detail in Sect. 10.4.

10.3 Classic Processing of Carrageenan Containing Seaweeds

Today, commercial carrageenans from *K. alvarezii* are produced by alkaline treatment of the algal biomass using different methods (Fig. 10.2). "Refined carrageenans (RC)" are recovered by extracting carrageenan from cleaned and washed seaweed using hot alkaline solutions at around 95–110 °C. The alkaline solution not only helps in dissolving the carrageenan but also achieves the transformation of the precursors μ – carrageenan and ν – carrageenan into κ – carrageenan and ι – carrageenan, respectively. Subsequent filtering removes the remaining seaweed matrix, and then carrageenan is precipitated by adding alcohol (mainly isopropanol), or by gel forming in potassiumchloride solution (i.e. "gel press process"). In the gel press pathway, the major part of water is removed after gelling by squeezing or freeze-thawing. In contrast to the alcohol precipitated product, the gel press product therefore usually contains higher amounts of potassiumchloride. After precipitation, the carrageenans are dried and milled. Semi-refining is the "younger" technology, which does not extract the carrageenan from the seaweed matrix, but treats the matrix with aqueous potassium hydroxide at around 75 °C to dissolve and remove soluble compounds other than carrageenan, i.e. mainly salts, soluble proteins, and soluble sugars. The embedded carrageenan is kept in the gel state and does not dissolve into the solution, since the process is operated with salt concentrations that keep the carrageenan below its gel melting temperature. Nevertheless, the transformation of the precursors from μ – carrageenan to κ – carrageenan is still taking place. The semi-refining technology is less expensive than the RC extraction process, since the costs for carrageenan precipitation or solvent recovery are avoided. The extracted primary

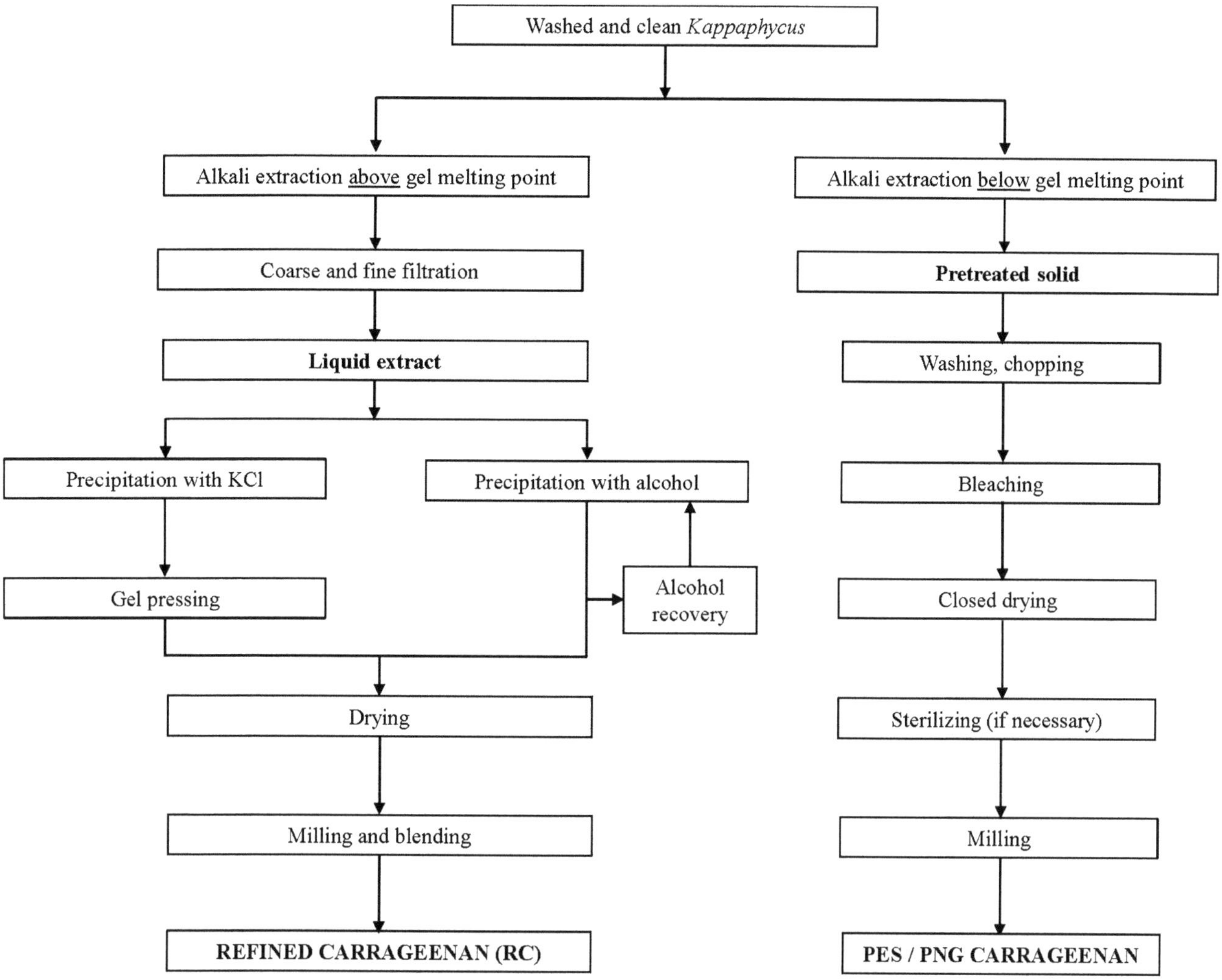

Fig. 10.2 Carrageenan production processes (Modified from McHugh 2003 and Phillips 1996)

product of the semi-refining process is dried and milled to yield a product of lower quality termed "seaweed flour", "alkali modified flour (AMF)", or "semi-refined carrageenan (SRC)", which is not suitable for human food application but is destined for pet food production. Higher qualities of SRC, termed "Processed *Eucheuma* Seaweed – PES" or "Philippine Natural Grade – PNG", which are suitable for food use, are achieved by an optional, additional bleaching or a drying step in a closed system to maintain low levels of microbial contamination. In contrast to RC, PES/PNG still contains the cellulose of the original seaweed. This distinction is also made in the European legislation for food, where PES is represented by its own number (e.g. E407a in contrast to E407 for RC), and is allowed to contain up to 15% cellulose (McHugh 2003; Phillips 1996; Rhein-Knudsen et al. 2015; McHugh 1987; Commission 1996).

RC has been, and is still mainly, produced in Europe and the US, while semi-refined products are mainly produced in Asia. In the decade 1999–2009 the major commercial production shifted slightly from alcohol precipitated to gel press process carrageenan due to lower production costs, and it shifted also considerably from low quality SRC to PES. The production of PES increased from 9,200 MT in 1999 to 20,500 MT in 2009, an increase from 22 to 41% of the global carrageenan production. The total global carrageenan production was estimated to be 50,000 MT in 2009 and 60,000 MT in 2015 (US$ 626 million) (Rhein-Knudsen et al. 2015; Bixler and Porse 2011).

10.4 Lessons from a Lignocellulose Bio-Refinery: Options for a Red Seaweed Bio-Refinery

Since the industrial production of carrageenan from *Kappaphycus* is a mature and well-established technology and the recovered carrageenan is well-established on the global market, *Kappaphycus* bio-refinery systems should be primarily based on carrageenan production. Yet, an integrated sustainable bio-refinery should aim to use as many compounds of the harvested algal biomass as possible, for production of valuable products.

Effluent solutions from carrageenan recovery containing mineral salts may be used for fertilizer. The use of seaweed for fertilizer and soil conditioner is generally well known and traditionally practiced (McHugh 2003). Seaweed extracts of various species are also well known to stimulate plant growth, a phenomenon which is attributed to growth regulators (Chapman and Chapman 1980; Crouch and van Staden 1993). In the "juice" of *K. alvarezii*, plant growth regulators such as indole 3-acetic acid, gibberellin GA3, kinetin and zeatin have been identified (Prasad et al. 2010; Zodape et al. 2009). A method was proposed for the recovery of seaweed juice for fertilizer and foliar spray application in addition to carrageenan production from the seaweed solid (Eswaran et al. 2005). For that, the juice, containing various minerals, mainly sodium, potassium, chloride and sulphate, as well as 1–1.4% organic matter (Zodape et al. 2009), is separated by crushing and filtering the freshly harvested seaweed. After separation of the juice, the remaining seaweed matter is dried and further processed by state-of-the-art carrageenan recovery methods (Eswaran et al. 2005). Amongst others, the main benefit of this method may lie in the fact that the juice is not subject to harsh conditions such as alkaline treatment, which might decompose parts of the dissolved organic compounds. Besides direct use for fertilizer, further valorization of the juice, e.g. by recovery of bioactives or pigments, may be an option.

Reddy et al. (2015) proposed a seaweed bio-refinery that starts with aqueous extraction of the fresh seaweed to get pigments from the extract by precipitation with ammonium sulphate. The algal matter is then subjected to lipid extraction with a chloroform-methanol solvent mixture. If *Kappaphycus* is processed, the extracted solid is treated with KOH solution to dissolve and precipitate the carrageenan while cellulose can be obtained from the residual solid of the carrageenan recovery (Reddy et al. 2015).

The major part of carrageenan in *K. alvarezii* is hydrosoluble, while the cellulose is mainly hydro-insoluble (Lechat et al. 1997). Insoluble residues after carrageenan extraction consequently contain mainly cellulose, small amounts of mannose and xylose-based polymers, as well as residual galactose and anhydro-galactose, depending on the efficiency of the carrageenan extraction process (Phillips 1996; Masarin et al. 2016). In native *Kappaphycus*, cellulose contributes to less than 15% to the dry matter (Phillips 1996). Cellulose can be enzymatically hydrolyzed to obtain glucose to be used as a fermentation substrate (Reddy et al. 2015), since there are different fermentation pathways using glucose as a carbon source, e.g. production of bioethanol, biobutanol, lactic acid, succinic acid, or poly-hydroxy-alcanoates (PHA). The cellulose fraction in the laboratory-scale production of SRC for instance, was reported to be nearly completely hydrolyzed by commercially available cellulases (Masarin et al. 2016).

Proteins could be another interesting by-product, since they make up a considerable share of the *Kappaphycus* dry matter (Table 10.1). The protein quality of seaweeds in terms of essential amino acids is reported to be comparable or even superior to traditional protein sources, such as soybean meal, or fish meal. However, the concentration of essential amino acids in the seaweed dry matter is lower, and the variations between different species are high, requiring development of suitable standardization technologies for protein concentration and extraction (Angell et al. 2016). During the alkaline treatments used for conventional carrageenan production, any soluble proteins are usually dissolved in the extraction solutions. In a recent semi-refining type laboratory-scale study, for example, a reduction in protein content of about 90% was reported for KOH (6% w/v aqueous solution) treated *K. alvarezii*strains (Masarin et al. 2016). Yet, alkaline conditions lead to denaturation and hydrolysis of proteins. The recovery of native proteins thus should be performed prior to alkaline treatment. An alternative valorization, which might be less affected by the pre-hydrolyzation of proteins, could be the conversion of amino acids to N-containing chemicals, e.g. glutamic acid conversion to products such as N-methypyrrolidine, succinonitril, or acetonitrile (Lammens et al. 2012).

Apart from the alkaline-based technologies, advanced alternative methods for the recovery of carrageenan have been proposed in recent years. Accordingly carrageenans can be successfully enzymatically extracted using proteases and cellulases. Even several decades ago this approach was considered as an alternative production route (Niwano 1970; Amat et al. 2000; Zamorano Palma et al. 2002). However, transfer to production-scale was not economical at that time. The strong decrease in enzyme costs for degradation of lignocellulose in the last decade, however, might now open this route for commercialization. In addition, sulphurylases, the enzymes that naturally catalyze the transformation of D- into DA-units, have been tested to convert the precursors to κ – or ι – carrageenan, which would allow the elimination of alkaline conditions (Genicot-Jancour et al. 2009). Unfortunately, no recombinant production of this enzyme has been described so far, which would be a requirement for it application at low cost. Rhein-Knudsen et al. (2015) recently summarized strategies for enzyme-based processes. Besides the recovery of carrageenan, enzyme-based technologies may also offer new perspectives for enhanced by-product separation and valorization.

Figure 10.3 outlines an integrated bio-refinery for advanced enzyme-based processing with respect to the state of knowledge discussed above. The particular steps proposed

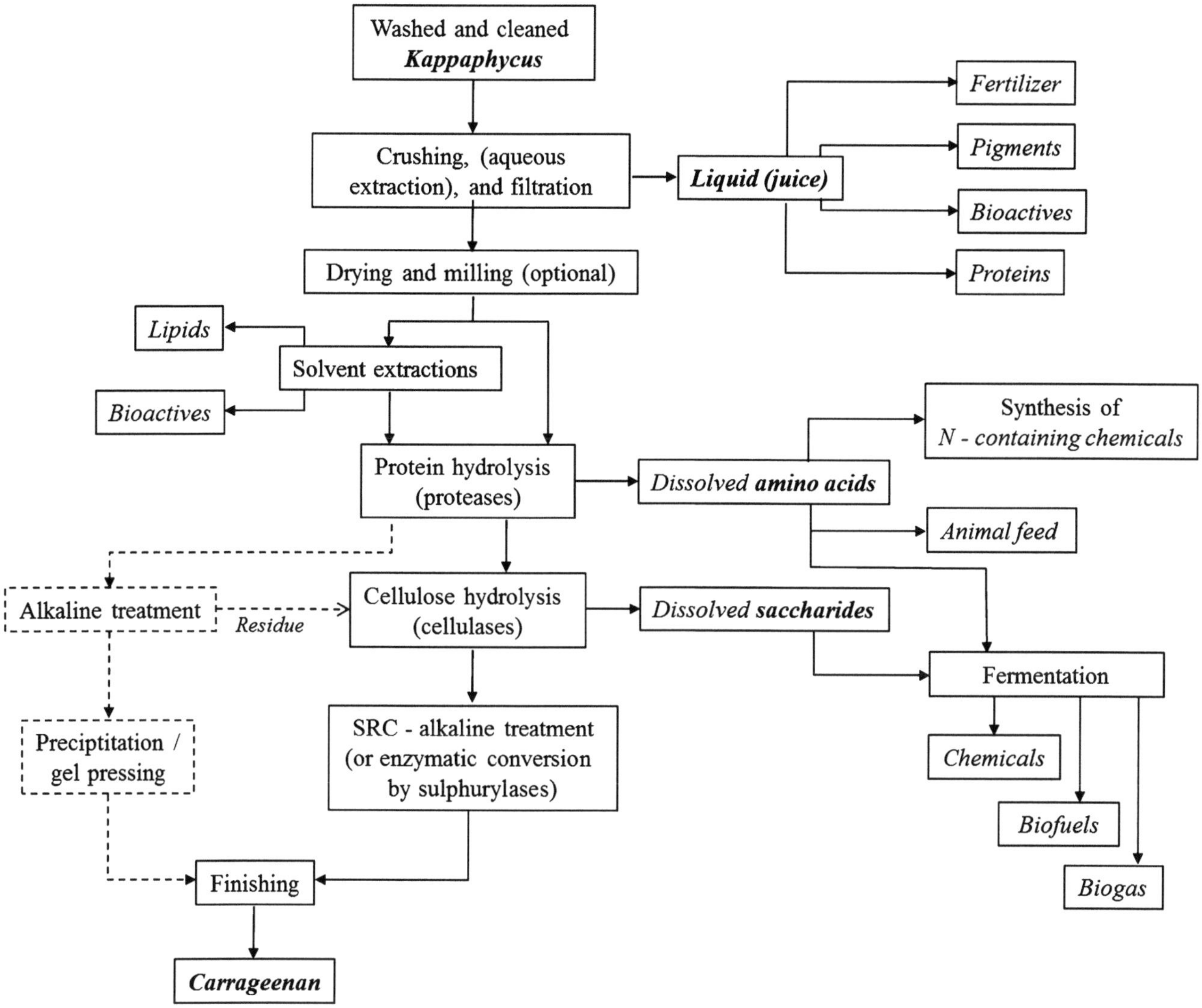

Fig. 10.3 Outline of *Kappaphycus* bio-refinery options applying advanced enzymatic processing. The *dashed lines* indicate the pathway for production of a refined carrageenan-like product

may be performed depending on technical and economic feasibilities. The biorefinery starts with crushing and filtration or aqueous extraction of the fresh, cleaned, wet algal biomass to separate the cell liquid (juice) from the solid plant matter, as proposed by Eswaran et al. (2005) and Reddy et al. (2015), respectively. The juice may be used for fertilizer or further processed to obtain pigments, native proteins or bioactives, such as plant growth regulators. The solid plant matter may then be extracted to recover lipids (Reddy et al. 2015) or bioactives. Drying and milling of the solid matter prior to solvent extraction may be applied, if the solid is meant to be transported or stored, or if hydro-insolublebioactives should be obtained by solvent extraction. For example, cardiatic glycosides for medical application can only be extracted with highly hydrophobic solvents (e.g. petrol ether), flavonoids can be extracted using methanol and anthraquinones, terpenoids or saponins by either solvents (Leelavathi and Prasad 2015). After extraction, the remaining solid may be subjected to enzymatic hydrolysis by proteases and cellulases. Hydrolysates from protein hydrolysis may be used for recovery of amino acids or for synthesis of chemicals, as discussed above. Saccharide streams from cellulosehydrolysis can be used for bioconversion to chemicals or biofuels. Therefore, protein and cellulosehydrolysis may as well be performed simultaneously to yield amino acids and saccharide containing substrates for subsequent fermentation. Residual streams from fermentation, as well as other waste streams of the bio-refinery containing organic matter, may be supplied to biogas production. The final conversion of the precursors to κ – and ι – carrageenan would be achieved by alkaline treatment, similar to the SRC process, or by advanced enzymatic conversion using sulphurylases.

Alternatively, carrageenan may be extracted after proteinhydrolysis by an alkaline solution and precipitated to get a refined carrageenan - like product. In that case, saccharides would be recovered from the solid residue of alkaline extraction, as proposed by Reddy et al. (2015).

An alternative to the afore-mentioned strategies for valorization of the red algal biomass could be the strategy of complete saccharification of polysaccharides (including carrageenan) to generate sugars for the subsequent production of bioethanol or commodity chemicals via fermentation, similar to lignocellulosic bioethanol production. The difference to lignocellulosic treatment lies in the milder conditions for hydrolysis needed to release reducing monosaccharides (0.2 M H_2SO_4, 15 min, 130 °C) from algae (Meinita et al. 2012). This is possible due to the absence of lignin and the lower concentrations of crystalline cellulose. Chemical heterogeneous catalysts such as Amberlyst™ -15 can also be used to hydrolyze carbohydrates, making it easier to recover, but with lower yields (Tan et al. 2013). In general, lower fermentation product yields are obtained from seaweeds for two reasons. Firstly, ordinary fermentative microorganisms (e.g. *Saccharomyces cerevisiae, Zymomonas mobilis*) are not capable of catabolizing 3,6-anhydro-galactose. However, the recent discovery and recombinant production of the two key catabolic enzymes of 3,6-anhydro-galactose from marine bacteria enabled ethanologenic strains of *E. coli* to grow and produce ethanol in agarose hydrolysates (Yun et al. 2015). Secondly, the galactose present in carrageenan is metabolized more slowly than glucose in classic fermentative organisms. This can be overcome by pre-culturing the yeasts in galactose medium, which diminishes the glucose inhibition and prepares the galactose metabolism in the cell (Hargreaves et al. 2013). Khambhaty et al. (2012) produced ethanol from the solid residues of pressed *Kappaphycus* with the yeast *S. cerevisiae*, reporting ethanol conversion of ca. 80% of reducing sugars after acid hydrolysis. Enzymatic hydrolysis using commercial enzymes for saccharification and subsequent fermentation has also been demonstrated. However, a modest hydrolysis yield of 62% was obtained, suggesting the need for specific carrageenases (Abd-Rahim et al. 2014). Other studies proved that simultaneous saccharification and fermentation of a cellulose containing stream separated from a liquid galactose stream could improve ethanol production, obtaining 105 L of ethanol per ton of seaweed (Hargreaves et al. 2013). More recently, this two-stream model was applied to generate concentrated sugar hydrolysates (98.7 g L^{-1}), by a combined approach, using a solid catalyst (Dowex™ Dr.-G8) and a commercial cellulase, yielding 11.6 g L^{-1}bioethanol from each fraction. After performing an economic analysis, the authors concluded that the fixed price of bioethanol produced from red seaweed was competitive with that obtained from other feedstocks (Tan and Lee 2016). However, despite the successful examples for the production of bioethanol from *Kappaphycus*, the economic viability of all these carrageenan hydrolyzing and fermenting strategies has to be assessed in contrast to the production of RC or SRC. With the carrageenan market prices (e.g. 6–15 US$ per kg in 2009, depending on quality grade (Bixler and Porse 2011); retail price of ca. 10.4 US$ per kg in 2015 (Rhein-Knudsen et al. 2015), similar profitable products from the hydrolysis and fermentation of carrageenan should be established that justify this process strategy.

References

Abd-Rahim F, Wasoh H, Zakaria MR, Ariff A, Kapri R, Ramli N, Siew-Ling L (2014) Production of high yield sugars from *Kappaphycus alvarezii* using combined methods of chemical and enzymatic hydrolysis. Food Hydrocoll 42:309–315

Amat M, Masoyer J, Demande F (2000) Method for purifying carrageenans FR 2794474 A1 20001208. FR 2794474 A1 2000120

Angell AR, Angell SF, de Nys R, Paul NA (2016) Seaweed as a protein source for mono-gastric livestock. Trends Food Sci Technol 54:74-84. doi:http://dx.doi.org/10.1016/j.tifs.2016.05.014

Azevedo G, Hilliou L, Bernardo G, Sousa-Pinto I, Adams RW, Nilsson M, Villanueva RD (2013) Tailoring kappa/iota-hybrid carrageenan from *Mastocarpus stellatus* with desired gel quality through pre-extraction alkali treatment. Food Hydrocoll 31(1):94–102

Azevedo G, Torres MD, Sousa-Pinto I, Hilliou L (2015) Effect of pre-extraction alkali treatment on the chemical structure and gelling properties of extracted hybrid carrageenan from *Chondrus crispus* and *Ahnfeltiopsis devoniensis*. Food Hydrocoll 50:150–158

Barahona T, Prado HJ, Bonelli PR, Cukierman AL, Fissore EL, Gerschenson LN, Matulewicz MC (2015) Cationization of kappa- and iota-carrageenan – characterization and properties of amphoteric polysaccharides. Carbohydr Polym 126:70–77

Bixler HJ, Porse H (2011) A decade of change in the seaweed hydrocolloids industry. J Appl Phycol 23(3):321–335

Bixler HJ, Johndro K, Falshaw R (2001) Kappa-2 carrageenan: structure and performance of commercial extracts: II. Performance in two simulated dairy applications. Food Hydrocoll 15(4–6):619–630

Campo VL, Kawano DF, Silva DB Jr, Carvalho I (2009) Carrageenans: biological properties, chemical modifications and structural analysis – a review. Carbohydr Polym 77(2):167–180

Chapman VJ, Chapman DJ (1980) Chapter 2. In: Seaweeds and their uses, 3rd edn. Chapman and Hall, London, pp 334–339

Commission Europena (1996) Commission directive 96/77/EC of laying down specific purity criteria on food additives other than colours and sweeteners. The European Commission

Crouch IJ, van Staden J (1993) Evidence for the presence of plant growth regulators in commercial seaweed products. Plant Growth Regul 13(1):21–29

Das AK, Sharma M, Mondal D, Prasad K (2016) Deep eutectic solvents as efficient solvent system for the extraction of κ-carrageenan from *Kappaphycus alvarezii*. Carbohydr Polym 136:930–935

de Araújo CA, Noseda MD, Cipriani TR, Gonçalves AG, Duarte MER, Ducatti DRB (2013) Selective sulfation of carrageenans and the influence of sulfate regiochemistry on anticoagulant properties. Carbohydr Polym 91(2):483–491

De Ruiter GA, Rudolph B (1997) Carrageenan biotechnology. Trends Food Sci Tech 8(12):389–395

Dul M, Paluch KJ, Kelly H, Healy AM, Sasse A, Tajber L (2015) Self-assembled carrageenan/protamine polyelectrolyte nanoplexes—investigation of critical parameters governing their formation and characteristics. Carbohydr Polym 123:339–349

Eswaran K, Ghosh PK, Siddhata AK, Patolia JS, Periyasamy C, Mehta AS, Mody KH, Ramavat BK, Prasad K, Rajyaguru MR, Reddy CRK, PAndya JB, Less AT (2005) Integrated method for production of carrageenan and liquid fertilizer from fresh seaweeds. USA Patent US6893479 B2

Fayaz M, Namitha KK, Murthy KNC, Swamy MM, Sarada R, Khanam S, Subbarao PV, Ravishankar GA (2005) Chemical composition, iron bioavailability, and antioxidant activity of *Kappaphycus alvarezii* (Doty). J Agric Food Chem 53(3):792–797

Gaweł K, Karewicz A, Bielska D, Szczubiałka K, Rysak K, Bonarek P, Nowakowska M (2013) A thermosensitive carrageenan-based polymer: synthesis, characterization and interactions with a cationic surfactant. Carbohydr Polym 96(1):211–217

Genicot SM, Groisillier A, Rogniaux H, Meslet-Cladière L, Barbeyron T, Helbert W (2014) Discovery of a novel iota carrageenan sulfatase isolated from the marine bacterium *Pseudoalteromonas carrageenovora*. Front Chem 26:1–15

Genicot-Jancour S, Poinas A, Richard O, Potin P, Rudolph B, Kloareg B, Helbert W (2009) The cyclization of the 3,6-anhydro-galactose ring of ι-carrageenan is catalyzed by two D-galactose-2,6-sulfurylases in the red alga *Chondrus crispus*. Plant Physiol 151(3):1609–1616

Guibet M, Boulenguer P, Mazoyer J, Kervarec N, Antonopoulos A, Lafosse M, Helbert W (2008) Composition and distribution of carrabiose moieties in hybrid κ-/ι-carrageenans using carrageenases. Biomacromolecules 9(1):408–415

Hargreaves PL, Barcelos CA, Costa ACA, Junior NP (2013) Production of ethanol 3G from *Kappaphycus alvarezii*: evaluation of different process strategies. Bioresoure Technol 134:257–263

Helbert W (2012) Analysis and modification of carrageenan structure using enzymes. In: Gums and stabilisers for the food industry 16. Royal Society of Chemistry, Cambridge, pp 19–25

Hilliou L (2014) Chapter two – hybrid carrageenans: isolation, chemical structure, and gel properties. In: Se-Kwon K (ed) Adv Food Nutr Res 72:17–43

Hilliou L, Sereno AM, Gonçalves MP (2014) Gel setting of hybrid carrageenan solutions under steady shear. Food Hydrocoll 35:531–538

Hu X, Jiang X, Aubree E, Boulenguer P, Critchley AT (2006) Preparation and *in vivo* antitumor activity of κ-carrageenan oligosaccharides. Pharm Biol 44(9):646–650

Jung KA, Lim S-R, Kim Y, Park JM (2013) Potentials of macroalgae as feedstocks for biorefinery. Bioresour Technol 135:182–190

Karthikeyan S, Selvasekarapandian S, Premalatha M, Monisha S, Boopathi G, Aristatil G, Arun A, Madeswaran S (2016) Proton-conducting I-Carrageenan-based biopolymer electrolyte for fuel cell application. Ionics:1–6. doi:10.1007/s11581-016-1901-0

Khambhaty Y, Mody K, Gandhi MR, Thampy S, Maiti P, Brahmbhatt H, Eswaran K, Ghosh PK (2012) Kappaphycus alvarezii as a source of bioetanol. Bioresour Technol 103

Lammens TM, Franssen MCR, Scott EL, Sanders JPM (2012) Availability of protein-derived amino acids as feedstock for the production of bio-based chemicals. Biomass Bioenergy 44:168–181

Larotonda FDS, Torres MD, Gonçalves MP, Sereno AM, Hilliou L (2016) Hybrid carrageenan-based formulations for edible film preparation: benchmarking with kappa carrageenan. J Appl Polym Sci 133 (2). doi:10.1002/app.42263

Lau TY, Vittal DF, Chew CSY, Yong WTL (2014) Antiproliferative potential of extracts from *Kappaphycus* seaweeds on HeLa cancer cell lines. Sains Malays 43(12):1895–1900

Lechat H, Amat M, Mazoyer J, Gallant DJ, Buléon A, Lahaye M (1997) Cell wall composition of the carrageenophyte *Kappaphycus alvarezii* (Gigartinales, Rhodophyta) partitioned by wet sieving. J Appl Phycol 9(6):565–572

Leelavathi MS, Prasad MP (2015) Comparative analysis of phytochemical compounds of marine algae isolated from Gulf of Mannar. World J Pharm Pharm Sci 4(5):640–654

Leiter A, Ludwig A, Gaukel V (2017) Influence of heating temperature, pH and ions on recrystallization inhibition activity of κ-carrageenan in sucrose solution. J Food Eng 195:14–20. doi:http://dx.doi.org/10.1016/j.jfoodeng.2016.09.016

Liang W, Mao X, Peng X, Tang S (2014) Effects of sulfate group in red seaweed polysaccharides on anticoagulant activity and cytotoxicity. Carbohydr Polym 101:776–785

Ling ALM, Yasir SM, Matanjun P, Abu Bakar MF (2013) Antioxidant activity, total phenolic and flavonoid contents of selected commercial seaweeds of Sabah, Malaysia. Int J Pharm Phytopharm Res 3(3):234–238

Liu J, Zhan X, Wan J, Wang Y, Wang C (2015) Review for carrageenan-based pharmaceutical biomaterials: favourable physical features versus adverse biological effects. Carbohydr Polym 121:27–36

Long J, Yu X, Xu E, Wu Z, Xu X, Jin Z, Jiao A (2015) In situ synthesis of new magnetite chitosan/carrageenan nanocomposites by electrostatic interactions for protein delivery applications. Carbohydr Polym 131:98–107

Masarin F, Cedeno FRP, Chavez EGS, de Oliveira LE, Gelli VC, Monti R (2016) Chemical analysis and biorefinery of red algae *Kappaphycus alvarezii* for efficient production of glucose from residue of carrageenan extraction process. Biotechnol Biofuels 9(1):1–12. doi:10.1186/s13068-016-0535-9

McHugh DJ (1987) Production and utilization of products from commercial seaweeds. FAO Fisheries Technical Paper No 288

McHugh DJ (2003) A guide to the seaweed industry. FAO Fisheries Technical Paper No 441:61–72

Meinita MDN, Yong YK, Kang GT, Jeong GT, Koo HM, Park SM (2012) Bioethanol production from the acid hydrolysate of the carrageenophyte *Kappaphycus alvarezii* (cottonii). J Appl Phycol 24:857–862

Mishra PC, Jayasankar R, Seema C (2006) Yield and quality of carrageenan from *Kappaphycus alvarezii* subjected to different physical and chemical treatments. Seaweed Res Util 28(1):113–117

Niwano S (1970) Carrageenan JP 1970-92570. JP:1970–92570

Phillips GO (1996) The chemical identification of PNG-carrageenan. In: Phillips GO, Williams PA, Wedlock DJ (eds) Gums and stabilisers for the food industry, vol 8. IRL, Oxford, p 403

Prasad K, Das AK, Oza MD, Brahmbhatt H, Siddhanta AK, Meena R, Eswaran K, Rajyaguru MR, Ghosh PK (2010) Detection and quantification of some plant growth regulators in a seaweed-based foliar spray employing a mass spectrometric technique sans chromatographic separation. J Agric Food Chem 58(8):4594–4601

Préchoux A, Genicot S, Rogniaux H, Helbert W (2013) Controlling carrageenan structure using a novel Formylglycine-dependent sulfatase, an Endo-4S-iota-carrageenan sulfatase. Mar Biotechnol 15(3):265–274. doi:10.1007/s10126-012-9483-y

Préchoux A, Genicot S, Rogniaux H, Helbert W (2016) Enzyme-assisted preparation of furcellaran-like κ-/β -carrageenan. Mar Biotechnol 18(1):133–143. doi:10.1007/s10126-015-9675-3

Rajasulochana P, Krishnamoorthy P, Dhamotharan R (2012) Experimental studies to determine various vitamins available in *Kappaphycus alvarezii*. J Chem Pharm Res 4(12):5176–5179

Rajasulochana P, Kumar S, Krishnamoorthy P (2015) A new antimicrobial bromophenol from *Kappaphycus alvarezii*. IN2014CH00602A

Reddy CRK, Baghel RS, Trivedi N, Kumari P, Gupta V, Prasad K, Meena R (2015) An integrated process to recover a spectrum of bioproducts from fresh seaweeds. Patents WO 2015102021 A1

Rhein-Knudsen N, Ale M, Meyer A (2015) Seaweed hydrocolloid production: an update on enzyme assisted extraction and modification technologies. Mar Drugs 13(6):3340

Seta L, Baldino N, Gabriele D, Lupi FR, de Cindio B (2013) The influence of carrageenan on interfacial properties and short-term stability of milk whey proteins emulsions. Food Hydrocoll 32(2):373–382

Sun W, Saldaña MDA, Zhao Y, Dong T, Jin Y, Zhang J (2016) New application of kappa-carrageenan: producing pH-sensitive lappaconitine-loaded kappa-carrageenan microparticle using two-step self-assembly. J Appl Phycol 28(3):2041–2050

Suresh Kumar K, Ganesan K, Subba Rao PV (2015) Seasonal variation in nutritional composition of *Kappaphycus alvarezii* (Doty) Doty—an edible seaweed. J Food Sci Tech 52(5):2751–2760

Tan IS, Lee KT (2016) Comparison of different process strategies for bioethanol production from *Eucheuma cottonii*: an economic study. Biores Tech 199:336–346. doi:http://dx.doi.org/10.1016/j.biortech.2015.08.008

Tan IS, Lam MK, Lee KT (2013) Hydrolysis of macroalgae using heterogeneous catalyst for bioethanol production. Carbohydr Polym 94(1):561–566

Tirtawijaya G, Mohibbullah M, Meinita MDN, Moon IS, Hong Y-K (2016) The ethanol extract of the rhodophyte *Kappaphycus alvarezii* promotes neurite outgrowth in hippocampal neurons. J Appl Phycol 28(4):2515–2522

Vairappan CS, Razalie R, Elias UM, Ramachandram T (2014) Effects of improved post-harvest handling on the chemical constituents and quality of carrageenan in red alga, *Kappaphycus alvarezii* Doty. J Appl Phycol 26(2):909–916

van de Velde F, Knutsen SH, Usov AI, Rollema HS, Cerezo AS (2002) 1H and 13C high resolution NMR spectroscopy of carrageenans: application in research and industry. Trends Food Sci Tech 13(3):73–92

van de Velde F, Antipova AS, Rollema HS, Burova TV, Grinberg NV, Pereira L, Gilsenan PM, Tromp RH, Rudolph B, Grinberg VY (2005) The structure of κ/ι-hybrid carrageenans II. Coil–helix transition as a function of chain composition. Carbohydr Res 340(6):1113–1129

Vu CHT, Won K (2014) Leaching-resistant carrageenan-based colorimetric oxygen indicator films for intelligent food packaging. J Agric Food Chem 62(29):7263–7267

Webber V, de Carvalho SM, Ogliari PJ, Hayashi L, Barreto PLM (2012) Optimization of the extraction of carrageenan from *Kappaphycus alvarezii* using response surface methodology. Cienc Tecnol Aliment 32(4):812–818

Yong YS, Yong WTL, Ng SE, Anton A, Yassir S (2015) Chemical composition of farmed and micropropagated *Kappaphycus alvarezii* (Rhodophyta, Gigartinales), a commercially important seaweed in Malaysia. J Appl Phycol 27(3):1271–1275

Yun EJ, Lee S, Kim HT, Pelton JG, Kim S, Ko H-J, Choi I-G, Kim KH (2015) The novel catabolic pathway of 3,6-anhydro-L-galactose, the main component of red macroalgae, in a marine bacterium. Environ Microbiol 17(5):1677–1688

Yun EJ, Kim HT, Cho KM, Yu S, Kim S, Cho I, Kim KH (2016) Pretreatment and saccharification of red macroalgae to produce fermentable sugars. Bioresoure Technol:199. doi:10.1016/j.biortech.2015.08.001

Zamorano Palma JHR, Andres R, Pierre-Etienne Bost (2002) Process for producing carrageenan with reduced amount of insoluble material WO 2002057477 A1 20020725. WO 2002057477 A1 20020725

Zodape ST, Mukherjee S, Reddy MP, Chaudhary DR (2009) Effect of *Kappaphycus alvarezii* (Doty) Doty ex silva. extract on grain quality, yield and some yield components of wheat (*Triticum aestivum* L.). Int J Plant Prod 3

Zuldin WH, Yassir S, Shapawi R (2016) Growth and biochemical composition of *Kappaphycus* (Rhodophyta) in customized tank culture system. J Appl Phycol 28(4):2453–2458

11 Applications of Carrageenan: With Special Reference to Iota and Kappa Forms as Derived from the Eucheumatoid Seaweeds

Rafael R. Loureiro, M.L. Cornish, and Iain C. Neish

Abstract

Carrageenan is a natural, long-chain carbohydrate (polysaccharide), obtained from a variety of red algae including the Eucheumatoid seaweeds, viz. *Kappaphycus* and *Eucheuma*. Carrageenan shows a tremendous diversity in its molecular structure, a feature that directly influences its commercial applications and extraction methods. Known applications of carrageenan include its use as a gelling, thickening and binding agent in the processed foods industry. A notable feature of its usage is that it is commonly blended in ingredient solutions with other gums where it exhibits synergistic or complementary characteristics. Carrageenan has also been used in experimental medicine, pharmaceutical formulations, cosmetics and several industrial applications.

11.1 Introduction

Carrageenan (Cg) is a term derived from the Irish word *carraigín*, meaning "little rock". Carrageenophytes (seaweeds containing carrageenan) were known to be exploited in Ireland (with no clear evidence of sustainable utilization) as long ago as 400 A.D. Irish moss (*Chondrus crispus*) was used mainly for animal feed and medicinal purposes; recipes date back as far as the eighteenth century. Furthermore, archeological evidence proved that the Chinese also used carrageenan-bearing macroalgae such as *Gigartina* spp., for similar purposes, around 600 B.C. (Mitchell and Guiry 1983).

Carrageenan is a generic name for a family of gel-forming and viscosifying polysaccharides, it comprises a linear polysaccharide of D-galactose and 3, 6-anhydro-D-galactose, units, with varying levels of sulphation which are obtained by chemical extraction from a number of seaweeds from within the class Rhodophyceae. These macroalgae are distributed globally (Table 11.1) showing the early development and distribution of the carrageenan-producing genera. Being an assemblage of naturally occurring polysaccharides, which can shift in properties depending on the amount and position of the SO_3^- groups in its main molecular structure, e.g. kappa, iota and lambda variations, the family of macromolecules has transitioned from providing structural fibre and perhaps a form of natural defense mechanism to the thalli of certain red macroalgae, to a commercial product. Extensive, open-water cultivation of *Eucheuma* and *Kappaphycus* species and their variants, the major forms of raw materials for carrageenan production, now makes significant contributions to the economy of several developing countries in East Africa, the Caribbean and South America, whilst being led by the world's largest producers, as of early 2017: Indonesia and The Philippines (Necas and Bartosikova 2013; FAO 2014; Prajapati et al. 2014).

This chapter focuses mainly on the products of farming *Eucheuma* and *Kappaphycus* species, which is now practiced on a large-scale in several countries. This apparent widespread distribution was enabled by the seaweeds' natural

R.R. Loureiro (✉)
Blue Marble Space Institute of Science,
1001 4th Ave, Suite 3201, Seattle, WA 98154, USA
e-mail: rafael@bmsis.org

M.L. Cornish
Acadian Seaplants Limited,
J.S. Craigie Research Centre, 4 Bren Street, Cornwallis, NS, Canada

I.C. Neish
PT Sea Six Energy Indonesia, Bali, Indonesia

A.Q. Hurtado et al. (eds.), *Tropical Seaweed Farming Trends, Problems and Opportunities*, Developments in Applied Phycology 9, DOI 10.1007/978-3-319-63498-2_11

Table 11.1 A list of the major carrageenan-bearing genera, with their type(s) of carrageenan and global distribution

Genus	Carrageenan	Global distribution
Chondrus spp.	Kappa, iota and lambda carrageenan	Worldwide -North Atlantic and Japan.
Gigartina spp.	Kappa, iota and lambda carrageenan	Worldwide -Northwestern North American coast Japan, New Zealand, Chile, and South Africa.
Gymnogongrus spp.	Kappa and iota carrageenan	Atlantic waters -Caribbean sea and Vietnam.
Eucheuma spp.	Iota carrageenan	Northern Indian Ocean and the western Pacific from southern Japan to northern Australia Mauritius, East Africa, and the Gulf of California.
Kappaphycus spp.	Kappa carrageenan	Sulu Sea and the Sulu Archipelago- East Africa, Indonesia, Malaysia, Hainan Island, the Philippines and Micronesia.
Furcellaria spp.	Carrageenan-like sulphated galactan	North Atlantic-Greenland, Iceland, Northern Russia, Ireland, France, the Baltic Sea, Norway and Canada.
Hypnea spp.	Kappa carrageenan	Worldwide tropical distribution -Africa, Brazil and the Caribbean.
Iridaea spp.	Lambda carrageenan and carrageenan-like sulphated galactan	Western South America along the North American coast, southern Alaska and northern Japan.

Adapted from Guiry & Guiry (2017)

resilience to abiotic impacts, a requirement for simple, yet effective, cultivation techniques and rapid vegetative propagation of cultivars. Such characteristics contributed to an increased and secured, stabilized supply of the hydrocolloid carrageenan, ensuring its availability for multiple applications, more recently, beyond the traditional, primary focus of the processed food industry (Hafting et al. 2015). Pure carrageenan has no proven nutritional value and is widely used for its rheological properties (Van de Velde et al.2002; Prajapati et al. 2014; Hafting et al.2015) as a gelling and thickening agent. It is interesting to note that more recent uses have been found in the pharmaceutical industry as, amongst others: elicitors to test anti-inflammatory agents, as a raw material for hydrogels, for the construction of drug carriers such as pills and capsules; other exciting applications will be further discussed later in this chapter (Zacharopoulos and Phillips 1997; Van de Velde et al. 2002; Campo et al. 2009; Liang et al. 2014; Prajapati et al.2014).

11.2 The Extraction and Refining Processes of Various Forms of Carrageenan

Carrageenans are water-soluble hydrocolloids displaying a highly viscous aspect in aqueous solutions. Viscosity and gel strength are the physical attributes traditionally most sought after by the food industry; such characteristics are directly associated with temperature and the presence of other solutes (also known as blending), together with the type of carrageenan and its molecular weight (Lai et al. 2000; Stanley 2011; Necas and Bartosikova 2013; Table 11.2).

As might be expected, different methods are used for the extraction of various types of raw material to provide various types or mixes of carrageenans, i.e. three similar process steps: (1) drying of the raw material; (2) an alkali treatment or alcoholic precipitation phase and (3) milling of the final product.

11.2.1 The Extraction of Kappa Carrageenan

Biomass from various species of *Kappaphycus* provides for over 85% of the world's entire kappa-carrageenan production. The process starts with the drying of the seaweed (usually sun dried), causing the pigments to slowly degrade, resulting in bleaching of the thalli. The pre-bleached and dried thalli are treated with 5–10% of NaOH, at 80–90 °C, followed by a KCl precipitation processes (see: semi-refined extraction) followed by a series of washes. The product is dehydrated by a pressing method, dried and milled to a kappa carrageenan powder (FAO 2014; Prajapati et al. 2014).

11.2.2 The Extraction of Lambda Carrageenan

Gigartina spp. and *Chondrus* spp. are the main sources of raw materials for the production of lambda carrageenan. Being more hydrophilic than kappa carrageenan, lambda lacks gel formation capabilities, but in turn is able to achieve high viscosities when in solution. A drum drying process,

Table 11.2 Types of commercial carrageenans, their gel characteristics and main methods of extraction

Iota	Clear, elastic (with calcium salts), freeze/thaw stable gels	Alcoholic precipitation
Kappa	Strong, rigid gel, formed with potassium salts, or a brittle gel formed with calcium salts.	5–10% of NaOH at 80–90 °C and KCl precipitation.
Lambda	No gel formation, forms high viscosity solutions.	Alcoholic precipitation

Adapted from FAO (2014)

where pureed raw macroalgae are dried, at relatively low temperatures over rotating steel drums, is the most common method used to obtain the dried seaweed which is usually followed by an alcohol precipitation process (i.e. propyl or ethyl alcohol) to further dehydrate the macroalgal fibre making it ready for a milling process (FAO 2014; Prajapati et al. 2014).

11.2.3 The Extraction of Iota Carrageenan

Members of the genera *Eucheuma* spp. and *Gymnogongrus* spp. are the main sources of iota carrageenan. This forms clear, elastic gels when combined. Extraction, utilising calcium salts, is somewhat similar to the lambda carrageenan extraction process, whereby after sun drying and bleaching, followed by hot extraction and filtration, with pH adjustment, alcoholic precipitation is used following milling of the product to powder (FAO 2014; Prajapati et al. 2014).

11.2.4 Refined and Semi-refined Extraction

Used only with *Eucheuma* spp. and *Kappaphycus* spp. this process is mainly distinguished by its low energy input to produce better gel quality, as well the additional care taken when removing salt and sand from the raw material thalli after being sun dried in preparation for extraction. Seaweed biomass is immersed in hot KCl solution and then soaked in fresh water in order to extract most of the residual alkali, thus further increasing the gel strength of the extract which further enhances its economic value. The refined process differs from the semi-refined process by adding extra steps required in order to dissolve and filter cell wall debris from the carrageenan, which is precipitated by either isopropanol or potassium chloride (FAO 2014).

11.3 Food Industry Applications

Carrageenan had its first formal introduction to humans through its identification as a constituent in a seaweed that was, at that time, called *Fucus crispus*. It was Turner (1809) who discovered the gelling properties of boiled *F. crispus*, which were largely responsible for the utilization of this seaweed for hundreds of years as food and medicine (Thorbjarnarson 1939; Allen and Hatfield 2004; Mouritsen 2013). Historically, in Ireland, carrageenan was extracted from certain red seaweeds by boiling their fronds in water or milk to be utilized as a traditional treatment for chest and lung ailments, appropriately mirroring some of the more modern day applications of carrageenan nasal sprays (Ludwig et al. 2013). Indeed, the relaxing effects of carrageenans on mucus membranes provided relief in tuberculosis, whooping cough, pneumonia, quinsy (abscessed tonsils) and for various gastro-intestinal complaints, but its convenient gelling characteristics also made it a useful kitchen ingredient (Collén et al. 2014). The distinctive utilization of the hydro- (phyco-) colloid for food processing was eventually popularized and commoditized; by the mid-1970s, cultivated "*Eucheuma*" spp. from the Philippines began to enter the market-place to meet with global demands for processed foods (Ricohermoso and Deveau 1979).

The commercial utilization of carrageenan, from the late nineteenth century to the present day, was almost exclusively dedicated to the food and pharmaceutical industries (Collén et al. 2014). In particular, Cg belongs to a structurally diverse group of hydrocolloid compounds, the sulphated polysaccharides (SP) which are commonly found in many marine macroalgae; these possess a wide range of functional gelling and emulsifying properties. Typically, food applications of Cg include: modifications of texture, reduction of fat and salt, enhancement of storage stability and flavour, improved fiber content, anti-microbial activities, protein flocculation (for the brewing industry) and particle suspension in dairy products (Bixler and Porse 2011; Menon 2012; Gurpilhares et al.2016).

Such versatile, industrial utility of refined carrageenans (in food processing applications) has inadvertently sidelined the importance of the fundamental nutritional value of the commercial carrageenan-bearing seaweeds, viz. *Kappaphycusalvarezii* (an industrial source of kappa Cg) and *Eucheumadenticulatum* (an industrial source of iota Cg). Whilst the modern literature reports considerable biomedical and therapeutic prospects associated with various SPs (Jiao et al. 2011; Patel 2012), it is important to recognize that diet also plays a significant role in the manifestation of many of the risk factors associated with animal and human illnesses and compromised health. Many global societies are experiencing an obesity epidemic (WHO Factsheet No. 311 2015) leading to a massive increase in cardiovascular diseases (CVD) and other chronic pathologies, while, alarmingly, neurological disorders are also on the increase (Cornish et al. 2015; Cornish et al. 2017). Carrageenan, with its unique complement of rheological properties, is also valuable for its dietary fiber (i.e. prebiotic) component, contributing to improvements in overall human health. The consumption of seaweed fibers helps promote satiety and delays glucose absorption, whilst also contributing to improved plasma lipid profiles, simultaneously promoting and sustaining populations of beneficial gut microbes (Burton-Freeman 2000; Aleixandre and Miguel 2008; Gómez-Ordóñez et al. 2010). In this respect, the whole family of Cg structures (i.e. kappa, iota, lambda, beta, etc.), contain various and some highly sulphated polysaccharides (galactans). This family of polysaccharides has demonstrated repeatedly, significant prebiotic potential, especially when in the form of depolymerized, low molecular weight, oligo-carrageenans.

Oligo-carrageenans are resistant to absorption and effects of enzymes in the upper, gastro-intestinal tract (GIT) and are only fermented by the (beneficial) microbial flora of the large intestine, thereby selectively enhancing populations of gut microbes (González et al. 2013; Zaporozhets et al. 2014; Jutur et al. 2016). The mechanisms as to why the effects should be so pronounced in favour of beneficial microbes is yet to be conclusively determined. Thus, dietary fibers from various carrageenophytes can play an important role in efforts required to combat obesity (of humans and companion animals) through provision of both soluble and insoluble dietary fibers, in addition to their processed food-favorable versatility (Warrand 2006).

As a fat substitute in certain food products (Bixler 1996; Mahungu et al. 2002), Cg reduces caloric content, while maintaining desirable organo-leptic characteristics; it provides significant antioxidant activities such that the health benefits of a food are also enhanced (Yuan et al. 2005). Furthermore, in aspects of food functionality *i*-carrageenan, in particular extends additional beneficial properties in the realm of antimicrobial activities against food-borne pathogens. This makes Cg a useful means of controlling microbes in poultry and meat products (Yamashita et al. 2001; Menon 2012). Indeed, the potential health benefits of selective consumption of various Cgs are only recently beginning to come to light; the immune-modulatory, anti-fungal and antioxidant activities identified in pharmacological research (Soares et al. 2016; Yermak et al. 2016) may also apply to dietary applications of whole seaweed ingredients. Furthermore, human food science technologies may yet build upon (production) animal studies regarding the prebiotic potential of red seaweeds, which have shown marked improvements in intestinal histo-morphology, short-chain fatty acid formation and a proliferation of beneficial gut microbes (Kulshreshtha et al. 2014, for an example using a poultry model).

The positive effects of specific applications of a variety of carrageenans (with varying levels of sulphation), to plants and their pathogens have been demonstrated as being mediated by oligo-carrageenans indicating increased cell cycle activity and thus increased cell division, as well as protective effects against pathogens, as determined by the accumulation of compounds with antimicrobial activities (González et al. 2013). Of particular agricultural interest, Cg has been shown to enhance crop plant resistance to viral infections, potentially eliminating, or reducing the need for synthetic pesticides (Ghannam et al. 2013; Sangha et al. 2015). Hence the circle is completed and judicious applications of Cg can be beneficial through indirect means due to integration into plant health and via its properties as a functional food, or food ingredient. It is widely accepted that *in vivo* research regarding dietary constituents and their effects in humans is both complicated and challenging. However, numerous published studies identifying the therapeutic, pharmacological functionality of Cg suggested its potential value to human society may well go beyond traditional, rheological contributions to the processed food industry. Clearly this is an area where a significant increase in natural product research in relation to human health and wellness would provide many societal and indeed financial benefits from the higher-value applications based upon biological activity of the long-chain polysaccharides.

11.4 Carrageenan Interactions with Other Hydrocolloids

Table 11.3 lists major food system hydrocolloids with their interactions with carrageenan. Much of the 'magic' of ingredient solutions is in the provider's ability to produce best-value performance by blending various combinations and proportions of hydrocolloid building blocks, along with ingredients to produce proprietary, complementary efficacies (Saha and Bhattacharya 2010; Hotchkiss et al. 2016; Porse and Rudolph 2017). The proportions and types of carrageenan in blends varies, depending on functionality and cost factors. For certain ingredient solutions the proportion of various types of carrageenan in blends may range from comprising most of a blend to comprising little. This fact is key and part of the value-proposition from leading manufacturers of carrageenan building-blocks (i.e. oligomers).

There are several hydrocolloids with which carrageenans can have either complementary, or competitive interactions, as they can be used singly or in combination (Hotchkiss et al. 2016). For example, modified starch can replace carrageenan in meat-pumping applications or xanthan can replace carrageenan in toothpaste. In many cases the decisions over which to use are based on supplier and price-performance. Synergistic, kappa carrageenan combinations with galactomannans and/or glucomannans are widely applied to several ingredient solutions. For example, they are used in table jellies and low sugar, low-acid fruit gels, in which such combinations have an advantage over pectin in that they dissolve well in sugar solutions, whereas pectin has to be predissolved in water. However, under acidic conditions ($pH < 4$), carrageenans tend to degrade while pectins are acid-stable.

11.5 Biological Activity and Pharmaceutical Applications

Sulphated polysaccharides, from numerous species of marine macroalgae have been identified to have diverse biological activities including anti-coagulant/anti-thrombotic, anti-viral and anti-tumor effects (Necas and Bartosikova 2013; Prajapati et al. 2014). Carrageenan, being an inert polymer is known to have the ability to form thermo-reversible gels in aqueous solutions and in the presence of cations. It is a supe-

Table 11.3 Major hydrocolloids and their interactions with carrageenan

Hydrocolloid	Characteristics	Examples of carrageenan interactions
Alginate	Thermo-stable gels do not melt on heating.	Can be complementary in dairy and brewery applications. May compete in textile and other applications.
Carboxymethyl cellulose (CMC)	High viscosity reduced by adding electrolytes and at low pH.	Can be complementary or may compete in salad dressings, fillings and sauces
Galactomannans including guar, locust bean (carob) gum, cassia and tara	High low-shear viscosity & shear thinning independent of electrolytes; degrades at extreme pH and high temperatures	Synergistic with kappa carrageenan in water gels, dairy desserts and meat products; also in dairy products sauces, juices, pudding and cake batter
Gellan gum	Thermo-reversible, transparent gels formed on cooling	Competes with kappa carrageenan and agar in water-based gelation applications.
Gum Arabic	Low viscosity gum; shear-thinning at low shear; near-Newtonian behaviour at higher shear	Complementary or competitive in fruit juice based beverages and soft drinks
Gum tragacanth	Swells rapidly in cold or hot water to form highly viscous dispersions	Complementary or competitive in salad dressings, bakery emulsions, fruit beverages and sauces
Konjac glucomannan	Forms highly viscous dispersions not influenced by salts; forms thermally stable gels with alkali	Synergistic with kappa carrageenan in gel formation in noodles, jelly desserts and meat products
Methyl celluloses (MC and HPMC)	Viscosity increases with temperature independent of pH; thermo-reversible gels	Complementary or competitive salad dressings, cake batters, beverages, whipped toppings
Modified starch	Thermo-irreversible opaque gels formed on cooling	Complementary or competitive in dairy desserts and meat-pumping
Pectin (high and low methoxy)	Thermo-irreversible gels on cooling at high acidic pH	Complementary or competitive in jams, jellies, glazes and dairy desserts
Xanthan	Highly shear thinning; maintains viscosity in the presence of electrolytes, high temperature and wide pH ranges	Complementary or competitive in soups, gravies, sauces, instant beverages, desserts, toppings and fillings; competitive in toothpaste.

After Saha and Bhattacharya (2010) and Hotchkiss et al. (2016)

rior agent for pharmaceutical formulations involving the stabilization of disperse systems, viscosity modifications, gel formation, encapsulation of solutions in oral extended-release tablets and as a novel extrusion-aid for the production of pellets and also a carrier/stabilizer in micro/nanoparticle systems. Moreover, selected carrageenans have shown promising results in tissue-regeneration therapy as bio-macromolecules and cell delivery systems (Liang et al.2014).

Whilst not derived from either *Kappaphycus* nor *Eucheuma*, it is worth mentioning that lambda carrageenan (i.e. primarily from *Chondrus* spp. and *Gigartina* spp.) has been proven to have four times the anti-thrombotic activity, as compared to kappa-carrageenan (from *Kappaphycus* spp. sources), being directly tied to thrombin inhibition, thus also corroborating with the anti-coagulant properties (Necas and Bartosikova 2013; Liang et al.2014). The higher level of sulphation in lambda carrageenan is the likely reason for its higher efficiency in inhibiting the actions of thrombin anti-via heparin co-factor II (HC-II) and/ or a direct anti-thrombin effect, thereby prolonging the anti-coagulant activity time. There is however, the drawback of being more toxic than kappa-carrageenan (Kindness et al. 1980a, b; Necas and Bartosikova 2013).

Various carrageenans are known to be selective inhibitors of several enveloped viruses, acting primarily by preventing binding or entry of viruses particles into cells (Buck et al.2006, Grassauer et al.2008; Stiles et al.2008; Necas and Bartosikova 2013). As with its anti-thrombotic/anti-coagulant properties, lambda-carrageenan (from *Gigartinaskottsbergii*) has more effective, anti-viral properties against different strains of herpes simplex virus (HSV) type 1 and type 2, when compared to other carrageenan types (Carlucci et al.1997, 2004). However, most carrageenan types (i.e. kappa, lambda, iota) were proven to be potent inhibitors of the common, sexually transmitted HPV types, proving that its inhibitory effects were able to act regardless of its molecular conformation (Necas and Bartosikova 2013).

For over a decade (as of 2017), anti-tumoral studies have been conducted using all types of carrageenan in order to observe their anti-proliferative activities in various cancer cell-lines. Inhibitory activities towards tumor growth in mice and anti-metastatic activity, by blocking interactions between cancer cells and the basement membrane (thus inhibiting tumor cell proliferation and tumor cell adhesion) have been reported (Yuan et al.2006; Zhou et al. 2006; Rocha de Souza et al. 2007) Most results showed significant reductions of cancerous cells in mice, especially when the proliferation of those cells was related to colo-rectal cancer (Necas and Bartosikova 2013).

Kappa-carrageenan (from *Kappaphycus* spp.) appears to be the most suitable source (amongst other carrageenan types), to be used as a support for 6-aminopenicillanic acid (6-APA) in the synthesis process for antibiotic production;

Nagalakshmi and Pai (1997) were able to obtain 90% efficiency of cell immobilization in *E. coli*, using penicillin-amidase activity on a kappa-carrageenan support.

The cell immobilization effects of kappa carrageenan can also be observed in the production of d-aspartic acid and l-alanine using *Pseudomonas dacunhae* cells. Industrial applications of immobilized, whole cells, by kappa carrageenan gels, for the preparation of l-aspartic acid was first developed by Chibata (Tanabe Seiyaku, Japan) in 1973. The industrial production of a number of other beneficial compounds (i.e. l-malicacid, l-alanine, l-tryptophan, 1, 5-dimethyl-2-piperidone) for use in food and medical applications, is based on the same principles (van de Velde et al. 2002; Necas and Bartosikova 2013).

11.6 Conclusions

Aside from the newly discovered applications of different blends of carrageenan in the food industry and the growing body of research related to their uses in crop yields, applications of benefits in the pharmaceutical field have increased exponentially. Various carrageenans are not only being employed for drug delivery, but also for cutting-edge applications such as tissue regeneration therapy. As a versatile polysaccharide, with an edge for being highly hydrophilic, many of the future uses for carrageenans are yet to be discovered. The potential for future, value-added applications is truly staggering, only held back by the need for further investments in applied research, testing and more especially providing human clinical trials to validate efficacy.

References

Aleixandre A, Miguel M (2008) Dietary fiber in the prevention and treatment of metabolic syndrome: a review. Crit Rev Food Sci Nutr 48:905–912

Allen DE, Hatfield G (2004) Medicinal plants in folk tradition – an ethnobotany of Britain & Ireland. Timber Press, Portland/Cambridge

Bixler HJ (1996) Recent developments in manufacturing and marketing carrageenan. Hydrobiol 326(327):35–57

Bixler HJ, Porse H (2011) A decade of change in the seaweed hydrocolloids industry. J Appl Phycol 23:321–335

Buck CB, Thompson CD, Roberts JN, Muller M, Lowy DR, Schiller JT (2006) Carrageenan is a potent inhibitor of papilloma virus infection. PLoS Pathog 2(7):671–680

Burton-Freeman B (2000) Dietary fiber and energy regulation. J Nutr 130:272S–275S

Campo VL, Kawano DF, da Silva DB, Carvalho I (2009) Carrageenans: biological properties, chemical modifications and structural analysis. Carbohydr Polym 77:167–180

Carlucci MJ, Pujol CA, Ciancia M, Noseda MD, Matulewicz MC, Damonte EB, Cerezo AS (1997) Antiherpetic and anticoagulant properties of carrageenans from the red seaweed *Gigartina skottsbergii* and their cyclized derivatives: correlation between structure and biological activity. Int J Biol Macromol 20:97–105

Carlucci MJ, Scolaro LA, Noseda MD, Cerezo AS, Damonte EB (2004) Protective effect of a natural carrageenan on genital herpes simplex virus infection in mice. Antivir Res 64:137–141

Collén J, Cornish ML, Craigie J, Ficko-Blean E, Hervé C, Krueger-Hadfield SA, Leblanc C, Michel G, Potin P, Tonon T, Boyen C (2014) *Chondrus crispus* – a present and historical model organism for red seaweeds. In: Bourgougnon N (ed) Advances in botanical research, Sea plants, vol 71. Academic, Amsterdam, pp 53–89

Cornish ML, Critchley AT, Mouritsen OG (2015) A role for dietary macroalgae in the amelioration of certain risk factors associated with cardiovascular disease. Phycologia 54:649–666

Cornish ML, Critchley AT, Mouritsen OG (2017) Consumption of seaweeds and the human brain. J Appl Phycol . doi:10.1007/s10811-016-1049-3

FAO (2014). FAO JECFA monographs, specifications: Carrageenan. http://wwwfaoorg/. Accessed 7 Jan 2017

Ghannam A, Abbas A, Alek H, Al-Waari Z, Al-Ktaifani M (2013) Enhancement of local plant immunity against tobacco mosaic virus infection after treatment with sulphated-carrageenan from red alga (*Hypnea musciformis*). Physiol Mol Plant Physiol 84:19–27

Gómez-Ordóñez E, Jiménez-Escrig A, Rupérez P (2010) Dietary fiber and physicochemical properties of several edible seaweeds from the northwestern Spanish coast. Food Res Int 43:2289–2294

González A, Castro J, Vera J, Moenne A (2013) Seaweed oligosaccharides stimulate plant growth by enhancing carbon and nitrogen assimilation, basal metabolism, and cell division. J Plant Growth Regul 32:443–448

Grassauer A, Weinmuellner R, Meier C, Pretsch A, Prieschl-Grassauer E, Unger H (2008) Iota-carrageenan is a potent inhibitor of rhinovirus infection. Virol J 5:1–13

Gurpilhares DB, Moreira TR, Bueno JL, Cinelli LP, Mazzola PG, Pessoa A, Sette LD (2016) Algae's sulfated polysaccharides modifications: potential use of microbial enzymes. Process Biochem 51:989–998

Guiry MD, Guiry GM (2017) AlgaeBase. World-wide electronic publication. National University of Ireland, Galway. http://www.algaebase.org; searched on 10 March 2017

Hafting JT, Craigie JS, Stengel DB, Loureiro RR, Bushmann AH, Yarish C, Edwards MD, Critchley AT (2015) Prospects and challenges for industrial production of seaweed bioactives. J Phycol 51:821–837

Hotchkiss S, Brooks M, Campbell R, Philip K, Trius A (2016) The use of carrageenan in food. In: Pereira L (ed) Carrageenans, sources and extraction methods, molecular structure, bioactive properties and health effects. Nova Science Publishers, New York, pp 142–150

Jiao G, Yu G, Zhang J, Ewart HS (2011) Chemical structures and bioactivities of sulfated polysaccharides from marine algae. Mar Drugs 9:196–223

Jutur PP, Nesamma AA, Shaikh KM (2016) Algae-derived marine oligosaccharides and their biological applications. Front Mar Sci 3:83

Kindness G, Long WF, Williamson FB (1980a) A anticoagulant effects of sulphated polysaccharides in normal and antithrombin III-deficient plasmas. Br J Pharmacol 69:675–677

Kindness G, Williamson FB, Long WF (1980b) Involvement of antithrombin III in anticoagulant effects of sulphated polysaccharides. Biochem Soc Trans 8:82–83

Kulshreshtha G, Rathgeber B, Stratton G, Thomas N, Evans F, Critchley A, Hafting J, Prithiviraj B (2014) Feed supplementation with red seaweeds, *Chondrus crispus* and *Sarcodiotheca gaudichaudii*, affects performance, egg quality, and gut microbiota of layer hens. Poult Sci 93:1–11

Lai VMF, Wong PA-L, Lii C-Y (2000) Effects of cation properties on sol-gel transition and gel properties of κ-carrageenan. J Food Sci 65:1332–1337

Liang L, Rui N, Yang S, Mao S (2014) Carrageenan and its application is drug delivery. Carbohydr Polym 103:1–11

Ludwig M, Enzenhofer E, Schneider S, Rauch M, Bodenteich A, Neumann K, Prieschl-Grassauer E, Grassauer A, Lion T, Mueller CA (2013) Efficacy of a carrageenan nasal spray in patients with common cold: a randomized controlled trial. Respir Res 14:124–127

Mahungu SM, Hansen SL, Artz WE (2002) Fat substitutes. In: Brown AL, Davidson PM, Salminen S, Thorngate JH III (eds) Food additives. Marcel Dekker, New York, p 311

Menon VV (2012) Seaweed polysaccharides – food applications. In: Se-Kwon K (ed) Handbook of marine macroalgae: biotechnology and applied phycology. Wiley, New York, pp 541–555

Mitchell ME, Guiry MD (1983) Carrageenan: a local habitation of a name? J Ethnopharmacol 9:347–351

Mouritsen OG (2013) Seaweeds edible, available & sustainable. University of Chicago Press, London

Nagalakshmi V, Pai JS (1997) Immobilization of penicillin acylase producing *E. coli* cells with carrageenan. Indian J Microbiol 37:17–20

Necas J, Bartosikova L (2013) Carrageenan: a review. Vet Med 58(4):187–205

Patel S (2012) Therapeutic importance of sulfated polysaccharides from seaweeds: updating the recent findings. Biotech 2:171–185

Porse H, Rudolph B (2017) The seaweed hydrocolloids industry: 2016 updates, requirements and outlook. J Appl Phycol. doi:10:1007/s10811-017-1144-0

Prajapati VD, Maheriya PM, Jani JK, Solanki HK (2014) Carrageenan: a natural seaweed polysaccharide and its applications. Carbohydr Polym 105:97–112

Ricohermoso MA, Deveau LE (1979) Review of commercial propagation of *Eucheuma* (Florideophyceae) clones in the Philippines. Proc Int Seaweed Symp 9:525–531

Rocha de Souza MC, Marques CT, Dore CMG, da Silva FRF, Rocha HAO, Leite EL (2007) Antioxidant activities of sulfated polysaccharides from brown and red seaweeds. J Appl Phycol 19:153–160

Saha D, Bhattacharya S (2010) Hydrocolloids as thickening and gelling agents in food: a critical review. J Food Sci Technol 47(6):587–597

Sangha JS, Kandasamy S, Khan W, Bahia NS, Singh RP, Critchley AT, Prithiviraj B (2015) λ-carrageenan suppresses tomato chlorotic dwarf viroid (TCDVd) replication and symptom expression in tomatoes. Mar Drugs 13:2875–2889

Soares F, Fernandes C, Silva P, Pereira L, Gonçalves T (2016) Antifungal activity of carrageenan extracts from the red alga *Chondracanthus teedei* var. *lusitanicus*. J Appl Phycol 28:2991–2998

Stanley N (2011) FAO Corporate document repository. Chapter 3: Production, properties and uses of carrageenan. FMC Corporation, Marine Colloids Division 5 Maple Street, Rockland Maine 04841, USA

Stiles J, Guptill-Yoran L, Moore GE, Pogranichniy RM (2008) Effects of carrageenan on *in vitro* replication of feline herpesvirus and on experimentally induced herpetic conjunctivitis in cats. Investig Ophthalmol Vis Sci 49:1496–1501

Thorbjarnarson T (1939) Nyting fjφrugrasa (utilization of Irish moss). Ægir 9:193

Turner D (1809) Fuci, sive plantarum fucorum generi a botanicis ascriptarum icons, descriptions et historia. John and Arthur Arch, London

Van de Velde F, Lourenco ND, Pinheiro HM, Bakkerd M (2002) Carrageenan: a food-grade and biocompatible support for immobilization techniques. Adv Synth Catal 344:815–835

van de Velde F, Rollema HS, Grinberg NV, Burova TV, Grinberg VY, Tromp RH (2002) Coil-helix transition of i-carrageenan as afunction of chain regularity. Biopol 65(4):299–312

Warrand J (2006) Healthy polysaccharides-the next chapter in food products. Food Technol Biotechnol 44:355–370

WHO Factsheet No. 311 (2015) Obesity and overweight. http://www.who.int/mediacentre/factsheets/fs311/en/. Accessed 18 Dec 2016

Yamashita S, Sugita KY, Shimizu M (2001) In vitro bacteriostatic effects of dietary polysaccharides. Food Sci Technol Res 7:262–264

Yuan H, Zhang W, Lu X, Li H, Geo X, Song J (2005) Preparation and in vitro antioxidant activity of k-carrageenan oligosaccharides and their over-sulfated, acetylated, and phosphorylated derivatives. Carbohydr Res 340:685–692

Yuan H, Song J, Li X, Li N, Dai J (2006) Immunomodulation and antitumoractivity of carrageenan oligosaccharides. Cancer Lett 243:228–234

Yermak IM, Sokolova EV, Davydova VN, Solov'eva TF, Aminin DL, Reunov AV, Lapshina LA (2016) Influence of red algal polysaccharides on biological activities and supramolecular structure of bacterial lipopolysaccharide. J Appl Phycol 28:619–627

Zacharopoulos VR, Phillips DM (1997) Vaginal formulations of carrageenan protect mice from *Herpes simplex* virus infection. Clin Diagn Lab Immunol 4:465–468

Zhou G, Sheng W, Yao W, Wang C (2006) Effect of low molecular carrageenan from *Chondrus ocellatus* on antitumor H-22 activity of 5-Fu. Pharmacol Res 53:129–134

Zaporozhets TS, Besednova NN, Kuznetsova TA, Zvyagintseva TN, Makarenkova ID, Kryzhanovsky SP, Melnikov VG (2014) The prebiotic potential of polysaccharides and extracts of seaweeds. Russ J Mar Biol 40:1–9

12 Development of Eucheumatoid Seaweed Value-Chains Through Carrageenan and Beyond

Iain C. Neish and Shrikumar Suryanarayan

Abstract

As of 2016, about 250,000 dry MT of *Kappaphycus* and *Eucheuma* (collectively referred to as the eucheumatoid seaweeds) entered value-chains as raw materials for single-stream processing that led to the production of carrageenan. Virtually all eucheumatoid seaweeds had been sold through such value-chains since the carrageenan industry developed farming, which began in the early 1970s. Over the ensuing decades, changes in value-chain (V-C) management and a general decline in industrial innovation, combined with over-capacity in the processing segment, created a situation where eucheumatoid seaweed production reached a plateau that was correlated with low-to-slow growth in global carrageenan markets and dominance of value networks by 'alpha' actors Indonesia (for seaweed supplies) and China (for processing to carrageenan). By the turn of the century, eucheumatoid seaweed farming was a source of livelihood for tens of thousands of coastal people in tropical to sub-tropical regions, so both private- and public-sector value-chain players had hopes for major expansion in seaweed production. Events have shown that such expansion cannot be supported, as long as the industry is dependent primarily on sales simply as raw material used for carrageenan production. Emerging, mechanized cultivation technologies and multi-stream, zero-effluent (MUZE) processing technologies have recently provided opportunistic pathways towards increased utilization of current and future volumes of cultivated, eucheumatoid seaweed biomass. Such technologies will enable production of a broad range of value-added agricultural, chemical and biofuel products that can support future value-chain development with market size and value beyond the present prospects for a scenario dominated solely by carrageenan.

12.1 Introduction

12.1.1 *Value-chain context*

As of 2016, about 250,000 dry MT of *Kappaphycus* and *Eucheuma* (the eucheumatoid seaweeds) served as raw materials for single-stream processing that led to production of various categories of carrageenans (Fig. 12.1). *Kappaphycus* spp. ('cottonii' of the trade) was a source of kappa-carrageenan and *Eucheuma* spp. ('spinosum' of the trade) was a source of iota carrageenan. Extracts were made either as semi-refined carrageenan (SRC) with E-number 407a, or as clarified, refined carrageenan with E-number 407. Minor quantities of eucheumatoid seaweeds have been consumed fresh as sea vegetables in the regions where the seaweed farms are located and about 10,000 tons per annum were dried for export from Indonesia to China, where they were consumed as sea vegetables (Baricuatro 2015).

Single-stream processing of carrageenan consequently resulted in substantial waste streams, because essentially all biomass components other than carrageenan (i.e. about 50%

I.C. Neish (✉)
PT Sea Six Energy Indonesia, Bali, Indonesia
e-mail: ineish@sea6energy.com

S. Suryanarayan
Sea6 Energy Pte. Ltd., Bangalore, India

A.Q. Hurtado et al. (eds.), *Tropical Seaweed Farming Trends, Problems and Opportunities*, Developments in Applied Phycology 9, DOI 10.1007/978-3-319-63498-2_12

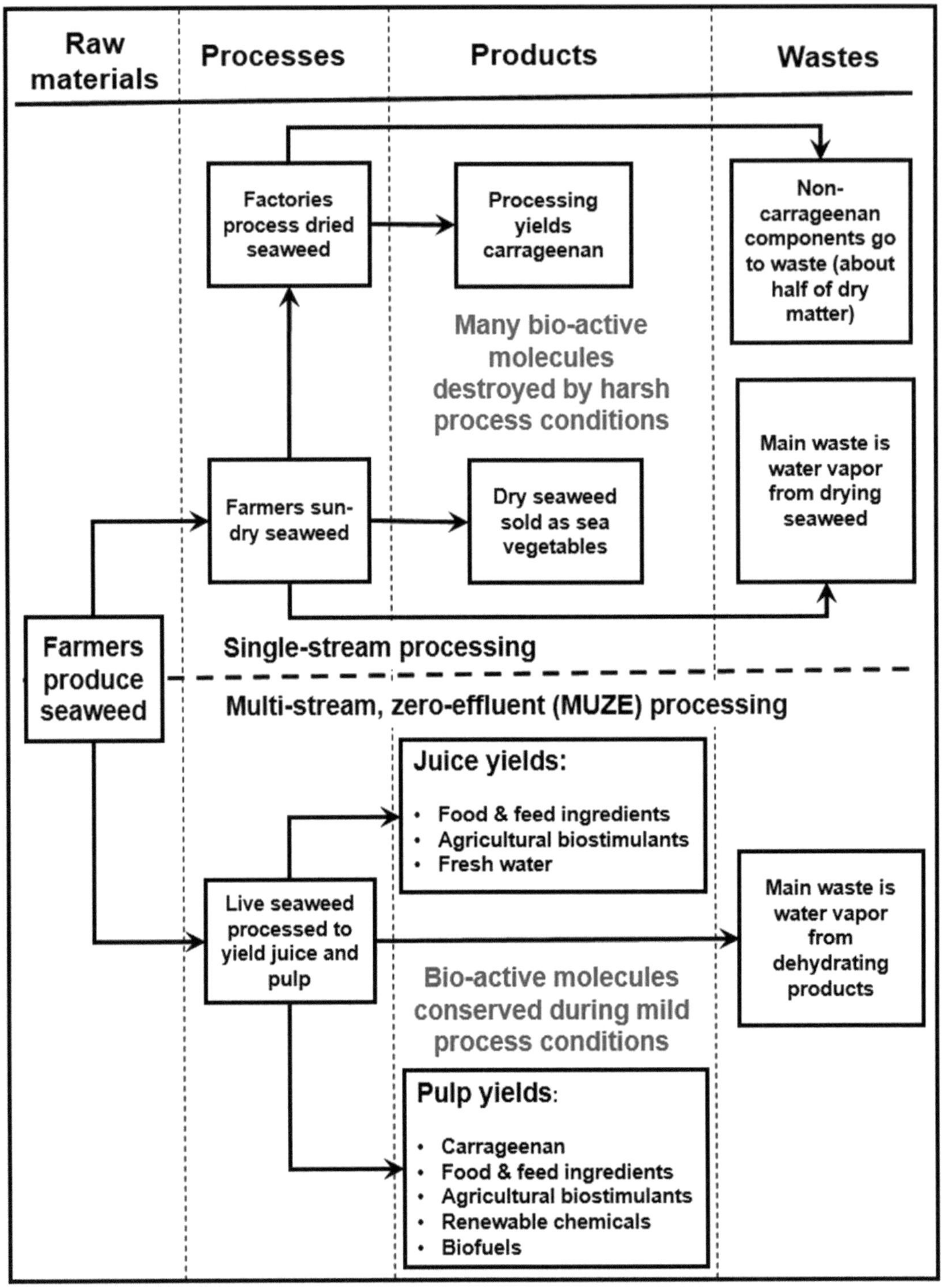

Fig. 12.1 Live eucheumatoid seaweeds enter value-chains where they are subjected either to single-stream processing or to multi-stream, zero-effluent (MUZE) processing

of the algal dry-matter) were discarded and many bio-active compounds were destroyed during harsh processing conditions. By 2016, MUZE processing was being undertaken in India and it had also spread to Indonesia. The application of MUZE-processing technology virtually eliminated waste and enabled production of a wide range of products from eucheumatoid seaweed biomass (see Figs. 12.1 and 12.10). In this case, bio-active compounds were conserved during mild, process conditions.

Agronomy systems used for eucheumatoid seaweeds are the foundation-link in value-chains that lead toward end-products entering global market places. Value-chains operate simultaneously within market systems and within the socio-economicproduction seascapes and landscapes where they function (Fig. 12.2).

Virtually all eucheumatoid seaweed farms operate within communities where farmers live and work, in the broader context of socio-economicproduction seascapes (SEPS) and socio-economicproduction landscapes (SEPL). SEPL are embraced by the Japanese term: 'satoyama' and SEPS by the term: 'satoumi'. A substantial literature has developed around these concepts (UNU 2010). They provide a model for sustainable environmental stewardship and resource management which in turn contributes to human well-being and, as such, they are useful concepts for defining the contexts of value-chain operations. The word 'satoumi' was coined by Professor Tetsuo Yanagi of Kyushu University in 1998. In Japanese 'sato' means village and 'umi' means the sea, so Yanagi defined 'satoumi' as: "***High productivity and biodiversity in the coastal sea area with human interaction***" (Yanagi 2008). The term 'satoumi' was derived from 'satoyama' which is a Japanese concept for long-standing traditions associated with land management practices. In the past such traditions fostered sustainable use of resources through symbiotic human interaction with ecosystems. The Japan Satoyama Satoumi Assessment project (JSSA 2010) employed satoyama and satoumi principles as useful heuristic

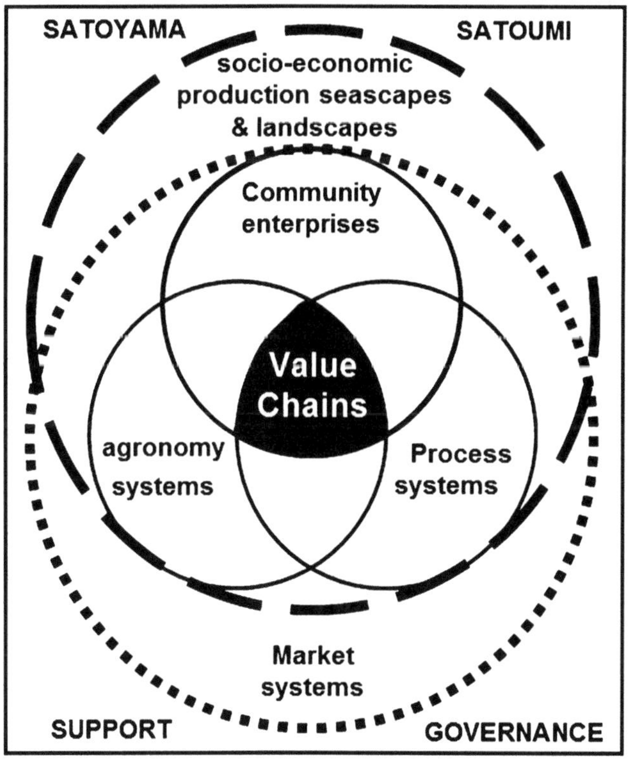

Fig. 12.2 A simplified heuristic diagram depicting the context of eucheumatoid seaweed value-chains, within the context of seascape communities and market systems

devices for framing and analyzing the relationship between ecosystem services and human well-being.

Eucheumatoid seaweed V-Cs are operated by community enterprises that own and manage the agronomy and process systems that produce seaweed biomass and also make value-added products from that biomass. Through value-chains, community enterprises link into local, regional and global market systems that provide support and governance functions (Fig. 12.3). Market systems are multi-player, multi-functional arrangements involving diverse stakeholders, all of whom have direct or indirect impacts on seaweed farming enterprises. The central set of market system exchanges is between the providers (i.e. the supply side) and the consumers (i.e. the demand side) of goods and services that are at the heart of a market system. The medium of exchange can be financial or non-financial (e.g. through accountability mechanisms). Support functions inform, provide and facilitate in the support of essential V-C functions. Six support functions that are crucial for the smooth operation of eucheumatoid seaweed value-chains include:

- Access to effective, affordable communication and logistics;
- Access to affordable, high quality education, science and technology;
- Provision of essential goods, services and infrastructure;
- Provision of affordable, fair finance;
- Access to fair trade in local, regional and global markets; and
- Facilitation for building strategic and tactical business alliances.

Governance functions impacting seaweed farms include:

- Government laws and regulations
- National and international standards
- Industry and buyer standards; and
- Traditional systems of norms, folkways and mores.

Herr and Muzira (2009) emphasized the importance of putting services at the core of market systems. They pointed out that key determinants of improved productivity and competitiveness amongst small and medium sized enterprises (SME) lay in improved knowledge, information and skills. Services are a critical means through which industry-specific information and knowledge is generated and transferred. However, the mechanisms by which these services are delivered and the market players who deliver them vary. One reason for poor performance of SMEs at the lower end of the value-chains is lack of access to information on available business development services (BDS). Since SMEs do not make use of such services, their performance is generally poor. Seaweed farm development therefore requires compi-

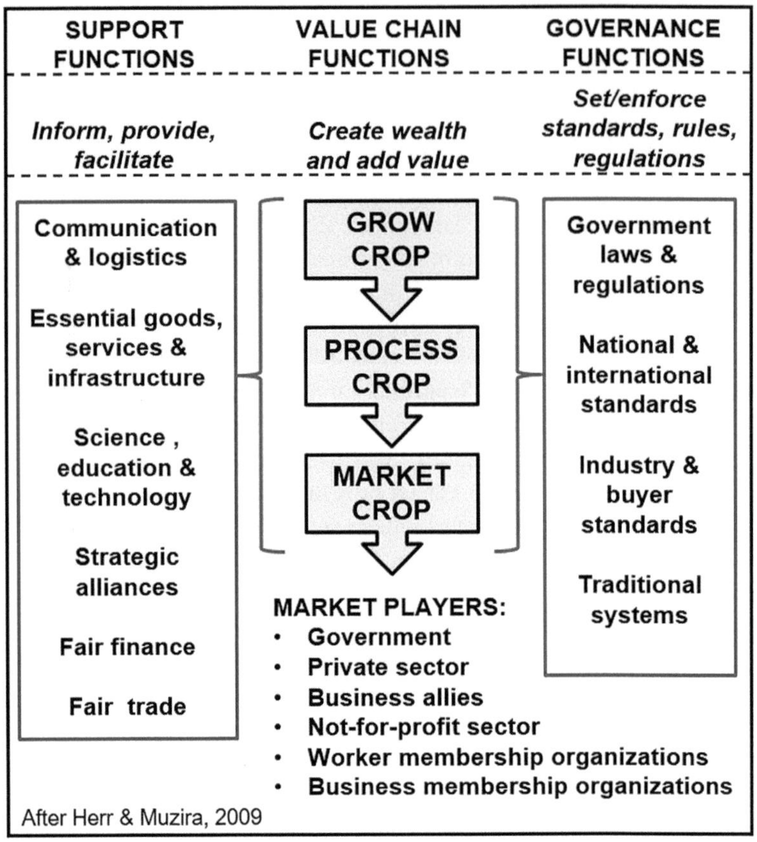

Fig. 12.3 Heuristic schematic that indicates how eucheumatoid seaweed value-chain functions fit within market systems (After Herr and Muzira 2009)

lation and publication of information on relevant BDS providers and input suppliers, their services and contact details. Unfortunately such services are commonly unavailable in many regions where eucheumatoid seaweeds are farmed (Neish 2013, 2015).

12.2 Mapping of Seaweed-to-Carrageenan Value-Chains

It is a fact that many eucheumatoid seaweed V-Cs have functional weaknesses and/or are weakly integrated into market systems (Neish 2013, 2015). This situation has attracted substantial attention from a variety of private sector and public sector organizations with the intent to effect interventions meant to correct V-C dysfunctions. A first step in the process is to undertake value-chain, diagnostic analyses. UNIDO (2011) provided a useful guide to V-C diagnostics which recommended seven diagnostic dimensions, as follows:

12.2.1 *Sourcing of Inputs and Supplies:* Emphasis is placed on understanding sources of goods and services that the seaweed industry players use in their production cycles. Their relationships with providers of primary materials (i.e. unprocessed, dried seaweed) and inputs must be examined and sourcing steps evaluated from the farm to the processors.

12.2.2 *ProductionCapacity and Technology*: The diagnosis must examine capabilities of seaweed industry players to manufacture and transform goods, including a means of production (e.g. machinery), human capital and the knowledge and technologies used in that production. Indicators of technical productivity, cost-efficiencies and profit margins are used in the analysis.

12.2.3 *End-Markets and Trade*: The examination of markets that ultimately absorb a value-chain's seaweed products and the seaweed quality demands of those markets is key. There should be emphasis on understanding existing capacity of V-Cs to meet demands and to access existing and potential markets.

12.2.4 *Value-Chain Governance*: Relationships amongst seaweed farmers and their buyers must be examined in light of the complex inter-dependencies between value-chain players. The focus on V-C governance should be on rules and regulations which determine functioning and coordination of seaweed value-chains, existing barriers to entry and the power of certain V-C players, such as carrageenan processors and exporters. See Chap. 1 of this book for a history of V-C governance evolution in seaweed-to-carrageenan value-chains.

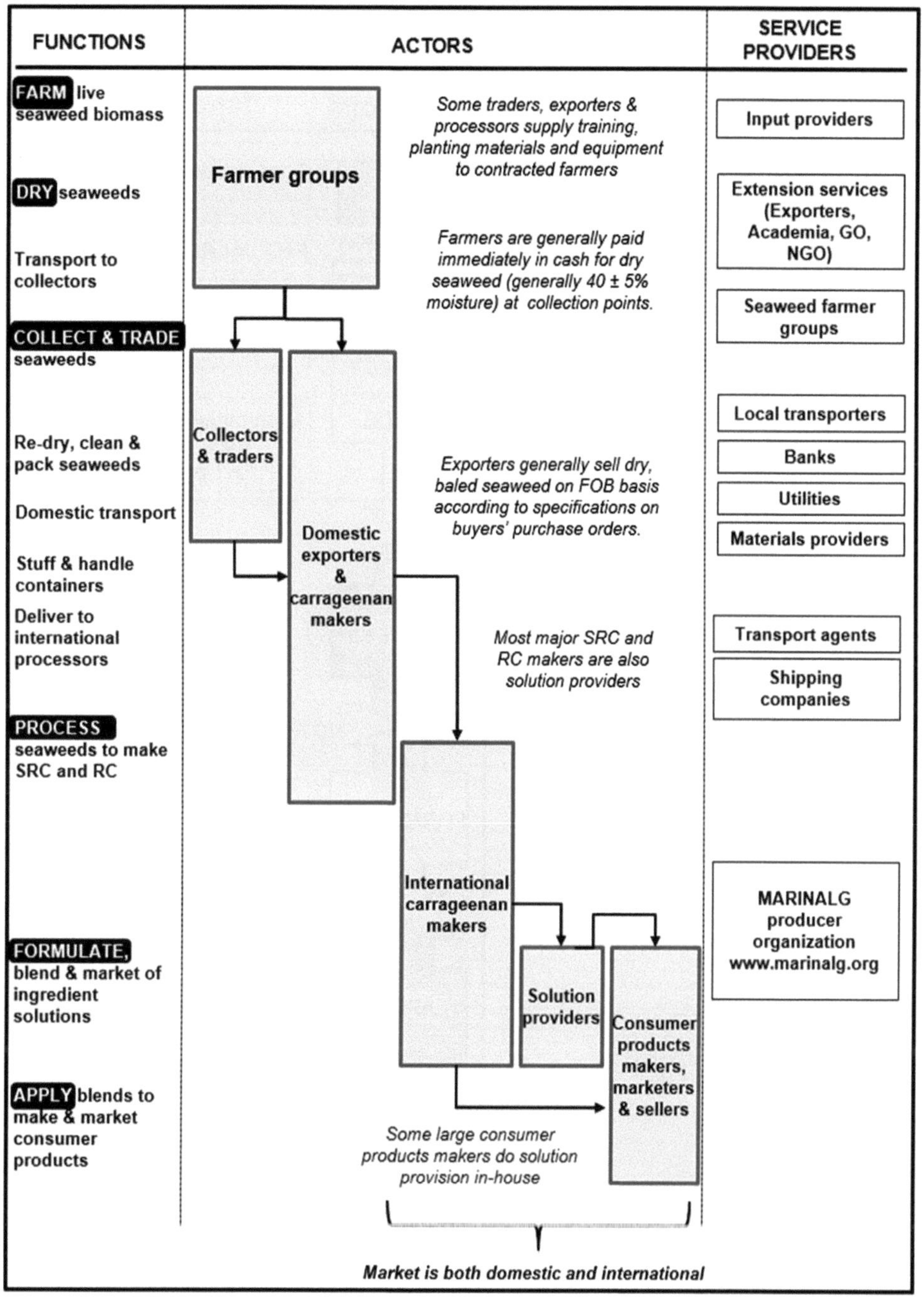

Fig. 12.4 A generic value-chain (V-C) map for eucheumatoid seaweeds (Format after UNIDO 2011)

12.2.5 *Sustainable Production and Energy Use*: The seaweed industry is regarded by many of its stakeholders as being sustainable and environment-friendly. Diagnostic analyses should examine how V-C actors comply with the standards of environmentally sustainableproduction.

12.2.6 *Finance of Value-Chains:* It is necessary to understand how various actors in the V-C finance their operations, the appropriateness and sufficiency of available finance mechanisms and how delivery can be made more efficient. Distinctions must be made between credits provided by formal financial institutions and informal financing through buyer-supplier relationships.

12.2.7 *Business EnvironmentandSocio-Political Context:* It is necessary to understand how policies and institutions have impacted V-Cs and what public institutions could do to support their development. The availability of public and private support services and the business culture of public and private actors should be examined (Fig. 12.4).

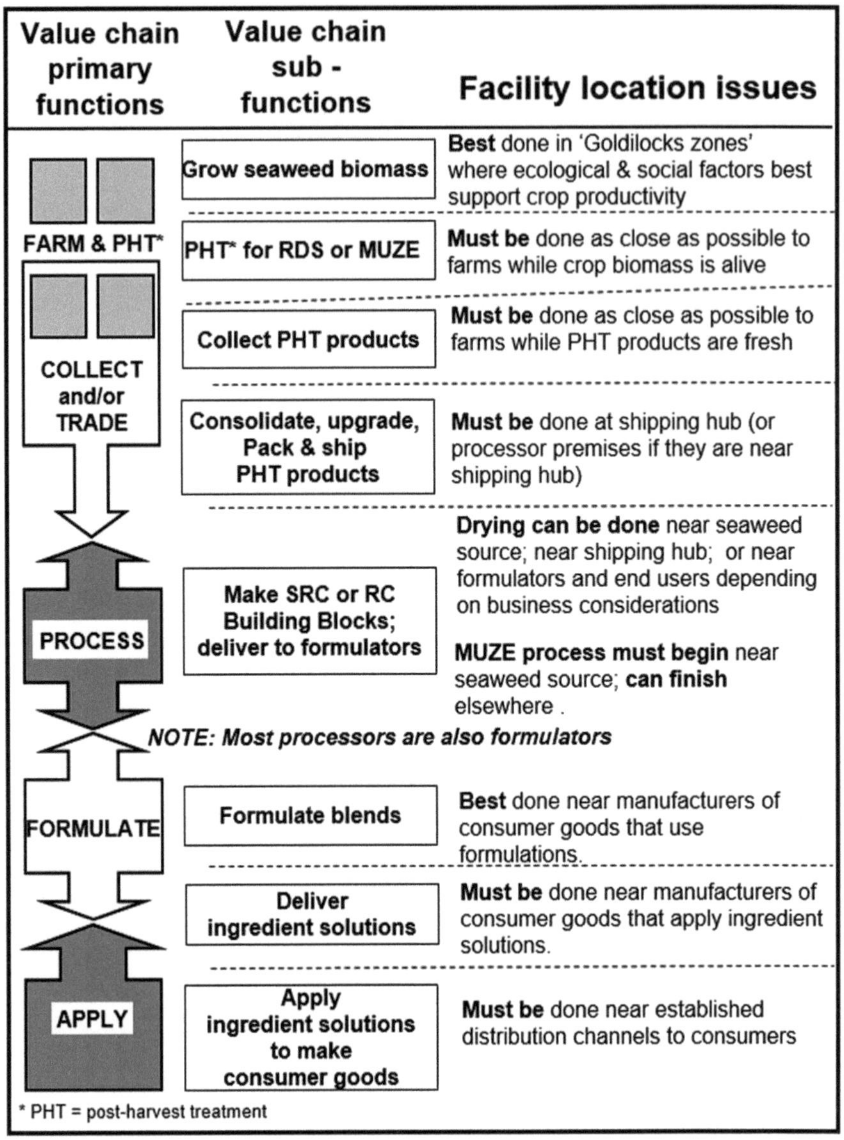

Fig. 12.5 Basic seaweed-to-carrageenan value-chain functions are shown in association with the products they make and facility location issues are indicated

A generic value-chain map for eucheumatoid seaweeds is shown in Fig. 12.1 and basic value-chain functions are shown in association with the products that are made from seaweed biomass in Fig. 12.2. The essential feature of V-C functions is that each step adds socio-political context value to products as they move geographically and economically from the original sources of raw materials to end-users.

It is an important feature of carrageenan that it is a product with a 'derived demand'. It is not sold directly to end users; rather, it is a 'building block', performance chemical that is blended with other ingredients as a component of 'ingredient solutions' that are applied in end products such as processed foods and personal care products.

Overlay notes on the value-chain map indicate how some actors can cover more than one function. For example, processors/exporters are commonly involved in functions extending all the way from picking up seaweed at farms to stuffing and handling export containers and many carrageenan makers are also solution providers. Arrows on the V-C map represent the flow of products from one actor to the next. The overlay notes include information on types of contractual arrangements. Arrows indicate market channels, with end-markets at the end of the map. Generic categories of support services are indicated on the right-hand side of the map including extension services, financial services, transport, materials, etc.

Issues around the location of facilities are indicated in Fig. 12.5 and location considerations are always a significant element of value-chain diagnostics. There are cost-effectiveness aspects to facility locations, but there are also political aspects that impact choices of facility location. There is often pressure to place as many value-adding steps as possible near to the seaweed source because it may be considered to be 'fair' for local economies to be value-adders and not simply purveyors of raw material. On the other hand, there are usually commercial pressures to have value-adding facilities placed in proximity to markets and within the same legal jurisdictions as those markets.

12.3 Overview of Seaweed-to-Carrageenan Value-Chains

Functions of the major actors in seaweed-to-carrageenan value-chains can be summarized as follows:

12.3.1 *Seaweed farmers*: Two broadly defined models for seaweed farm operation are common in major producing regions. Various combinations of the two models are employed in actual practice:

(a) *The 'lead farmer'model*: is one where one person or a small team of people own the enterprise; are actively involved in day-to-day operations; assume responsibility for managing the farm enterprise; and undertake marketing and selling of the crops produced. Much of farm labor is supplied by piece-work laborers who perform tasks including attaching cuttings to lines and drying the seaweed. Most of this labor tends to be supplied by farmer-group members, relatives and friends of farm-owning households.
(b) *The 'nuclear family'model*: is where spouses and their immediate relatives tend to share work and also share the derived income (in cash or in-kind) among themselves, their children, their parents and/or other first-degree blood relations.

12.3.2 *Seaweed collectorsandintermediate traders*: Most seaweed farmers sell directly to collectors who, in turn, sell to domestic traders or processors. Traders then sell to national or international carrageenan processors. Seaweed 'quality' is generally defined in terms of clean, dry weight and color; not based on rheological properties of the carrageenan contained in the seaweeds. Buyers select for rheological properties through selection of supplies amongst known sources. Collectors commonly finance seaweed purchases using funds provided by their buyers. Several collectors may operate in close proximity to a given seaweed farming area.

12.3.3 *National seaweed traders/exporters*: Local collectors generally sell seaweeds with little or no processing to national traders or processors. Some national processors are also exporters of raw seaweeds. Traders sell to national processors or export to international processors. Collectors often use the funds from national traders or processors to finance seaweed purchases. It is significant to note that traders and exporters/processors are generally not just 'middlemen' who simply buy and sell seaweed. Trader/exporters generally bear several risks and cost burdens. For example, exporters normally buy seaweed in local currency and sell in US dollars. Many buyers withhold as much as 20% of the purchase price pending testing at their factories. Final payment is then adjusted according to moisture content (MC) on a *pro-rata* basis to take a penalty or award a bonus. Payment of the balance can take several months. Trader/exporters also absorb processing and transportation costs and they absorb shrinkage losses (moisture and solids lost during transit and processing) that are often in the range of 15–20%. In summary the key value-chain functions of traders/exporters include:

(a) Provision of capital to farmers and collectors in the form of cash advances or rapid payments.
(b) Logistic arrangements for transporting seaweeds from their sources to processors.
(c) Technical and information exchange with farmers and collectors.
(d) Simple processing needed to make farmer-delivered seaweed meet buyer specifications including sorting and cleaning; drying to a uniform, low moisture content (e.g. about 35–38%); and pressing seaweed into compact, wrapped bales for export shipment.
(e) Absorption of foreign exchange risk or rewards and provision of working capital to cover holding of inventories and work-in-progress.

12.3.4 *Seaweed processing companies around the world*, in the dynamic seaweed industry, are irregularly opening, closing, merging and otherwise changing their identities. The list in Table 12.1, whilst not totally comprehensive, includes most major processors known to the authors to be in business, at the time of writing. Plant capacities were company confidential so the numbers shown are speculative and mainly based on companies' promotional literature or estimates by interviewed seaweed suppliers. The industry seemed to have a large, under-utilized processing capacity. The total estimated plant capacity of almost 120,000 MT year^{-1} was approximately double the known total carrageenan market of approximately 50–60,000 MT per annum (Porse and Ladenburg 2015; Porse and Rudolph 2017). The magnitude of over-capacity was inflated further by the fact that several alkali-treated chips (ATC) and semi-refined carrageenan (SRC) factories were known to be inactive at the time of writing (e.g. Neish 2015).

Table 12.1 Prominent carrageenan processors worldwide (2016) with estimates of capacity (MT)

Country	Company	Alcohol	Hybrid	Gel press	SRC FG	SRC TG	Total
Brazil	Algasbras				2,000		2,000
	Griffiths				750		750
Chile	Dupont		2,000		1,500		3,500
	Gelymar	1,000	2,000	1,000			4,000
China	Longrun Newstar			2,000	2,000		4,000
	Lubao bio (Greenfresh)			3,000			3,000
	Shanghai brilliant gum (BLG)	1,000		4,000	8,000		13,000
	Sheli foods			2,000	500		2,500
	Other China			2,500	2,500	1,000	6000
Denmark	CP Kelco	4,000					4,000
France	Cargill	4,500					4,500
India	Aquagri processing Pvt. Ltd.				600	600	1,200
Indonesia	PT Algalindo Perdana				1,000		1,000
	PT Amarta Carragenan Indonesia				600	1,200	1,800
	PT Bantimurung Indah				600	600	1,200
	PT Cahaya Cemerlang				600	1,200	1,800
	PT Centram			1,000			1,000
	PT Galic Arthabahari				1,000	1,000	2,000
	PT Giwang Citra Laut				1,000		,1000
	PT Gumindo Perkasa Indistri				500	1,000	1,500
	PT hydrocolloid Indonesia				1,200		1,200
	PT Indonusa Algaemas prima				3,000		3,000
	CV. Karaginan Indonesia				1,000		1,000
	PT Sansiwita				500		500
	PT phoenix Mas				180		180
	PT Seamatec			1,000			1,000
	PT Wahyu Putra Bimasakti					600	600
Japan	Mitsubishi	1,250					1,250
	Nippon Carrageenan			500			500
Korea	MSC co. ltd.	1,500					1,500
Malaysia	Omnigel				600	600	1,200
	Tawau Carrageenan (Takara)					600	600
Philippines	CP Kelco			2,000	1,000		3,000
	FMC biopolymer				4,000		4,000
	Kerry				2,000		2,000
	Marcel				1,500	4,000	5,500
	Martsons food Corpn.				1,200		1,200
	MCPI/D atingbayan				600	600	1,200
	Philippine bioindustries			1,800			1,800
	Polysaccharide (Shemberg)					2,400	2,400
	W hydrocolloids co.			500	1,500		2,000
	Shemberg marketing			1,220	5,400		6,620
	Shemberg biotech	1,200					1,200

(continued)

Table 12.1 (continued)

Country	Company	Alcohol	Hybrid	Gel press	SRC FG	SRC TG	Total
	TBK				1,000	500	1,500
	CEAMSA				2,000		2,000
Spain	CEAMSA	1,750			1,000		2,750
USA	FMC	6,000					6,000
Other	Other producers worldwide			2,000	1,000	1,500	4,500
		22,200	**4,000**	**24,520**	**49,830**	**17,400**	**119,950**

Source: PT Jaringan Sumber Daya [Jasuda]

12.3.5 *Solution providers:* Dozens of solution providers around the world take delivery of processed carrageenan products and other gums; then blend them with other ingredients to reliably and predictably satisfy the requirements of hundreds of applications based on rheological (and other) properties. In the marketplace, carrageenan competes with many products and solutions so it must be cost-effective and acceptable to end-consumers before it is formulated into end-products. The development and sustained production of ingredients solutions for long-term commercial applications is generally a combination of art and science that is based on long years of experience. Initial product development can be the result of several years of joint R&D between a solution provider and an application implementer. Consequently, particular ingredient solutions, once established in the market place, may have considerable long-term endurance. Quality parameters that define ingredient solutions involve strict specifications and specialized performance tests (Lustre and Bixler 2006).

12.3.6 *End-users*: Some large, end-users undertake the solution provision function, in-house but most of the thousands of end-user companies around the world rely on solution providers in order to supply them with functional blends. Major carrageenan end-uses include meat and pet-food processing, confectionaries, dessert gels, stabilization of dairy products and texturizing of personal care products.

12.4 Investment and Intervention Impacts on Value-Chains

The value-chain map shown in Fig. 12.6 is based on the legacy carrageenan V-C depicted in Fig. 12.4. It indicates the roles of two sectors that impact on most eucheumatoid seaweed value-chains, namely:

12.4.1 *Investment-sector actors* who directly invest their funds and other assets as debt and/or equity in enterprises that undertake value-adding functions.

12.4.2 *Intervention sector actors* that are market system support and governance entities including governmental and non-governmental agencies (GO and NGO); United Nations Organizations (UN); and International Financial Institutions (IFIs). Interventions tend to fall into two general categories:

(a) *Selective interventions* that are generally short-term (one to five years), with initiatives that strive to impact targeted segments of value-chains in market systems. They may target perceived imbalances in system dynamics (e.g. in farmers' share of V-C revenues). They may focus on building market share of one region or segment versus another region or segment. They may also seek 'rents' such as taxes and levies on seaweed products passing through the value-chain. Agendas may have social and/or political dimensions, as well as having market dimensions.

(b) *Comprehensive interventions* that strive to benefit the entire V-C by catalyzing and supporting innovations in process, product and market development and/or other technologies impacting overall economics. Relations between investment and intervention actors can be harmonious, but sometimes they are not. Tension or conflicts can develop especially when selective interventions are undertaken in value-chains that are poorly understood and with a goal of altering perceived 'imbalances' among V-C players.

12.5 Slow Growth of Carrageenan Value-Chains

A history of eucheumatoid seaweed V-C governance, pertinent to farming activities, was presented in Chap. 1 of this book, entitled *"Reflections on the commercial development of eucheumatoid seaweed farming"*. That chapter also reviewed trends that led to carrageenan V-Cs having slow growth in the order of 2–3%, per annum, since the turn of the century (Bixler and Porse 2011; Porse and Ladenburg 2015; Porse and Rudolph 2017). Based on personal experience and

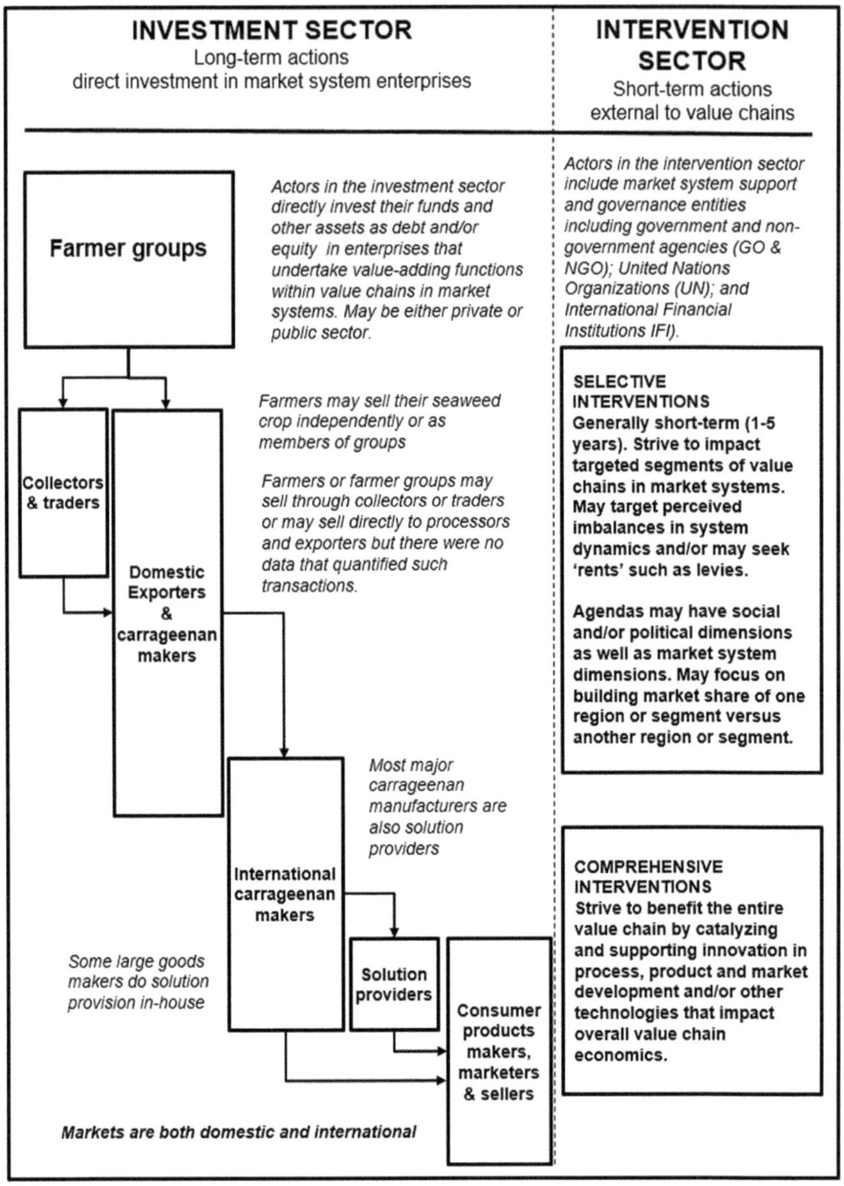

Fig. 12.6 The value-chain of Fig. 12.4 is presented here in the context of investment- and intervention-sector interests

on numerous discussions with other industry actors, the authors postulated the following reasons for sluggish growth in carrageenan markets:

12.5.1 *Research and development stagnated* in the carrageenan industry as the innovative SMEs that once dominated the carrageenan business were purchased by large multi-national owners during the late 1970s and into the 1980s. Those events coincided with proliferation of SRC producers, first in the Philippines and later in Indonesia, China, Chile and Malaysia. Since the advent of SRC technology, much process capacity was developed on the basis of technology obtained from former employees, consultants and equipment suppliers of the previously established manufacturers. This process was facilitated by multinational owners of formerly innovative carrageenan enterprises who cut back on R&D and farm development activities and also discharged many senior technical and management staff. In carrageenan value-chains there has been a paucity of innovation leading to new applications and markets for two to three decades. The last major new applications were developed about 30 years ago, in the form of iota carrageenan as used in dental products and kappa carrageenan in meat processing.

12.5.2 *Anti-carrageenan activistsand their impacts on market development.* Bixler (2017) traced the development of negative attitudes toward carrageenan to a research group at the University of Illinois that cited results of *in vitro* studies in support of claims that carrageenan up-regulated genes involved in inflammation of the intestinal epithelium. Despite a growing body of research that found no evidence of this phenomenon during *in vivo* studies; and in the face of decades of evidence that carrageenan was safe for human consumption, including in infant formula, anti-carrageenan sentiments went viral on the internet blogosphere. One result of such controversy was the recommendation by the United States National Organic Standards Board (NOSB) that carrageenan should be delisted for use in organically registered products in the USA (Cornucopia 2016). Anti-carrageenan campaigns created uncertainty that caused potential customers to be wary of carrageenan as an ingredient, regardless of how stoutly the hydrocolloids industry defended its products. At the time of writing, several major food processors had adopted policies of formulating carrageenan out of their products and/or avoiding the use of carrageenan in developing products. Bixler (2017) provided information on studies that supported carrageenan food safety and also proposed actions that could be taken to reduce impacts from the controversy. The NOSB recommendation had yet to be ratified by the FDA at time of writing (April, 2017)

12.5.3 *Softening of Chinese, Eastern European and other markets* due to a general economic recession, appeared to be a significant factor which dampened development of the carrageenan market. This trend was especially noticeable during 2014–2015. China became a major force in the business and was also a major driver for expansion of markets for eucheumatoid seaweeds cultivated in Indonesia. Softening of demand from China, therefore had a clear, dampening impact on that market.

12.5.4 *Industrial over-capacity for theproductionof undifferentiated products stifled profits* in the face of keen competition. Especially with respect to kappa-SRC and gel-press, kappa carrageenan, plant capacity seemed to be able to manufacture about two to three times what the market could absorb. Buyers were able to pick and choose amongst dozens of companies which sold similar, commoditized products. Producers could only develop markets or increase their market-share by taking it away from others, so the carrageenan business had virtually become a zero-sum game.

12.5.5 *Proliferation of entrants into carrageenan value-chains resulted in successive shifts* in V-C dynamics from captive governance in the mid-1980s, to modular governance by the mid-1990s. This was superceded by market governance by about 2001 (see Chap. 1 for more information). The disintegration of exclusive V-C relationships, caused 'side-selling' to become a major problem, both for the original MARINALG companies and also for subsequent entrants. Investments in process development and farm development virtually ceased by the mid-1990s because investments could no longer be protected and internalized by private investors. Investment in farm development also diminished because the larger hydrocolloid producers decided that seaweed farming comprised aquaculture and they wanted to focus on their core businesses rather than digressing into the aquaculture business.

12.5.6 *Carrageenan is a 'performance' chemical that is generally marketed as a component ofingredient solutions*. Product performance and production costs cannot be optimized unless seaweed raw material sources are transparently linked from farms through processors to solution providers and end-users. Market-governance value-chains do not result in such links.

12.5.7 *Farming is still dependent onrudimentary,labor-intensive technology* such as individually tying cuttings on to lines; often using contract labor systems that stifle productivity.

12.5.8 *Inefficientproductiontechnology predominates.* The obvious example is the fact that most manufacturers only extract the carrageenan portion of seaweed biomass using a single-stream processes. This wastes about 50% of seaweed dry matter and creates high-chloride, waste streams. Much of the value of the seaweed biomass that could be recovered and sold as products is wasted unless a multi-stream, zero-effluent (MUZE)-process is used.

12.6 Current Single-Stream and MUZE Value-Networks

The sum-total of eucheumatoid value-chains operating within market systems results in value-networks that provide an overview of the global situation. The value-network for single-stream carrageenan production as of 2017, is depicted in the heuristic of Fig. 12.7 and in Tables 12.2 and 12.3. An emerging scenario for MUZE-based value networks is depicted in Fig. 12.8.

12.6.1 *Single-stream carrageenan value- network per 2017*

Figure 12.7 depicts the authors' opinions on a heuristic model of the value-network based on single-stream production of carrageenan, from eucheumatoid seaweeds, as of 2017. Tables 12.6.1 and 12.6.2 explain classifications and categories relevant to the model. The defining characteristics of this model are dominance of seaweed supplies by Indonesia, the 'alpha-source' and dominance of the processing-sector by China, where domestic 'alpha'

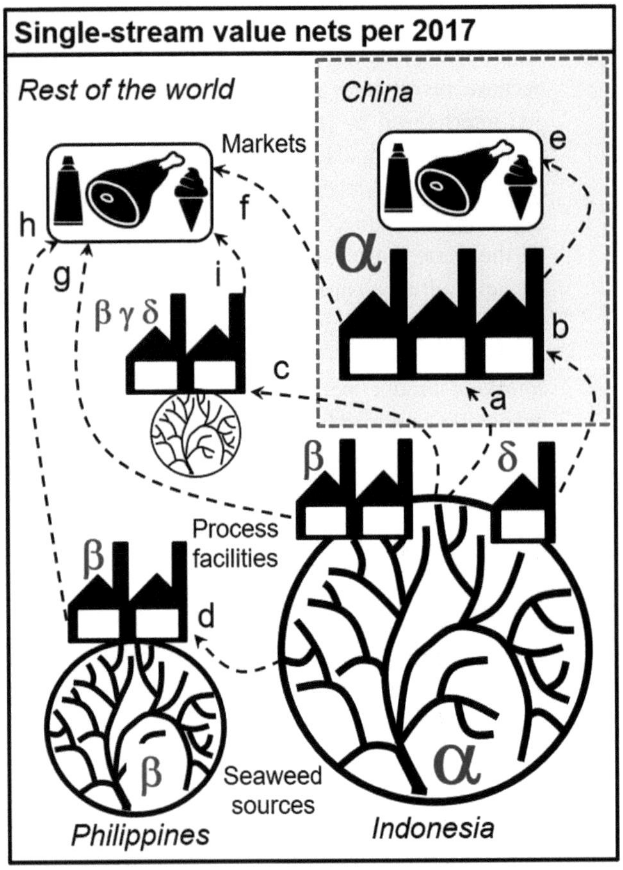

Fig. 12.7 A heuristic model of the state of the value-network that is based on single-stream production of carrageenan from eucheumatoid seaweeds, as of 2017. See text for explanations of arrow paths (Based on the authors' own observations and conclusions)

Table 12.2 Classification of seaweed sources, process facility development and markets for the value-network depicted in Figure 12.6.1

Symbol		Seaweed sources	Process facilities	Markets
α	Alpha	Indonesia (ID)	China (CN)	China (CN)
β	Beta	Philippines (PH)	Denmark (DK)	Canada + USA
			France (DK)	EU + UK
			Indonesia (ID)	
			Philippines (PH)	
			Spain (ES)	
			USA (US)	
γ	Gamma	Brazil (BR)	Brazil (BR)	Russia
		India (IN)	Chile (CL)	(RU)/FSR
		Malaysia (MY)	India (IN)	Latin America
		Tanzania (TZ)	Japan (JP)	East Asia
			Korea (KR)	SEAsia
δ	Delta	At least 30 countries around tropical & sub-tropical regions	Malaysia (MY)	West Asia
				N. Africa
				Sub-Saharan
				Africa

Table 12.3 Descriptions of the classifications depicted in Table 12.2

	Seaweed sources
α	Dominant source that can conceivably supply entire global demand
β	Significant source that can supply a substantial portion of global demand
γ	Currently active but minor source
δ	Formerly active and could revive as a minor source
	Process facility locations
α	Dominant in process capacity for finished building blocks
β	Secondary in process capacity for finished building blocks
γ	Formerly active or developing capacity for finished building blocks
δ	Has capacity to produce intermediate building blocks
	Market locations
α	Biggest market in the world
β	Major market
γ	Minor market
δ	Minimal market

carrageenan manufacturers have virtual, exclusive access to the domestic Chinese 'alpha market'. The authors postulate that for carrageenan value-chains Indonesia is fully capable of providing all the eucheumatoid seaweeds that the global market can bear and that China is on the verge of being capable of manufacturing all the carrageenan-building blocks that the world could utilize. Other sources are relegated to 'beta' or 'gamma' or 'delta' status (Table 12.3).

There is a virtual carrageenan 'Great Wall of China' that lets seaweed raw materials in from tropical-source countries, but which effectively bars entry for carrageenan made outside of China. The major carrageenan producer in the world, as of 2017, was Shanghai Brilliant Gum Co. Ltd. (BLG) with Chinese factories in Shanghai and Zhejiang and a large factory in Indonesia which was about to commence production of partially-processed carrageenan feed-stocks. On their website, index page (http://www.sh-blg.com/en/) *circa* 5th April, 2015, BLG identified themselves as the world's biggest carrageenan manufacturer, with a global, market-share of nearly 30% and consumption of 'Eucheuma Seaweed' volumes equal to more than 40% of Indonesian production (Neish 2015). Such claims did not appear in the BLG website at the time of writing, but as of 2017, interviewed Indonesian, value-network actors estimated that about 4000–6000 MT of dried eucheumatoid seaweeds, per month, were being exported from Indonesia to BLG, which had penetrated world markets with carrageenans at prices which competing processors claimed were at or below their production cost.

The strong performance by BLG was likely to be emulated by other Chinese, carrageenan processors including Longrun Newstar, Lubao Bio (Greenfresh) and Sheili. It is a sign of 21st century trends that FMC Health and Nutrition (sold to DuPont at the time of writing) closed its European kappa carrageenan factories (previously the largest in the world) by 2004 and commenced buying carrageenan building-blocks from China. By 2017, China appeared to be on a path to dominate global carrageenan production (Porse and Rudolph 2017).

Figure 12.6.1 includes arrows highlighting various value-network paths which can be described as follows:

(a) As the 'alpha' eucheumatoid seaweed source to the world Indonesia, was reputed by industry sources to export more than half of its annual production to China as raw, dried seaweeds (RDS).
(b) A developing trend was for Chinese processors to make intermediate building blocks, e.g. ATS and ATC in Indonesia, then undertake further processing in China. This can be a cost-effective strategy and it also guards against any future bans on raw seaweed exports.
(c) Indonesia exported RDS to carrageenan manufacturers all over the world.
(d) Philippines-based processors imported a substantial proportion of their eucheumatoid seaweed needs from Indonesia, since the commencement of commercial farming there. Meanwhile, essentially all of the Philippines' production was made into carrageenan in Philippines-based facilities, so little was exported as RDS.
(e) Much of the Chinese-made carrageenan was sold to the domestic 'alpha' markets, effectively protecting them from sales by processors outside of China (see above).
(f) Chinese processors also sold carrageenan in global markets at highly competitive prices (see above).
(g) Indonesian carrageenan processors sold into both domestic and global markets, but domestic markets were not protected.
(h) Likewise, Philippines' carrageenan processors sold into both domestic and global markets, but domestic markets were not protected.
(i) 'Beta', 'gamma' and 'delta' processors around the world purchased RDS from various seaweed sources but many were dependent on Indonesia, the 'alpha' source, for most of their RDS needs. They then sold carrageenan on the global market.

12.6.2 *Developing MUZE-value networks*

Figure 12.8 depicts the authors' opinions on a heuristic model of a developing value-network which is based on multi-stream, zero-effluent (MUZE)production of a wide range of products from eucheumatoid seaweeds.

A defining characteristic of MUZE value networks is that processing starts with live seaweed, so value-addition necessarily commences in process facilities near to farms. Biomass is effectively 100% utilized to make marketable products and markets are globally dispersed, including in local and regional areas near to seaweed farming and processing locations.

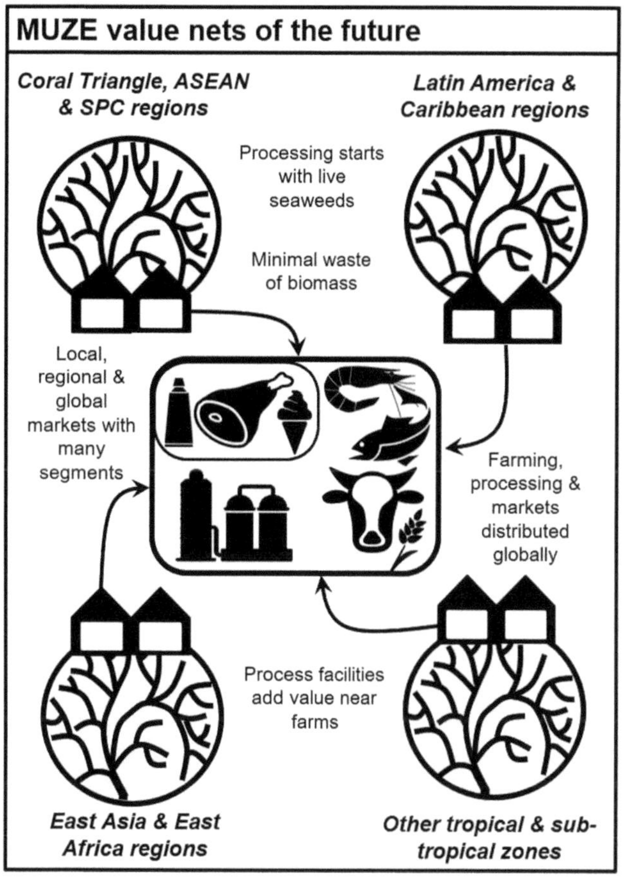

Fig. 12.8 A heuristic model of a developing value-network that is based on multi-stream, zero-effluent (MUZE)production of a wide range of products from eucheumatoid seaweeds (Based on the authors' own observations and conclusions)

With MUZE-value networks in place there should be a basis for the further development and growth of eucheumatoid seaweed farming enterprises globally, in a way that is not possible with carrageenan networks dominated by alpha players. Features of MUZE-value networks are elucidated in following sections of this chapter.

12.7 Mapping and Overview of Emerging MUZE Value-Chains

The V-C as mapped in Figs. 12.4 and 12.5 and reviewed above, described how virtually all eucheumatoid seaweeds were sold through seaweed-to-carrageenan V-Cs since the industry developed farming in the early 1970s. During the ensuing decades, changes in V-C management and a general decline in industry innovation, combined with processing over-capacity, created a situation where eucheumatoid seaweed production reached a plateau that was correlated with low-to-slow growth in the carrageenan markets as a whole. In effect, as described in Sect. 12.5 above, carrageenan markets and markets for the seaweeds that supplied them had become zero-sum games.

Eucheumatoid seaweed farming has become a source of livelihood for tens of thousands of coastal people in tropical to sub-tropical regions (Valderrama et al. 2013), so both private and public V-C players have hopes for major expansions in seaweed production. Clearly, such growth cannot be supported as long as the industry is dependent primarily on sales as simply raw material for carrageenan extraction. Fortunately, there are emerging alternative value-chains that provide hope for future development. These involve development of mechanized cultivation and multi-stream, zero-effluent (MUZE) processing technologies (Fig. 12.1). An example of a MUZE V-C is mapped in Fig. 12.9.

The V-C, as mapped in Fig. 12.9 depicts a business model that is being applied by enterprises such as those with which the authors are engaged. Substantial interest in MUZE-processing of eucheumatoid seaweeds has been developing even since the early 1970s, as companies such as Marine Colloids Inc. (where the senior author was Process Development Manager for a time) sought product diversification to follow the successful commencement of commercial seaweed farming. An outstanding example of such interest was research by the Indian Government which led to the granting of USA Patent number 6,893,479 B2 to Eswaran et al. (2005).

As of 2016, eucheumatoid seaweed value-chains that employed MUZE technology were at early stages of development and commercial production was underway in facilities of at least two companies in India (known to the authors), namely: Aquagri Processing Pvt. Ltd. and Sea6 Energy Pte. Ltd. Both had developed production facilities in Tamilnadu, southern India and both were engaged in efforts to expand their production in other eucheumatoid seaweed-source countries, including Indonesia. On the other side of the world, in Brazil, similar value-chains were being independently developed by Algasbras Biorrefinaria Ltda. (see Chaps. 1 and 4 in this book).

At least four key elements differentiate MUZE value-chains from legacy, carrageenan industry value-chains:

1. Processing commences with live seaweeds near to the farm sites so value-addition moves towards farmers and the seaweed must be delivered reliably, on a daily basis.
2. Essentially 100% of biomass is extracted as products for sale.
3. A wide range of products can be made (Fig. 12.10) and most of these products are sold into growing markets; not into zero-sum markets such as carrageenan.
4. Management of relational governance applied within strategic business alliances is a core competency of enterprises engaged in V-Cs of this type. The market governance, typical of legacy carrageenan value-chains, is not compatible with MUZE V-Cs (see Chap. 1).

Prominent features of the single-stream and MUZE-process options that are relevant to V-C differences can be compared and contrasted as follows (Suryanarayan et al.2016a, b):

12.7.1 *Both single stream and MUZE-processing commence with live seaweed*, but in the case of single-stream processing, seaweeds are generally sun-dried and packed before shipment to distant factories for further processing. In the case of MUZE-processing, live seaweeds enter the first stages of processing near to the farming sites, so the value is added close to those farming communities and multiplier effects can accrue there.

12.7.2 *The first step of MUZE-processing is generally to extract seaweed 'juice' and separate it from seaweed 'pulp'* using equipment typically found in the fruit and vegetable, juice industries. Mild processing conditions are used in order to conserve bioactive compounds. Typically, juice is concentrated under reduced pressure to minimize the need to transport low-solids liquid and to effect preservation of bioactive components. Pulp can be liquefied using a variety of methods and then enter liquid process streams or it can be dried for use as process feedstock. Juice and pulp can then be subjected to a wide range of further process options.

12.7.3 *Duringsingle-stream processingraw, dried seaweeds are generally subjected to cooking in alkali* prior to various combinations of recovery and dehydration. Refined hydrocolloids are dissolved, clarified and recovered through precipitation in alcohol or potassium prior to drying. Semi-refined hydrocolloids are retained in a gelled state during processing and are dried following a washing step. During the harsh conditions typical of single-stream processing, many bioactive compounds are destroyed and/or lost.

12.7.4 *The hydrocolloids produced bysingle-stream processingare milled to powder form* and then blended into ingredient solution products. A great deal of fresh water may be used during processing, especially in the case of semi-refined carrageenan (SRC) which is the highest-volume carrageenan product on the market. Several MT of high-chloride, high BOD

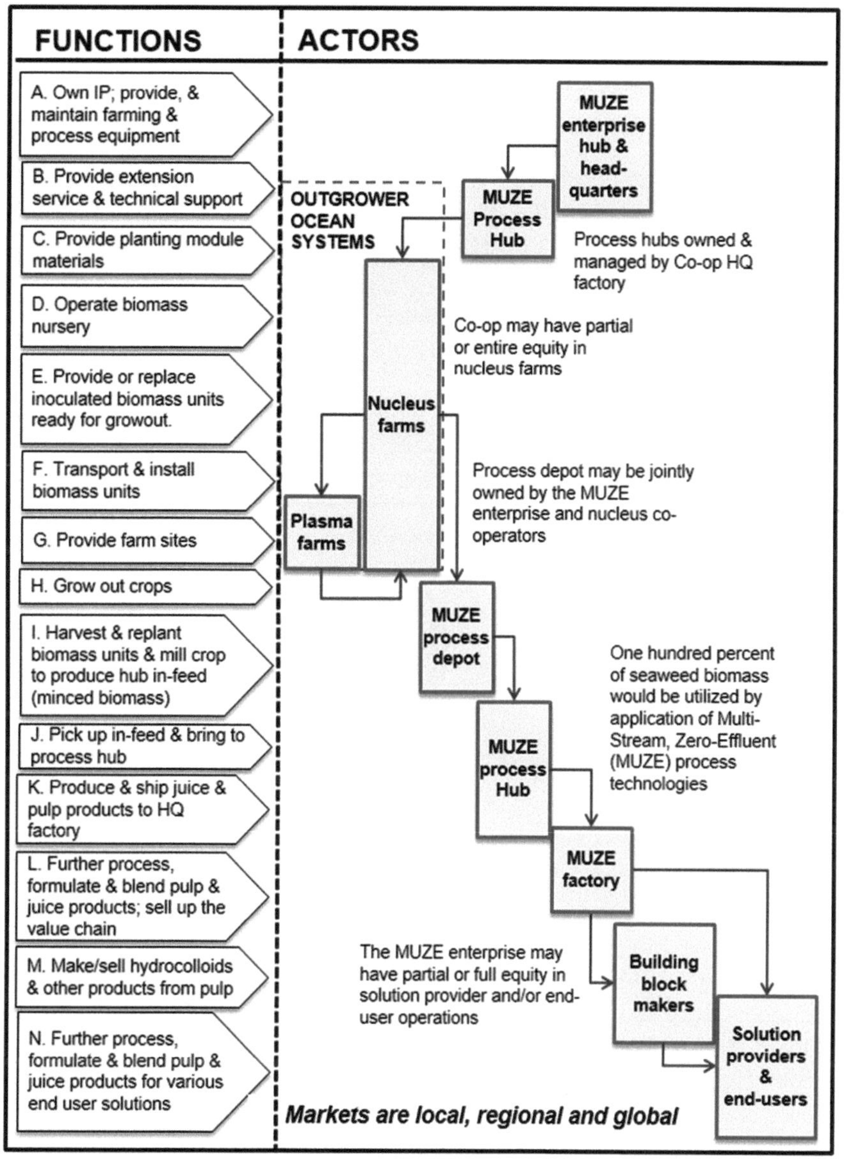

Fig. 12.9 A value-chain map depicting a vertically-integrated, MUZE-system operated by an enterprise

and COD, alkaline wastewater can result from the production of every MT of SRC. In addition, waste solids resulting from clarification of wastewater and clarification of the refined carrageenan, result in waste filter-cake which can be produced in substantial quantities. An example is the ex-Marine Colloids/FMC factory in Rockland, Maine, USA (owned by DuPont at the time of writing) where large, abandoned quarry systems were filled with filter-cake over several decades of operation.

12.7.5 *The only significant waste products from a well-designed, MUZE-processing system should be water vapor* removed during liquid concentration and drying and combustion wastes from fuel consumption. Fresh water produced during juice concentration can be also recycled to processing or can be sold as a product in its own right.

The authors postulate that MUZE V-Cs for eucheumatoid seaweeds are being made possible, largely due to the fact that their farming is a proven technology, providing numerous options for improved agronomic practices (Vadassery et al.2016 and Chap. 1 of this book). Furthermore, the fact that eucheumatoid seaweeds can be juiced using readily available, food industry, standard equipment facilitates low-cost processing.

12.8 Developing Innovations in Agronomy

At the time of writing, market prices and volumes indicated that present farm systems were more than capable of supplying the needs of current markets when prices are attractive to farmers (see Chap. 1 in this book). However, major developing markets that can be supplied by MUZE-production sys-

Fig. 12.10 An exemplary array of the types of products that can be made from eucheumatoid seaweeds through various MUZE-processes

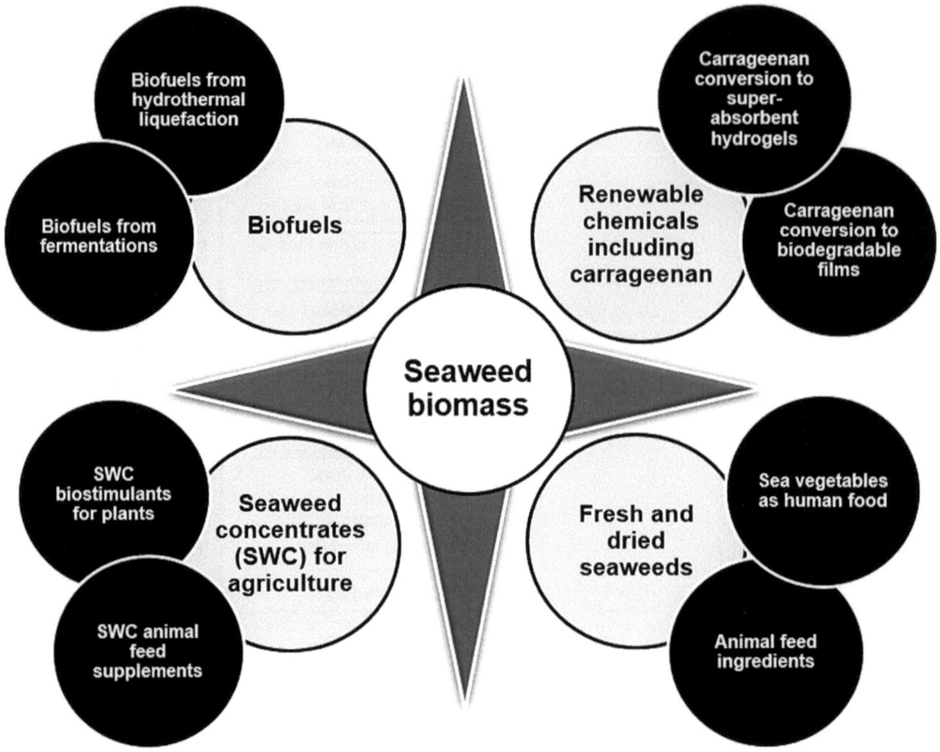

tems, provide a many-fold increase in production volumes that can be achieved without creating inflated unit-costs.

The vast majority of farm systems, current at the time of writing, were small-holder operations; many of which had low productivity and almost all of which involved a large factor of 'drudge-labor' comprising most of the seaweed production cost (Vadassery et al.2016). An outstanding example of such drudge-labor is the individual, manual attachment of seaweed cuttings to ropes. Production of one ton of raw, dried seaweed can involve tens of thousands of such attachments and an increasing proportion of potential seaweed farmers decline to participate in the industry because they are averse to such drudgery. Methods of seaweed farming are therefore being developed that enable simple mechanization of tasks that previously involved drudge-labor and hence enable farmers to increase farm productivity per unit-of-effort, several fold (Vadassery et al. 2016 and Chaps. 1 and 4 of this book). Key features of such systems are:

1. Biomass is inoculated into tubular nets, rather than being fastened manually on to ropes.
2. Planting, crop-tending, harvesting and crop handling are mechanized using simple machinery that can be operated either on shore, or at sea; thus eliminating most labor-intensive farm chores.
3. Losses of biomass caused by frond breakage is virtually eliminated and impacts by grazers are reduced.
4. Farming is done within contract farming systems known as 'outgrower' or 'nucleus-plasma' systems, as depicted in Figure 12.6.1 and the systems are managed so that there is security in the flow of fresh seaweed biomass to process facilities on an even, reliable and daily basis.
5. Farming is based on ecosystem approaches to aquaculture (FAO 2010).
6. Systems for marine agronomy are designed to operate in deeper and more turbulent waters than current systems can withstand; thus expanding the areas of ocean that can support seaweed cultivation.

Tube-net systems, of the type described, were pioneered in Baja California, Mexico in the 1990s when Zertuche-Gonzalez et al. (1999, 2001) used them to farm *Chondrus crispus*. Tube-nets were also used successfully to farm *Kappaphycus* in Brazil (de Góes and Perpetuo Reis 2011 and Chaps. 1 and 4 in this book). Tube-net farming systems were tested by several innovative farmers around the world, but published accounts of such work were unavailable to the authors at the time of writing. The only commercial development of such systems, known to the authors, was farm development by Sea6 Energy Pte. Ltd. in India and Indonesia (Vadassery et al.2016) and developments in Brazil (Chaps. 1 and 4 in this book). These developing agronomic operations were being developed to link directly into MUZE-processing systems.

12.9 Value-Chain Growth Driven by MUZE-Products

12.9.1 *Seaweeds as food for people*

Kappaphycus and *Eucheuma* have been traditional dietary items for people living in areas where these seaweeds are indigenous, however, they have not served as major contributors to macro-nutrition. There are still modest, localized markets for these and the eucheumatoid seaweeds are also dried for export to China, where they are consumed as sea vegetables. Baricuatro (2015) estimated that, as of 2014–2015, as many as 10,000 MT of dried *Eucheuma* were exported from Indonesia to China, for use as human food. Much of this was sun-bleached to make it white in color.

12.9.2 *Functional fiberbeyond carrageenan*

Section 12.5 addressed issues pertaining to slow growth of carrageenan value-chains and an important factor for slow growth was the issue addressed in Sect. 12.5.2, which was anti-carrageenan activists and their impacts on market development. Recent publications by Bixler (2017) and Tobacman (2017) presented opposite sides of anti-carrageenan arguments and regardless of the merits of either side it is arguably the case that growth of carrageenan markets for human food applications has been damaged in the wake of "The Carrageenan Controversy". Any role of carrageenan in future growth of eucheumatoid seaweed markets must depend, at least in part, on the development of carrageenan markets that are not for human food. Insofar as human food ingredient opportunities exist, MUZE processing may open them up partly by enabling production of functional fibers from eucheumatoid seaweed biomass that do not share negative attributes ascribed to carrageenan.

12.9.3 *Seaweed concentrates (SWC)for agriculture*

Liquid and **s**olid seaweed **c**oncentrates (SWC) are produced for agricultural applications by several companies in countries around the world (Craigie 2011; Suryanarayan et al. 2016a). Agricultural SWC markets have major potential for utilization of eucheumatoid seaweed extracts, but as of 2016 most SWC was manufactured from cold-water, temperate zone, brown seaweeds (Phaeophyta) such as the kelp genera (e.g. *Laminaria*, *Saccharina*, *Ecklonia* and *Durvillaea*) and rockweed genera (e.g. *Ascophyllum* and *Fucus*). These have been used as soil treatments and animal feed additives for many decades; if not for centuries (Craigie 2011; Suryanarayan et al. 2016a).

During most of the history of SWC production it was thought that their efficacious benefits for plants and animals were quite different; until 1970, only about 500 scientific papers related to the agricultural uses of seaweeds had been published. Since 1970 however there has been an almost exponential growth in scientific attention paid to SWC so by 2010, almost 2500 papers had been published (Craigie 2011). By the twenty-first century, knowledge gained from SWC research had been published in generally accessible scientific and technical literature.

Thorough treatment of this topic is beyond the scope of the present V-C discussion, but the reader can find especially useful reviews that have been published by Stirk and van Staden (2006), Khan et al. (2009), Craigie (2011) and Evans and Critchley (2013). Latest developments have been reviewed by Prithiviraj and Critchley (2016) and Evans et al. (2016). Recently, a: 'review of reviews' with special reference to cultivated tropical red seaweeds was included in a book chapter by Suryanarayan et al. (2016a).

By 2011, research had been focusing on SWC modes of action, specific health benefits and importantly, the underlying mechanisms (or modes) of action common to both plants and animals. Mechanisms of SWC action were being unraveled through the use of techniques developed for molecular biology, metabolomics, and genomics with bioassays used as a crucial means for assaying impacts on plants and animals. Observed effects seemed to result from the combined action of several SWC constituents including, not only small organic or inorganic molecules, but also larger organic molecules such as the biopolymers contained in SWC. It seemed that large organic compounds could influence the expression of genes to bring about systemic resistance and to modulate immune systems in plants and animals. Craigie summarized the state of knowledge, as of 2011, along the following lines:

1. Diverse physiological and biochemical processes are elicited in both plants and animals when SWC is administered to them.
2. Observed effects from both foliar- and soil-applications to plants could be either direct or indirect.
3. SWC application elicited changes in the rhizosphere community for plants and for the gut microbiome of animals grazing on fodder containing inclusions of seaweed products.
4. SWC elicited improvements in gastro-intestinal health, and immune system responses in animals as well as enhancing their ability to withstand abiotic stress such as heat.
5. Various bacterial, fungal, insect and parasite depredations could be attenuated in both plants and animals.

Evans and Critchley (2013) and Prithiviraj and Critchley (2016) reviewed research that focused on whole seaweed applications and SWC from *Ascophyllumnodosum* (perhaps one of the most studied species in the field) and updated elucidation of SWC bioactive mechanisms. By 2016, similar biological pathways appeared to be impacted across multiple groups of organisms such that research could be directed toward unifying models of extract bio-activity. Compounds isolated from *A. nodosum* commonly elicited activity against abiotic stresses in both plants and animals (e.g. temperature and osmotic stress) and also commonly induced up-regulation of genes associated with immune responses. *A. nodosum* extracts also exerted differential effects on microbes by favoring beneficial species that supported health, while regularly acting to the detriment of pathogenic microbes. Such effects have been characterized as 'synbiotic' or 'prebiotic' phenomena and they influenced gene expression and the tendencies for communications amongst pathogenic microbes.

As SWC products have entered mainstream agricultural and horticultural applications, it has become increasingly important that the bio-activity of branded products be uniform and predictable. Leading products must therefore be manufactured using processes that consistently yield SWC that can be blended into batches which are consistent in terms of their efficacy for both plants and animals. Bioassays are of crucial importance because they remain the most effective techniques for the rapid screening and quantification of SWC bioactivity. Important bioassay techniques can be built around 'model' organisms with well-understood genomes. Prominent examples include: *Arabidopsis thaliana* for plant systems (Rayorath et al.2008) and *Caenorhabditis elegans* for animal systems (Kandasamy et al.2011, 2012).

As of 2016, the state of knowledge of SWC was such that beneficial impacts on crops, livestock and other animals had been well documented. Even by 2011, Craigie could say: "The evidence at hand establishes that benefits derived from the use of seaweed products can be indisputable in agriculture." Craigie also stated that: "There are exciting discoveries yet to be made and we can be assured that some will be unanticipated."

In as much as this statement originally applied to SWC from the brown seaweeds it also applies, perhaps even more, to SWC products to be manufactured from cultivated, eucheumatoid seaweeds, which have received far less attention from researchers than temperate, brown seaweeds. *Kappaphycus* and *Eucheuma*, are amenable to juice extraction that employs readily available equipment such that extraction costs are minimized and SWC can be supplied at cost-effective prices which are affordable, even to poor farmers in developing countries. Furthermore, eucheumatoid seaweeds of commerce, are cultivated and, as more efficient agronomy systems are introduced, biomass can be available to processors in massive quantities, far beyond what was available, as of 2017. Within the past few years, pioneering efforts in India have already resulted in commercial production, sales and distribution of SWC-products from eucheumatoid seaweeds that were affordable by all segments of the terrestrial agriculture industry, including small-holders. From its origins in India, technology for the production of SWC from several species of cultivated, tropical and subtropical red seaweeds was spreading to other areas, including: Indonesia, Brazil and East Africa.

12.9.4 *Renewable chemicals from MUZE-processing*

Carrageenan can be manufactured in MUZE-processing systems and options for making it using mild-processing conditions are being explored by several companies. Besides being used as a food ingredient in legacy applications, carrageenan can be utilized as a substrate for making a substantial array of products through the applications of numerous biotechnology processes. One prospect for renewable chemical products showing promise for development is the conversion of carrageenans to biodegradable films. Since the early years of carrageenan utilization, it was known that edible and biodegradable, carrageenan-based films could be formed. Subsequent research has demonstrated that films can be formed with carrageenan, as the sole building-block (Briones et al. 2004; Karbowiak et al. 2006) and with carrageenan in combination with a variety of other organic polymers including starch (Abdou and Sorour 2014; Lafargue et al. 2007), pectin (Alves et al. 2006), konjac (Briones et al. 2004) and chitosan (Park et al. 2001). In film applications carrageenan tends to add mechanical strength and water vapor barrier properties. The potential applications are huge, particularly as societies move to replace non-degradable plastic containers and wrappings which do so much damage in the marine environment.

Another prospect for the further development of added-value applications for carrageenans include their conversion to super-absorbent hydrogels as cross-linked, hydrophilic polymers capable of absorbing large quantities of water, saline, or body fluids, including blood. They are widely used in applications including feminine hygiene products, disposable diapers, cleansing pads, cosmetics and culture media in agriculture. Pourjavadi et al. (2004, 2005) described the means by which hydrogels could be synthesized to yield anti-salt, super-absorbents with high-swelling capacities in salt solutions.

12.9.5 *Biofuelsfrom eucheumatoid seaweeds*

Carbohydrates present in eucheumatoid seaweeds can be converted into biofuels by a variety of thermo-chemical and bio-chemical methods, but as of 2017, the authors knew of no commercial production of biofuels from them; probably because there was insufficient biomassproduction to cost-effectively meet the considerable volume required by the biofuels market.

Anyone who works with raw extracts of eucheumatoid seaweeds quickly discovers that they commence the processes of autolysis and fermentation, as soon they are harvested from the water. These processes can result in the production of combustible gases (e.g. methane) and combustible alcohols (e.g. ethanol). The potential for seaweed ethanol production has attracted substantial attention and reasons for this focus were summarized by Khambhaty et al. (2012a), who pointed out that the market for bio-ethanol grew from less than a billion liters in 1975, to more than 39 billion liters in 2006 and about 100 billion liters by 2015. During that time frame, nearly all bio-ethanol was produced by the fermentation of sugars from food crops, including beets, corn and sugar cane. Khambhaty et al. suggested that limitations in the availability of agricultural land provided a potent argument for considering marine macroalgae as sources of biomass for bio-ethanol production (Khambhaty et al. 2012a, b). These authors described the production of bio-ethanol from freshly harvested, cultivated *Kappaphycusalvarezii* and also undertook a study of marine yeasts required for bio-degradation and ethanol production from *K. alvarezii*biomass.

Tan et al. (2013) demonstrated that heterogeneous-catalyzed hydrolysis could be combined with *Saccharomyces cerevisiae*fermentation to produce bio-ethanol from *Kappaphycus;* several Indonesian researchers published the results of ethanol production work that utilized *Kappaphycus* spp. as the feed-stock (Candra et al. 2011; Fakhrudin et al. 2014; Sekar et al. 2015; Wagiman et al. 2015).

Hydrothermal liquefaction is another means of making biofuels from eucheumatoid seaweed biomass. It has been demonstrated that light, sweet, crude oil (a.k.a. 'green crude' or 'bio-crude') can be produced from several species of macroalgae using hydrothermal liquefaction and much of this developing technology will probably be relevant to further eucheumatoid seaweed developments. For example, Anastasakis and Ross (2011) produced bio-crude from the brown macroalga*Laminariasaccharina* using hydrothermal liquefaction in a batch reactor and Zhou et al. (2010) made bio-crude using the green macroalga*Enteromorpha prolifera* as the biomass source. The highest bio-crude yield obtained was 23.0% on a weight basis. The bio-crude was a complex mixture of ketones, aldehydes, phenols, alkenes, fatty acids, esters, aromatics, and nitrogen-containing heterocyclic compounds.

In 2014, Sea6 Energy Pte. Ltd., India, working in collaboration with Aban Ltd. successfully produced bio-crude from *K. alvarezii*biomass through hydrothermal liquefaction undertaken in a batch reactor, under alkaline conditions. The results of this study indicated that *Kappaphycus* was useful as biomass feedstock for bio-crudeproduction in the presence of alkaline catalysts and yields of 33–40% (ash-free, dry weight) were obtained. Further studies are required to improve the carbon and energy recovery in the bio-crude fraction (Suryanarayan et al. 2016a).

Acknowledgements The authors have had the privilege of learning about value-chains from experience working with hundreds of seaweed farmers and with numerous colleagues from both the private and public sectors. The term 'MUZE' appears to have been coined by David Myslabodski and numerous discussions with him around the turn of the century contributed to the concept as we apply it today. For helping the senior author to learn the economic basics of value-chain operations, he owes a special debt of gratitude to Frank Hartwich of UNIDO and Bruce Wise of IFC. His colleagues Boedi Julianto and Irsyadi Sirajuddin in Jasuda enabled him to keep living and working in Indonesia and supplied data that were available from no other source. Finally, the authors owe a major debt of gratitude to our colleagues at Sea6. It is because of them that we are able to move forward as we develop the step-change, value-chain development that mechanical farming and MUZE-processing are bringing to eucheumatoid seaweed-based, industrial developments.

References

Abdou ES, Sorour MA (2014) Preparation and characterization of starch/carrageenan edible films. Int Food Res J 21:189–193

Alves V, Costa N, Hilliou L, Larotonda F, Gonçalves M, Sereno A, Coelhoso I (2006) Design of biodegradable composite films for food packaging. Desalination 199:331–333

Anastasakis K, Ross AB (2011) Hydrothermal liquefaction of the brown macro-alga *Laminaria saccharina*: effect of reaction conditions on product distribution and composition. Bioresour Technol 102:4876–4883

Baricuatro F (2015) Challenges in increasing Indonesia seaweed export. Paper presented at the 7th meeting of the ASEAN Seaweed Industry Club, Novotel Hotel, Makassar, Indonesia, 1–3 August 2015

Bixler HJ (2017) The carrageenan controversy. J Appl Phycol. doi:10.1007/s10811-017-1132-4

Bixler HJ, Porse H (2011) A decade of change in the seaweed hydrocolloids industry. J Appl Phycol 23:321–335

Briones AV, Ambal WO, Estrella RR, Pangilinan R, De Vera CJ, Pacis RL, Rodriguez N, Villanueva MA (2004) Tensile and tear strength of carrageenan film from Philippine *Eucheuma* species. Mar Biotechnol 6:148–151

Candra KP, Sarwono, Sarinah (2011) Study on bioethanol production using red seaweed *Eucheuma cottonii* from Bontang seawater. J Coast Dev 15:45–50

Cornucopia (2016) Carrageenan, new studies reinforce link to inflammation, cancer and diabetes. Report by The Cornucopia Institute

Craigie JS (2011) Seaweed extract stimuli in plant science and agriculture. J Appl Phycol 23:371–393

de Góes HG, Perpetuo Reis RP (2011) An initial comparison of tubular netting versus tie–tie methods of cultivation for *Kappaphycus alvarezii* (Rhodophyta, Solieriaceae) on the south coast of Rio de Janeiro State, Brazil. J Appl Phycol 23:607–613

Eswaran K, Ghosh PK, Siddhanta AK, Patolla S, Periasamy C, Mehta AS, Mody KH, Ramavat K, Prasad K, Rajyaguru MR, Reddy SKCR, Pandya JB, Tewari A (2005) Integrated method for production of carrageenan and liquid fertilizer from fresh seaweeds. USA Patent number 6, 893, 479 B2

Evans FD, Critchley AT (2013) Seaweeds for animal production use. J Appl Phycol 26:891–899

Evans F, Kandasamy S, Prithiviraj B, Critchley AT (2016) Symbiotic effects of Tasco, an *Ascophyllum nodosum* product used for animal feed applications. Paper presented at the 22nd International Seaweed Symposium, Copenhagen, Denmark, 19–24 June 2016

Fakhrudin J, Setyaningsih D, Rahayuningsih M (2014) Bioethanol production from seaweed *Eucheuma cottonii* by neutralization and detoxification of acidic catalyzed hydrolysate. Int J Environ Sci Dev 5:455–458

FAO (2010) Aquaculture development 4. Ecosystem approach to aquaculture. ISBN 978-92-5-106650-8. Publishing Policy and Support Branch Office of Knowledge Exchange, Research and Extension FAO

Herr ML, Muzira T (2009) Value chain development for decent work: a guide for private sector initiatives, governments and development organizations. ISBN 978-92-2-122488-4, International Labour Office, Geneva

JSSA (2010) Japan Satoyama Satoumi Assessment, 2010. Satoyama-Satoumi ecosystems and human well-being: socio-ecological production landscapes of Japan – Summary for Decision Makers. United Nations University, Tokyo, Japan. [SDM-EN_24Feb2011, PDF]

Kandasamy S, Fan D, Sangha JS, Khan W, Evans F, Critchley AT, Prithiviraj B (2011) Tasco®, a product of *Ascophyllum nodosum*, imparts thermal stress tolerance in *Caenorhabditis elegans*. Mar Drugs 9:2256–2282

Kandasamy S, Fan D, Khan W, Evans F, Critchley AT, Prithiviraj B (2012) Tasco®, a product of *Ascophyllum nodosum*, enhances immune response of *Caenorhabditis elegans* against *Pseudomonas aeruginosa* infection. Mar Drugs 10:84–105

Karbowiak T, Debeaufort F, Champion D, Voilley A (2006) Wetting properties at the surface of iota-carrageenan-based edible films. J Colloid Interface Sci 294:400–410

Khambhaty Y, Mody K, Gandhi MR, Thampy S, Brahmbhatt H, Eswaran K, Ghosh PK, Maiti P (2012a) *Kappaphycus alvarezii* as a source of bioethanol. Bioresour Technol 103:180–185

Khambhaty Y, Upadhyay D, Kriplani Y, Joshi N, Mody K, Gandhi MR (2012b) Bioethanol from macroalgal biomass: utilization of marine yeast for production of the same. Bioenergy Res 6:188–195

Khan W, Rayirath UP, Subramanian S, Jithesh MN, Rayorath P, Hodges DM, Critchley AT, Craigie JS, Norrie J, Prithiviraj B (2009) Seaweed extracts as biostimulants of plant growth and development. J Plant Growth Regul 28:386–399

Lafargue D, Lourdin D, Doublier J (2007) Film-forming properties of a modified starch/j-carrageenan mixture in relation to its rheological behavior. Carbohydr Polym 70:101–111

Lustre AO, Bixler HJ (2006) Quality assurance of carrageenan for the world, market. Paper given at the 6th Asia-Pacific conference on Algal Biotechnology held from October 12–15, 2006 at the Dusit Hotel Nikko, Makati City, the Philippines

Neish IC (2013) Some social and economic dimensions of seaweed aquaculture in Indonesia. In: Valderrama D, Cai J, Hishamunda N, Ridler N (eds) Social and economic dimensions of carrageenan seaweed farming. Fisheries and Aquaculture Technical Paper No. 580. Rome, FAO

Neish IC (2015) A diagnostic analysis of seaweed value chains in Sumenep Regency, Madura Indonesia. Report submitted for UNIDO Project no 140140 "Preparatory assistance to develop a seaweed value chain development project for Madura"

Park SY, Lee BI, Jung ST, Park HJ (2001) Biopolymer composite films based on k-carrageenan and chitosan. Mat Res Bull 36:511–519

Porse H, Ladenburg S (2015) Seaweed value chain programme: final report. Submitted to Smart-Fish Indonesia

Porse H, Rudolph B (2017) The seaweed hydrocolloid industry: 2016 updates, requirements, and outlook. J Appl Phycol. doi:10.1007/s10811-017-1144-0

Pourjavadi A, Harzandi AM, Hosseinzadeh H (2004) Modified carrageenan 3. Synthesis of a novel polysaccharide-based superabsorbent hydrogel via graft copolymerization of acrylic acid onto kappa-carrageenan in air. Eur Polym J 40:1363–1370

Pourjavadi A, Hosseinzadeh H, Mazidi R (2005) Modified Carrageenan. 4. Synthesis and swelling behavior of crosslinked C-g-AMPS superabsorbent hydrogel with antisalt and pH-responsiveness properties. J Appl Polym Sci 98:255–263

Prithiviraj B, Critchley A (2016) Convergence of mechanism(s) of action of bioactives of *Ascophyllum nodosum*. Paper presented at the 22nd International Seaweed Symposium, Copenhagen, Denmark, 19–24 June 2016

Rayorath P, Jithesh MN, Farid A, Khan W, Palanisamy R, Hankins SD, Critchley AT, Prithiviraj B (2008) Rapid bioassays to evaluate the plant growth promoting activity of *Ascophyllum nodosum* (L.) LeJol. using a model plant, *Arabidopsis thaliana*. (L.) Heynh. J Appl Phycol 20:423–429

Sekar P, Wagimana P, Ainuri M, Darmawan A, Haslianti N (2015) The production of bioethanol fermentation substrate from *Eucheuma cottonii* seaweed through hydrolysis by cellulose enzyme. Agric Agric Sci Procedia 3:200–205

Stirk WA, van Staden J (2006) Seaweed products as biostimulants in agriculture. In: Critchley AT, Ohno M, Largo DB (eds) World seaweed resources, an authoritative reference system. ETI Bioinformatics, Amsterdam

Suryanarayan S, Neish IC, Kumar S, Nori S (2016a) Beyond carrageenan: development of technologies for full utilization of *Kappaphycus* biomass. Paper presented at the 22nd international seaweed symposium, Copenhagen, Denmark, 19–24 June 2016

Suryanarayan S, Neish IC, Nori S, Vadassery N (2016b) Cultivation and conversion of tropical red seaweed into food and feed ingredients, agricultural biostimulants, renewable chemicals and biofuel. In: La Barre S, Bates SS (eds) Blue biotechnology – production and use of marine molecules. Wiley. (in press)

Tan IS, Lam MK, Lee KT (2013) Hydrolysis of macroalgae using heterogeneous catalyst for bioethanol production. Carbohydr Polym 94:561–566

Tobacman JK (2017) Reply to comments regarding "the Carrageenan controversy". J Appl Phycol. doi:10.1007/s10811-017-1145-z

UNIDO (2011) Industrial value chain diagnostics: an integrated tool. United Nations Industrial Development Organization (UNIDO)

UNU (2010) Satoyama–Satoumi ecosystems and human well-being: assessing trends to rethink a sustainable future. Policy Brief No.7, 2010. United Nations University. [unu_policybrief_10-07.pdf]

Vadassery N, Neish IC, Suryanarayan S (2016) Systems for achieving high productivity during cultivation of *Kappaphycus* and *Eucheuma*. Paper presented at the 22nd International Seaweed Symposium, Copenhagen, Denmark, 19–24 June 2016

Valderrama D, Cai J, Hishamunda N, Ridler N (2013) Social and economic dimensions of carrageenan seaweed farming. Fisheries and Aquaculture Technical Paper No. 580. Rome, FAO

Wagiman MA, Rinda G, Jumeri (2015) Design process of hydrolysis and fermentation bioethanol production from seaweed *Eucheuma* cottonii to renewable sovereignty: KnE Life Sci Int Conf Agr (ICoA) 2015

Yanagi T (2008) "Sato-Umi" – a new concept for sustainable fisheries. In: Tsukamoto K, Kawamura T, Takeuchi T, Beard D Jr, Kaiser MJ (eds) Fisheries for global welfare and environment, 5th World Fisheries Congress 2008, pp 351–358. © by TERRAPUB 2008. [wfcbk_351.pdf]

Zertuche-Gonzalez JA, Garcia-Lepe G, Pacheco-Ruiz I, Chee-Barragan A, Gendrop-Funes V (1999) A new approach to seaweed cultivation in Mexico. World Aquacult Mag 30(2):57–61

Zertuche-Gonzalez JA, Garcia-Lepe G, Chee PRA, Gendrop V, Guzman JM (2001) Open water *Chondrus crispus* Stackhouse cultivation. J Appl Phycol 13:249–253

Zhou D, Zhang L, Zhang S, Fu H, Chen J (2010) Hydrothermal liquefaction of macroalgae *Enteromorpha prolifera* to bio-oil. Energy Fuel 24:4054–4061

Carrageenan Industry Market Overview

13

Ross Campbell and Sarah Hotchkiss

Abstract

Carrageenan is the leading seaweed-derived food hydrocolloid and is used widely for its textural functionality. Carrageenan has a specific set of rheological properties that differentiates it from other hydrocolloids and renders it useful in certain processed foods in addition to a small number of other niche markets. Growing demand for processed foods is a global trend that is being driven in general by an increase in population but also by opposing changes in the world's economies. On one side, recession is forcing consumers to seek value for money, on the other, improved living standards in formerly disadvantaged economies are enabling consumers to look for more diverse and luxurious product offerings. From either perspective, carrageenan is a winner and the global market is expected to grow albeit, there are some indications that growth in certain regions and sectors has slowed.

This chapter provides a general overview of the global carrageenan industry with a particular focus on the market dynamics for *Kappaphycus* and *Eucheuma.* Below we present information on: production of *Kappaphycus* and *Eucheuma* seaweeds as raw materials; carrageenan production and use in end-markets. We also present data on pricing, key players and capacities and round off with a general discussion of the current and potential trends that are influencing the industry. In all of this, we take an industry-based approach.

13.1 Introduction

The carrageenan market is the largest of the seaweed-derived, food hydrocolloid markets and currently holds the fourth largest share (in $ value) of the wider global hydrocolloids market, behind gelatin, starch and pectin (Seisun 2016). In 2014, the global hydrocolloids market was valued at US$ 5.4 Billion but projected to reach US$ 7.6 Billion by 2020 (Markets and Markets 2016). Industry estimates for 2015/2016 put a value of US$ 600–700 Million on the global carrageenan market (Hotchkiss et al. 2016b; Seisun 2016) with a projected average growth of ≈ 3% per year (CAGR). Most recent market research predicted that the global market for carrageenan would approach US$ 1 Billion by 2024 (FMI 2017).

R. Campbell • S. Hotchkiss (✉)
CyberColloids Ltd, Carrigaline Industrial Park, Carrigaline, Co. Cork, Ireland
e-mail: sarah@cybercolloids.net

Current industry estimates for the global production capacity of carrageenan range from about 70,000–80,000 MT dwt year^{-1} (Porse and Ladenburg 2015; CyberColloids own data) to over 110,000 MT year^{-1} (Neish 2015). CyberColloids' estimate for current global usage, as based on end-sales data, is for 68,000 MT dwt or approximately 80% of the estimated production capacity (CyberColloids own data). Most of this carrageenan (≈90%) comes from species and strains of *Kappaphycus* and *Eucheuma* or "cottonii" and "spinosum" respectively as they were referred to historically and still are by the industry. In 2016, CyberColloids estimated the global production of *Kappaphycus* for the carrageenan industry to be in the region of 1.5 Million MT wwt or 200,000 MT dwt (Hotchkiss et al. 2016b) with most cultivation occurring in Indonesia and the Philippines. *Eucheuma* is not cultivated on such a large-scale

A.Q. Hurtado et al. (eds.), *Tropical Seaweed Farming Trends, Problems and Opportunities*, Developments in Applied Phycology 9, DOI 10.1007/978-3-319-63498-2_13

as *Kappaphycus*, current estimates put production for the carrageenan industry in the range of 0.3 Million MT wwt or 40,000 MT dwt (Hotchkiss et al. 2016b). It must be noted that a substantial market for sea vegetables also exists which is not included in these estimates. The remainder of the carrageenan produced comes from cold-water species that are harvested mostly in South America, approximately 15,000 MT dwt and from the dwindling remnants of a small, traditional N. Atlantic industry which produces only a few hundred dry MT annually from *Chondrus crispus* (CyberColloids own data).

A note on information sources: much of the information and data presented here is CyberColloids' own data that have been obtained by permission and interview with carrageenan industry contacts including processors, end users and other hydrocolloid experts. Where relevant, we have included our interpretation of "industry gossip" and review of activity as reported in the industry media. All other data sources are appropriately accredited.

Building an accurate picture for the dynamics of any global market such as carrageenan is a difficult task and one that is reliant on obtaining realistic and representative data across a world-wide, value-chain. Consequently, very few industry based reports for carrageenan are available. The approach we use here draws heavily on the combined expertise of CyberColloids and our network of industry contacts. The sales and production data we present are based on farm-gate statistics and industry intelligence and most importantly, bottom-up estimates of seaweed production are corroborated by top-down estimates of carrageenan production and end-use markets.

Where possible, we have put our data into context with that of other industry experts. It must be noted that direct comparisons are difficult given that industry market intelligence varies significantly and is often regionally biased (i.e. both over- and under-emphasis on certain markets and sectors). Extrapolation of data sets is inevitable and all players utilise different conversion rates/yield ratios based on their own particular experiences. However, in accepting all of the above it appears that most recent analyses are acceptably close.

13.2 Global Production of *Kappaphycus* and *Eucheuma* Seaweeds

Most of the world's carrageenan (≈90%) comes from *Kappaphycus* and *Eucheuma* seaweeds that are cultivated mainly in Indonesia and the Philippines. Significant production for export markets has also been achieved in Malaysia, Zanzibar in the United Republic of Tanzania and Madagascar (Fig. 13.1). Commercial production is growing in other parts of south-east Asia such as Vietnam, Cambodia, Myanmar, East Timor and southern China; in some Pacific Islands including Hawaii, Fiji, Kiribati and Solomon Islands; in mainland East Africa, Mozambique and Kenya and in the South and Central Americas including Argentina, Colombia, Brazil, Ecuador, Venezuela, Mexico, Peru and Panama.

Recent industry estimates put the global production of *Kappaphycus* in the region of 170,000–200,000 MT dwt and *Eucheuma* in the region of 40,000–45,000 MT dwt (CyberColloids own data; Neish 2015; Porse and Ladenburg 2015). Since the late 2000s Indonesia has been the world's largest producer of *Kappaphycus,* surpassing the Philippines, which had been the major producer since the 1970s, when large-scale cultivation of these seaweeds began (Hurtado et al. 2015). Today, Indonesia produces an estimated 110,000 MT dwt of *Kappaphycus* (Fig. 13.2) of which, about 70% was exported (Porse and Ladenburg 2015). The Philippines produced an estimated 60,000 dry MT (Fig. 13.2). Neish

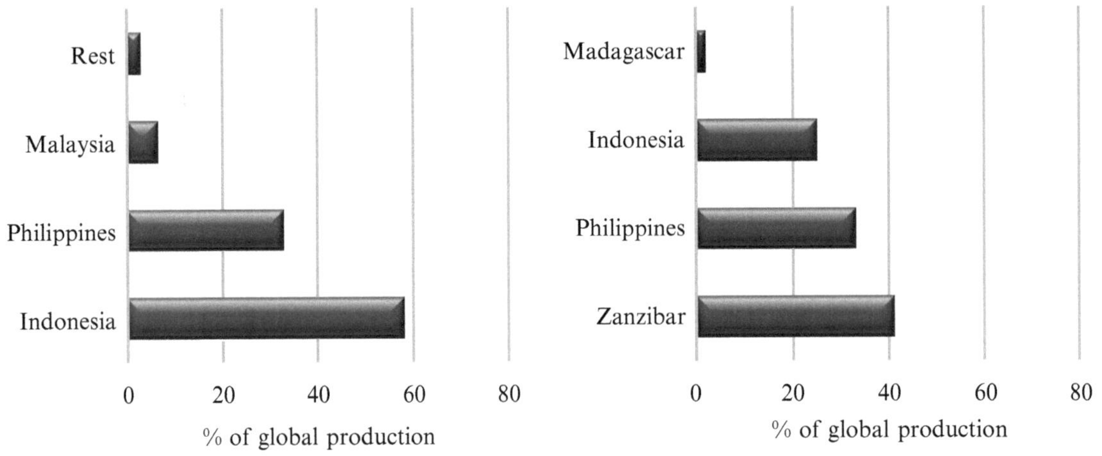

Fig. 13.1 Global production of *Kappaphycus* (*left*) and *Eucheuma* (*right*) in 2012 showing the relative split between major producers. NB: For *Kappaphycus* "Rest" includes Vietnam, Madagascar, India, Rep. of Tanzania, Central/South Pacific Islands and East Timor (CyberColloids 2012)

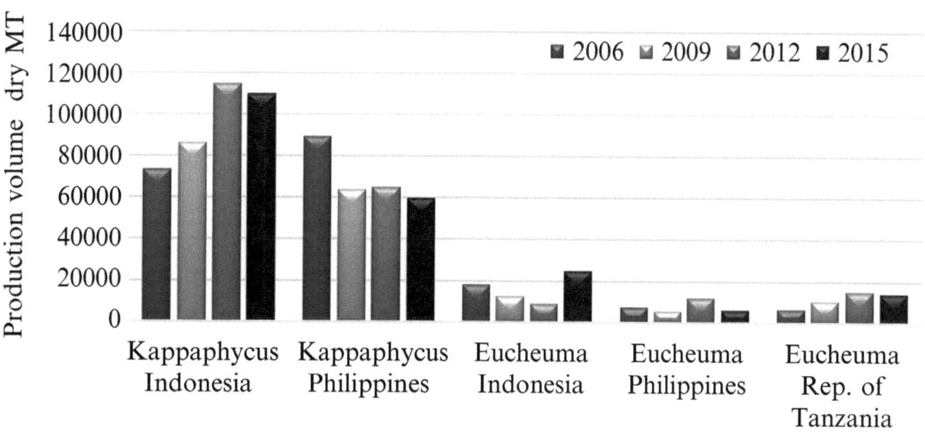

Fig. 13.2 Average annual production of *Kappaphycus* and *Eucheuma* (in dry MT) in the major producing regions from 2006–2015 (Based on the following sources: CyberColloids own data; Dakay 2010; Bixler and Porse 2011; CyberColloids 2012; Porse and Ladenburg 2015)

(2015) reported that the Philippines also imported around 50,000 MT dwt of seaweed from Indonesia for processing.

Kappaphycus has been the primary carrageenophyte seaweed used to establish commercial farming enterprises outside of Indonesia and the Philippines. These industries are relatively small on a global level. Production in Vietnam, Madagascar and India was typically in the range of 1000–2000 MT dwt year^{-1}, whereas in the Republic of Tanzania, Central/South Pacific Islands and East Timor volumes of 100–500 dry MT were reported (CyberColloids 2012).

In the last 10 years (i.e. 2010–2017), growth in production of *Kappaphycus* in Indonesia has been considerable, with a 50% increase in production during this period (Fig. 13.2). Since 2009, production of *Kappaphycus* in the Philippines appeared to have remained fairly stable (Fig. 13.2) but this trend followed a significant decline in production (in the region of 25%) after the 2008 industry "crisis", which initiated a re-balancing of global supply and demand for carrageenan (CyberColloids 2012). The combined outcome was that, as of 2017, Indonesia produced around 58% of the global *Kappaphycus* crop (CyberColloids own data, Porse and Ladenburg 2015).

Eucheuma is not cultivated on such a large scale as *Kappaphycus* and in addition to Indonesia and the Philippines, much of the world's *Eucheuma* is cultivated in Zanzibar, in the Republic of Tanzania (Figs. 13.1 and 13.2). Production in Zanzibar had been steadily increasing since the 1990s and in 2012, over 15,000 dry MT of *Eucheuma* were exported from there. This represented over one third of the global harvest (CyberColloids 2012; Neish and Msuya 2013). There is no processing of seaweed for carrageenan in Zanzibar and virtually all *Eucheuma* is exported for the production of **a**lcohol **p**recipitation (AP)iota carrageenan for use in toothpaste and food (CyberColloids own data, Neish and Msuya 2013). Although *Kappaphycus* is also grown in the Zanzibar region, crops have generally not flourished and have been subject to repeated failures, unlike *Eucheuma.* In 2012, Zanzibar produced less than 100 dry MT of *Kappaphycus* (Neish and Msuya 2013).

Markets for *Eucheuma* seaweed have been reported as showing "sluggish growth" in recent years. Consequently, there was an over-supply of seaweed and high competition between Indonesia, the Philippines and Zanzibar (Neish and Msuya 2013). In 2012, CyberColloids reported a declining trend for *Eucheuma* production in Indonesia with annual production being approximately half of that in 2006 (CyberColloids 2012). More recent estimates for *Eucheuma* production (Porse and Ladenburg (2015) showed production in Indonesia to have risen to about 25,000 MT dwt in 2015 whilst production in the Philippines appeared to have dropped. Zanzibar however, seems to have remained stable during that period.

13.3 Pricing Dynamics for *Kappaphycus* and *Eucheuma* Seaweeds

With Indonesia dominating the supply of *Kappaphycus,* as of 2017, global pricing strategies typically track the situation in Indonesia. Figure 13.3 shows the average price of *Kappaphycus* (exports in US$ MT^{-1} dwt) from Indonesia over the period 2007–2016. For simplicity, only the changes in pricing have been plotted to show the extremes in price and a trend-line added which indicates a general fall in price, post the price spike in 2008. Prior to 2008, the price of raw material was low, generally due to an over-capacity which kept prices down. This situation was not sustainable and consequentially resulted in the industry re-balancing itself. However, this new balance, between supply and demand and capacity, was too tight, hence the price spike in 2008. Since 2008, global capacity has risen, as has demand (detailed below) and industry prices have not returned to unsustainable levels.

In recent years, the average price for *Kappaphycus* has been dropping gradually, as of March 2017, it was around US$ 900 MT^{-1} dwt (CyberColloids own data). This is partially as a reflection of exchange rates, in particular a strong US$, but also due to a surplus in the supply chain. In 2016,

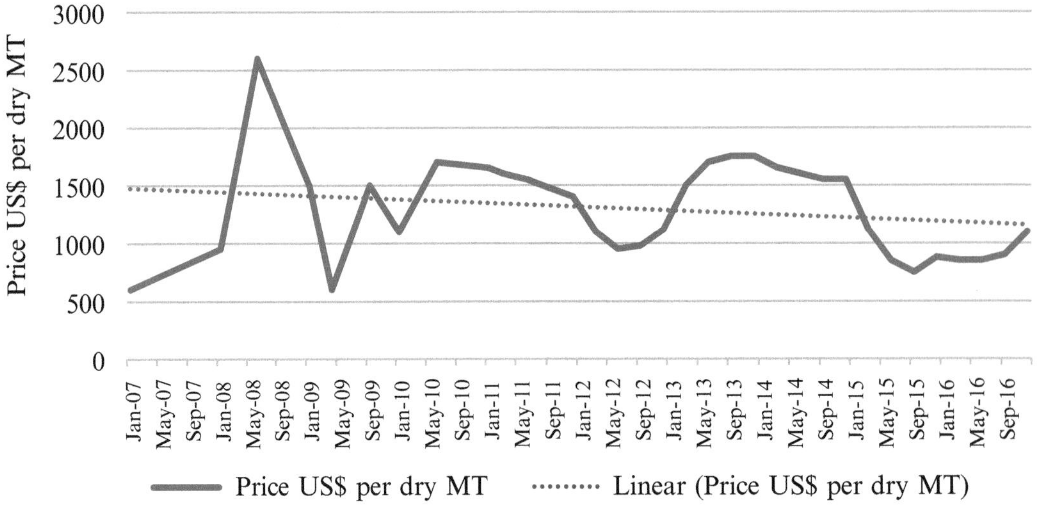

Fig. 13.3 Average price (US$ MT^{-1} dwt) for *Kappaphycus* seaweed from Indonesia from 2007–2016 based on export prices. This graph shows the price spike in 2008, followed by a gradual decline in price. NB: only significant changes in pricing were plotted (CyberColloids own data)

Fig. 13.4 Average price (US$ MT^{-1} dwt) for Eucheumatoid seaweed from the major growing regions from 2006–2012 based on export prices (CyberColloids own data)

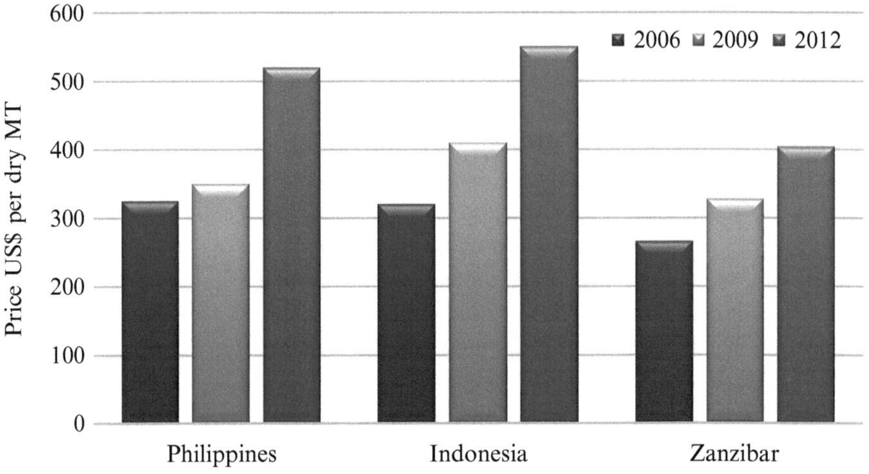

there were exceptionally good harvests in south-east Asia, Indonesia in particular, which caused the price of the cultivated seaweed to drop. South-east Asia is currently (April 2017) coming out of a very bad wet season which may further drive prices up, but this remains to be seen.

Currently, most of Indonesia's exported seaweed goes to China (Porse and Ladenburg 2015). The industry is waiting to see whether the Indonesian government will restrict exports in order to create a local 'added value' industry (as currently being discussed within Indonesia) and if so, what this will mean for the future of the carrageenan industry, in China in particular. How the Indonesian situation plays out will undoubtedly have a significant impact and most likely create pricing turmoil if current exports of their seaweeds diminish.

The price for *Eucheuma* seaweeds has historically been much lower than that of *Kappaphycus* (Fig. 13.4). Demand for *Eucheuma* is substantially lower (since there were fewer value applications for iota carrageenan) and there is generally a surplus of raw material supply which maintained the price at low levels. In 2012, farm gate prices for Zanzibar grown *Eucheuma* were in the range US$ 190–250 MT^{-1} dwt. These prices were reported as being lower on average than equivalent prices for Indonesian and Philippines grown *Eucheuma*. However, prices on the open market, at that time, i.e. with freight costs factored in, were more similar due to the higher cost of shipping from Zanzibar (Neish and Msuya 2013). These higher export prices are reflected in Figure 13.4 which shows that the average market price of *Eucheuma* from the major growing regions ranged from US$ 400–550 in 2012. The average market price for *Eucheuma* exported from Indonesia was reported to be about US$ 400 per dry MT in March 2017 (CyberColloids own data).

13.4 Overview of Global Carrageenan Production

Within the food industry, carrageenan is recognised as either "Carrageenan" or "PES – **P**rocessed **E**ucheuma **S**eaweed" however, within the carrageenan industry the terms "refined" and "SRC" (**s**emi-**r**efined **c**arrageenan) are more typically used.

Kappaphycus and *Eucheuma* are used to produce both refined and semi-refined grades of carrageenan. **R**efined **c**arrageenan (RC) is extracted using **a**lcohol **p**recipitation (AP) or **g**el **p**ress (GP) techniques to produce refined carrageenans that have had the cellulosic component (≈15–20% dry weight) filtered out. Alcohol precipitated, refined iota and kappa carrageenan are obtained from *Eucheuma* and *Kappaphycus* spp. respectively. Gel-press, refined carrageenan (kappa) is only extracted from *Kappaphycus* species. SRC is produced using a process that produces a semi-refined grade that still contains the cellulose component of the seaweed tissue. Semi-refined grades of both kappa and iota carrageenan are produced from *Kappaphycus* and *Eucheuma* respectively. For more detail on extraction methods and carrageenan sources see Ortiz-Tena et al. Chap. 12 in this book and Hotchkiss et al. (2016b).

Refined carrageenans are also extracted from various cold-water species of seaweed. AP lambda carrageenan is produced from *Chondrus crispus* (Irish Moss),*Mastocarpus stellatus* (False Irish Moss) and some of the South American seaweeds that contain a relatively high lambda carrageenan content.

An important group of hybrid grades of carrageenan are also extracted from the cold-water species using a range of proprietary processes that are based on alkali extraction, in an alcohol:water slurry and/or selective extraction techniques. These grades, also known as kappa-iota hybrids; kappa-lambda hybrids; "kappa-2-hybrids"; weak-gelling hybrids and heterogenous grades, are a mix of kappa, lambda and iota carrageenan, the exact composition of which depends on the source material and process (see Hotchkiss et al. 2016b for more detail). They are collectively referred to here as: "hybrid grades".

For the purposes of this overview, the following abbreviations will be used to denote the different types of carrageenan: RC = **r**efined **c**arrageenan; SRC = **s**emi-**r**efined **c**arrageenan; AP = **a**lcohol **p**recipitated, GP = **g**el **p**ress. The initial steps in a SRC process produce alkali-treated chips, either ATC (**a**lkali-**t**reated **c**ottonii) or ATS (**a**lkali-**t**reated **s**pinosum) that are then taken through a series of further purification and milling steps to produce SRC. Some processors take ATC as a preferential feedstock for the production of GP carrageenan.

Carrageenan is produced by companies around the world, Table 13.1 shows the key players in the industry by region and also the technology employed. It clearly shows the regional split between SRC and GP production in Asia and AP and hybrid production in Europe and the Americas. Historically, Asia was the supplier of seaweedraw material which was exported to the west for processing. However, the situation today is quite different. There are currently 16 processors in Indonesia and 14 in the Philippines. Foreign operators, CP Kelco, FMC, Cargill, Kerry and CEAMSA have production plants in the Philippines but until recently there were no foreign operators in Indonesia. In 2015 however, the Chinese company Shanghai Brilliant announced that it would build a large ATC/SRC plant in Indonesia which (as of mid 2017) was due to start production "soon". Most of the seaweed grown in Indonesia is exported for processing (mostly to China). Only a third of Indonesia's *Kappaphycus* is processed domestically, mostly by small producers (<15,000 MT) and mostly industrial grades for pet food and ATC (Porse and Ladenburg 2015).

The Chinese companies, Shanghai Brilliant and Greenfresh are currently the number one and two carrageenan producers in the world respectively; they were reported to be taking <4,000 MT dwt mo^{-1} and <2,000 MT dwt mo^{-1} (respectively) of *Kappaphycus* from Indonesia for the production of GP and SRC (Neish 2015; CyberColloids own data). Neish (2015) reported that Chinese companies were producing carrageenan products that could be US$ 1–2 kg^{-1} cheaper than Indonesian and Philippines grades. Many Indonesian and Philippines companies were finding it hard to compete with China on a cost-basis. For Indonesian companies in particular, competition with China was also proving difficult in terms of end-product quality (Neish 2015). However, CyberColloids is aware that some smaller Chinese processors are now toll-manufacturing in Indonesia and the Philippines. Outside of Indonesia, the Philippines and China, carrageenan was also processed in Malaysia and Japan.

In Europe and the Americas, a small number of companies are perceived to be the key players with all predominantly focusing on the production of AP grades (Table 13.1). Albeit, most of these companies have production facilities in the Philippines which produced RC GP and/or SRC grades. In 2013, Neish and Msuya reported that most buyers were taking roughly equal splits of seaweed raw material from Indonesia, the Philippines and Zanzibar for their AP iota production. Certainly, in 2012, *Eucheuma* from Zanzibar was exported to key players in Europe, the Philippines, North America and Chile (Neish and Msuya 2013). Gelymar is an interesting exception to the western norm, with a wider range of production technologies (e.g. RC AP, RC GP and hybrid) all based in Chile.

Production in Europe and North America, in general, appeared to be steady with no closures or investment/new plants reported in recent years. With concerns over consumer attitudes, future investments may become difficult to justify. However, whilst production appears to remain steady, the market continues to restructure. DuPont recently announced (March 2017) that it would acquire FMC Health and Nutrition which included the carrageenan production facility in Rockland, Maine, USA.

In 2015/2016, the estimated production capacity for Europe and the Americas accounted for about 12% (10,500 MT dwt) and 17% (15,000 MT dwt) respectively of total global carrageenan production*c.f.* 45% (or just under 40,000 MT dwt) and 25% (about 22,000 MT dwt) for Asia-Pacific and China (estimated averages based on CyberColloids own data, Porse and Ladenburg 2015; Neish 2015 and summarized in Table 13.1 below).

CyberColloids currently estimates the global production capacity of carrageenan to be in the region of 86,000 MT dwt; Porse and Ladenburg's estimates (2015) are within range at 73,000 MT dwt whereas Neish (2015) reported much higher figures of over 110,000 MT dwt year^{-1} (Fig. 13.5 and Table 13.2 below). Neish (2015) does however state that the estimates given included data for operations that were not necessarily active at the time of writing. Porse and Ladenburg (2015) did not give estimates for the hybrid grades separately but those data were distributed across other categories.

When considering the breakdown of global production by carrageenan type, more SRC grades (kappa and iota) are produced than other type of carrageenan. In 2015/2016 SRC accounted for about 51% of the total global carrageenan production and most of this SRC (≈80%) came from *Kappaphycus*. In fact, kappa carrageenan in general, had a higher market demand and accounted for about 70% of all

Table 13.1 Key players by region showing types of carrageenan produced

		RC AP	RC GP	Hybrid	SRC
Europe	CPKelco (Denmark)	✓			
	Cargill (France)	✓			
	CEAMSA (Spain)	✓			
Americas	FMC (USA)	✓			
	Gelymar (Chile)	✓	✓	✓	✓
	Du Pont (Danisco)			✓	
Asia Pacific	CPKelco (Philippines)		✓		
	Cargill/Phil.Bioindustries		✓		
	Kerry (Philippines)				✓
	Shemberg (Philippines)	✓	✓		✓
	CEAMSA (Philippines)				✓
	Other Philippines		✓		✓
	Indonesia		✓		✓
China	Shanghai Brilliant (China)		✓		✓
	Greenfresh (China)	✓	✓		✓
	Other China		✓		✓

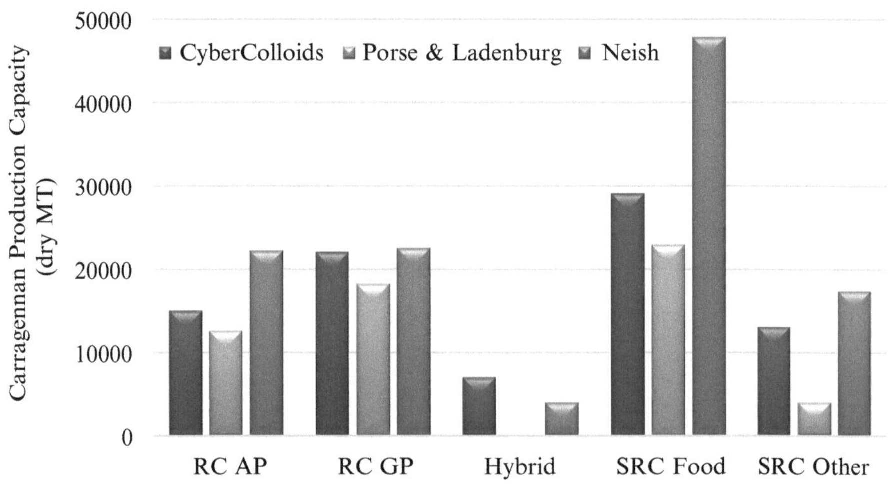

Fig. 13.5 Overview of global carrageenan production capacity in 2015/2016 (dry MT) by type. Based on recent industry estimates: CyberColloids own data for 2016; Porse and Ladenburg data for 2015 and Neish data for 2015. NB: RC = refined carrageenan; SRC = semi-refined carrageenan; AP = alcohol precipitated, GP = gel press; "Other" includes ATC/s, used in pet food, toothpaste and beer

Table 13.2 Estimated carrageenan production capacity (dry MT) by type and region

	RC AP	GP	Hybrid	SRC food	SRC other
Europe	7,000a	–	–	–	–
	6,900b	1,400b	–	–	–
	10,250c	–	–	1,000c	–
Americas	5,000a	1,000a	7,000a	2,000a	–
	5,000b	2,600b	–	1,500b	–
	7,000c	1,000c	4,000c	2,250c	–
Asia Pacific	3,000a	10,000a	–	22,000a	12,000a
	1,900b	6,200b	–	20,000b	7,000b
	3,950c	8,000c	–	30,700c	9,950c
China	–	11,000a	–	5,000a	1,000a
	–	15,000b	–	6,000b	–
	1,000c	11,500c	–	11,000c	1,000c
Total	12,500–22,500	18,200–22,520	4,000–7,000	22,885–47,830	4,000–17,400

2015/2016 data based on (a) CyberColloids own data (2016); (b) Porse and Ladenburg (2015); (c) Neish (2015)

carrageenan produced globally *c.f.* 20% iota (CyberColloids own data, Porse and Ladenburg 2015). SRC production capacity in 2006 was estimated to be in the order of 16,000 MT dwt (CyberColloids 2012) and thus, on average, doubled in the last 10 years. Current estimates were quite wide-ranging, viz. 23,000–48,000 MT dwt and reflected differences in industry intelligence.

SRC is by nature slightly cloudy (due to the cellulose content) but for many applications high clarity and gel strength are not as important as other functional properties. For this reason SRC is used as a cheaper alternative to RC. Market prices for SRC (as of March 2017) were in the range: US$ 6–8 kg^{-1} (CyberColloids own data). Semi-refined kappa carrageenan in particular, is used in processed meat products, to extend the meat and other sources of protein and thus add value. In some cases, SRC also gives some additional technical functionality, e.g. reduces "tiger striping" or visual defects in processed meat products. Current growth in the carrageenan market, as a whole, is being driven, to a large extent, by increased demand for cheaper and better value meat and meat-based products. SR iota carrageenan is also used in meat-based products (to a much lesser extent than kappa), in various carrageenan blends in order to add specific functionality and also has applications for beer refining.

A key, non-food application for SRC (both kappa and iota) is in pet food. SRC appeared to be finding renewed uses in pet food (from discussions with pet food industry contacts). When dry food formulations became popular, the demand for pet food-grade carrageenan fell but a new trend for high diversification in new, wet/moist food formulations is driving the re-use of carrageenan.

The global production capacity for refined carrageenans in 2015/2016 was in the range of 18,000–22,500 MT dwt of GP and 12,500–22,500 MT dwt of AP. Note that estimates for AP were not exclusively for iota carrageenan but also included other AP grades. Refined GP carrageenan has high gel strength and high clarity. It typically has a slightly higher cellulose component than AP carrageenan (≈0.5%) and also contains residual levels (1–2%) of potassium chloride from the extraction process. The residual potassium chloride means that GP carrageenan always gels on the addition of water and without the addition of ions, unlike AP carrageenan grades. These properties render refined GP carrageenan ideal for use in jellies and confectionery, particularly in Asia. In fact, most GP carrageenan is used in the Asian market for jellies and confectionery and this particular application is one of the key drivers in the global market at present.

Refined AP iota carrageenan from *Eucheuma* has particularly high purity and high clarity. Consequently it has a higher market value than refined GP and the market price is higher than that of refined kappa. Consequently, its main application is in high value products such as toothpaste. Current market prices for AP and GP (as of March 2017) were US$ 12–14 kg^{-1} and US$ 9–11 kg^{-1}, respectively (CyberColloids own data). Iota carrageenan is also used in dairy desserts as it forms soft gels and has good freeze/thaw stability. In 2016, CyberColloids reported that the production of AP carrageenan had fallen during the 1990–2000s, we attributed this decline to a rise in SRC and GP technology and the increased demand for SRC and GP kappa grades.

Since 2006, production of hybrid grades of carrageenans from South American red seaweeds has grown. Recent estimates (2015/2016) for production capacity ranged from: 4,000–7,000 MT dwt (Fig. 13.5 and Table 13.2). The hybrid carrageenans have particular applications in dairy products as they provide unique mouth-feel attributes which are highly desirable to food formulators. As such, these carrageenans command a higher market price, about US$ 14–17 kg^{-1} (as of March, 2017; CyberColloids own data). There has been strong growth in the Asian dairy market where these grades are being utilised. The sustainability of the South American hybrid production is a long running debate within the carrageenan industry but, as of 2017, a crisis in raw material supply had not happened. However, Chilean producers, like Indonesian, appear to be shifting towards the export of locally produced carrageenan, whilst exports of seaweedraw material dropped. There were no apparent restrictions in the availability of raw materials as of early 2017 (CyberColloids own data).

CyberColloids' current (March 2017) estimate for global carrageenan usage, as based on end-sales data, was for 68,000 MT dwt which equated to approx. 80% utilization of the estimated global production capacity. Percentage utilization across different technology categories varied from 60–90% (CyberColloids own data). Table 13.2 above provides a regional breakdown of carrageenan production capacity by type, based on our 2016 estimates and those given in the most recent industry reports (Neish 2015; Porse and Ladenburg 2015). To the best of our knowledge, and based on discussions with industry contacts, it appeared that production capacity and demand for the refined AP were tightly matched, whereas there was an over-capacity for the SRC grades and a possible over-capacity for GP. The magnitude of this over-capacity is debatable as it is difficult to confirm production figures for Indonesia, in particular. As of March 2017, we believed the utilization of current processing capacity in Indonesia to be in the range of 50–70%, although some industry contacts reported this to be even lower. The demand for hybrid grades remained strong, whilst production was rising.

13.5 End-Use Markets

Carrageenan is used in an extremely diverse range of food products but most typically those that are based on, or contain, ice cream, meat and poultry, dairy products (including cheese and cream, dairy drinks), non-dairy (e.g. nuts and seeds) drinks and water-based jellies (see Hotchkiss et al. 2016a for more detail). Currently, over 80% of the global carrageenan production is utilised by only three major application sectors: (i) processed meats; (ii) dairy and (iii) desserts and jellies (Fig. 13.6). Growth in demand for these three, end-use markets was steady over the past 10 years (i.e, from 2006–2015, with an estimated growth of 1.5–3%, per year, in each sector (CyberColloids own data). This growth reflected the global trend for increased demand for processed/convenience foods and was in response to both population growth and changes in regional economies. Convenience foods are a target for consumers either forced to seek value for money during recession, or those that are experiencing improved living standards and are able to buy more diverse food offerings.

The demand for processed meat products that offer value to consumers, i.e. those made from comminuted meat (burgers, nuggets), injected and reformed meats (hams, chicken rolls) and emulsions (frankfurters and sausages) had risen substantially and remains a key driver for the global production of SRC (kappa in particular and iota to a much lesser extent). In terms of the number of new products appearing in the market place, it appeared that little changed between 2007–2016, with the level of innovation remaining static at 10–11% of new products launched globally (Fig. 13.7).

The dairy sector was still dominated by the use of RC AP grades and SRC grades did not appear to have made any significant inroads, except in chocolate milk and ice cream. However, despite significant growth in the Asian dairy market and high global innovation in terms of new products in the market, the production of RC AP fell from >20,000 MT dwt in 2006 to about 12,500 MT dwt in 2015. The question as to how long the RC AP grades could defend themselves against 'cleaned up' SRC grades and more cost-effective hybrid grades still stands.

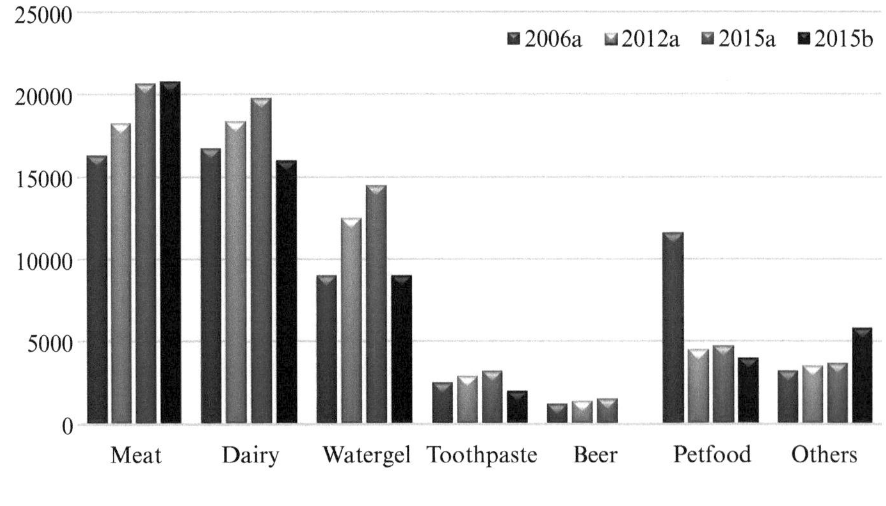

Fig. 13.6 Carrageenan sales by end-use showing major categories; estimated volumes (dry MT) for 2006, 2012 and 2015. NB: Dairy also included ice cream and milk based drinks, water gels included jellies, confectionery and shelf stable desserts (a = CyberColloids own data, b = Porse and Ladenburg 2015)

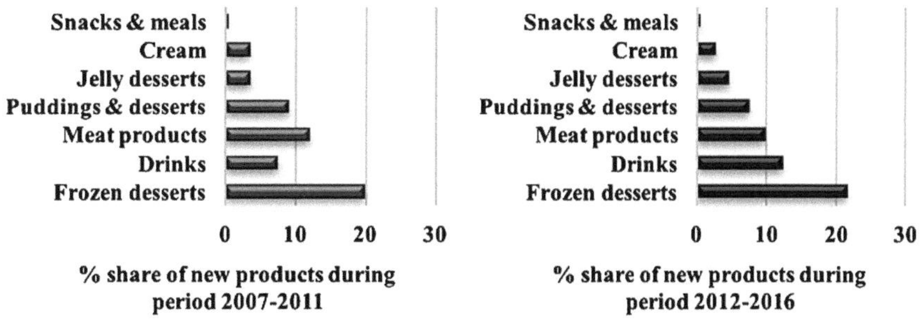

Fig. 13.7 New food products appearing in the global market containing carrageenan (refined and semi-refined grades) for the periods 2007–2011 and 2012–2016. The diagram shows the percentage share in key application sectors. NB: Frozen desserts included ice creams and frozen yogurts (CyberColloids 2017)

SR kappa is used as a thickening and suspension agent in flavoured milk drinks, e.g. chocolate milk and iced coffee. From 2007–2016, there appeared to be some growth in the use of carrageenan in new, milk-based drinks but note that the data presented here also includes use in non-dairy milks (e.g. nut and seed-based) and meal replacement drinks which were a globally growing sector.

The rise in demand for RC GP for water jellies and confectionery was seen to be very much an Asian market phenomenon, being linked to population growth and rising standards of living. Jelly-based convenience foods like snacks and shelf-stable desserts (as opposed to chilled/frozen dairy based desserts) are extremely popular. As the RC GP technology was very reliant on seaweeds which were cultivated in south-east Asia, any future demand would likely be satisfied by regional suppliers.

Significant smaller market sectors for carrageenan included: toothpaste, pet food and beer. Refined AP iota carrageenan has a niche application in toothpaste. It is used as a thickener, but has other key attributes, most importantly, stability against enzymatic degradation, good flavour release and rinse-ability making it ideal for use in toothpaste. The use of SR grades (mostly kappa with some iota blends) in beer had generally been steady to the time of writing however, there was also a noted move towards using less processing aids in beer, the outcome of which could mean a decline in the use of carrageenan. This awaits to be seen.

As previously mentioned, a revival in the use of carrageenan in pet food has been reported. Traditionally, this was one of the significant, non-food sectors which utilised SRC. Carrageenan is used as a retort-stable, gelling agent and stabiliser in jelly and gravy-based, canned and pouched products and more recently in chewable snacks as a binder and texturant. It is difficult to judge growth in this sector in terms of new products in the market, as in many regions, carrageenan is not declared on the ingredients list. However, in recognising this renewed global growth, changes in consumer attitudes in North America appeared to be having an impact. As of 2016/2017, North American pet food manufacturers were seeking replacements for carrageenan, however, since cost-effective alternatives were not readily available, sales of carrageenan appeared to be unaffected.

13.6 Regional End-Use

In 2016, the global carrageenan market was reported by Markets and Markets (2016) to be worth an estimated US$ 758 Million, based on a global market size of approximately 57,000 MT dwt (Fig. 13.8). This is in line with hydrocolloid industry estimates which put a value of US$ 600–700 Million on the market for which the demand was in the order of 60,000 MT dwt year^{-1} (Hotchkiss et al. 2016b; Seisun 2016; Porse and Ladenburg 2015). Markets and Markets (2016) predicted that the market volume would reach 70,723 MT dwt with a value of US$ 1047 Million by 2020 (CAGR of 4.41% and 6.66%, respectively).

Markets and Markets (2016) reported Europe and North America to be the major markets for carrageenan. The only regional data that CyberColloids has been able to verify is that for the North American markets (including USA, Canada and Mexico). CyberColloids estimated this to be around 20,000 MT dwt year^{-1} for 2016, as based on import data. It seems reasonable therefore that the Markets and Markets estimates of end-use in other regions were credible (CyberColloids own data).

In terms of the number of new products appearing in the market which contained carrageenan (refined and semi-refined), there was a substantial increase 2007–2016, which indicated that there was probably growth in the use of carrageenan in all global regions (Fig. 13.9). Europe remained the region with the highest level of innovation with >40% of the global share. There appeared to have been a shift in relative positions of the North American and Asia Pacific markets with the most products launched in Asia Pacific during the 2012–2016 period (Fig. 13.9). These observations could be illustrative of changed consumer attitudes towards carrageenan in North America, or simply the rise in carrageenan usage in Asia.

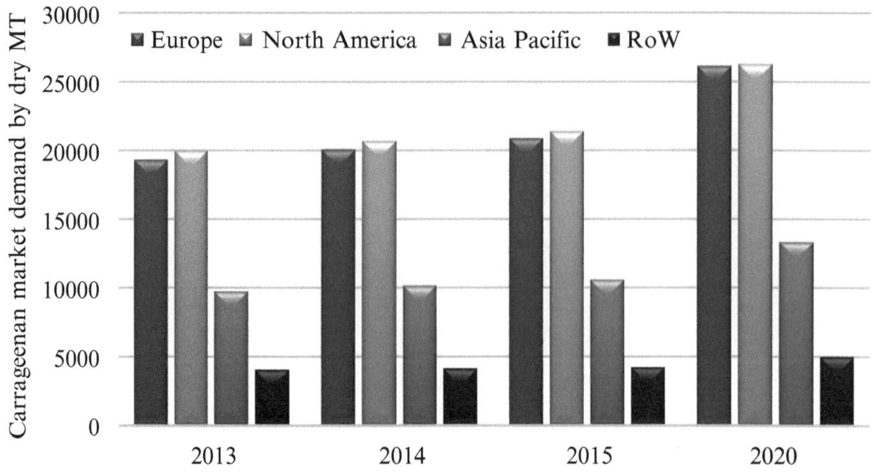

Fig. 13.8 Global carrageenan market by region from 2013–215 with projected demand for 2020. NB: RoW = Rest of World (Data by permission Markets and Markets 2016)

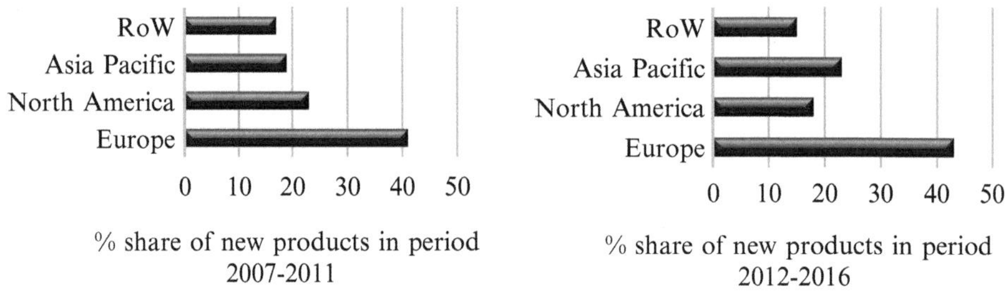

Fig. 13.9 New food products appearing in the global market place that contained carrageenan (refined and semi-refined grades) by region. For the periods 2007–2011 and 2012–2016. Showing % share by region. NB: Row = Rest of World (CyberColloids 2017)

13.7 Looking Forward – An Industry Perspective

In very broad terms, today's carrageenan industry could be seen as being divided into three major sectors based on production technology. Each satisfying its own end-markets and each being subject to particular fluctuations in supply and demand. These sectors are, the AP and hybrid sector, the GP sector and the SRC and ATC/S sector (Fig. 13.10). However, as already highlighted, there was also a clear geographical split between the production of AP and hybrid grades in the west and that of SRC and GP grades in Asia which was essentially linked to raw material supply.

Today (2017), the AP and hybrid sectors are characterized by a relatively small number of players that service high value, niche markets, including toothpaste and pharmaceutical, but this sector was clearly dominated by the dairy market. The AP iota market was generally considered to be mature and production of RC AP had fallen over the last decade. There appeared to have been little investment in the west, despite growth and high innovation in the dairy market (particularly in Asia). Demographic growth in middle-class consumers in Asia (in particular) was one of the key drivers for the dairy sector which was expected to continue growing. The use of Chilean seaweeds to produce carrageenan for the dairy sector had grown over the past 5 years (CyberColloids own data) and given this, concern remained within the industry for the future, sustainable supply of these Chilean seaweeds.

Key players in the AP and hybrid space appeared to be adopting different strategies in order to deal with any potential threat to raw material supply from Chile. Since 2013, Cargill openly announced their strategy to diversify their product offering and to utilize more "sustainable" seaweeds that are farmed in Asia. Chilean-based Gelymar however, were proactive in gaining a better understanding of the naturally available resource and invested in the development of resource management strategies and cultivation programmes for some of the key species (from discussion with Gelymar).

Thirty years ago, there was virtually no left hand side (GP and SRC) to the schematic illustrated in Fig. 13.10. There has been a relative explosion in seaweed cultivation and in carrageenan production capacity in South-east Asia. Despite growing markets and demographic changes, there was undoubtedly a significant over-capacity for SRC and probably GP. We have to ask what is going to happen with all of this over capacity and whether it will force the cut back or closure of some facilities?

The government roadmap for the Indonesian seaweed industry was reported to be driving the construction of several new carrageenan processing facilities, with the aim for Indonesia to become the largest seaweed processor in the world. We have been informed that 20 new plants were planned, eight of which had been built so far (as of March

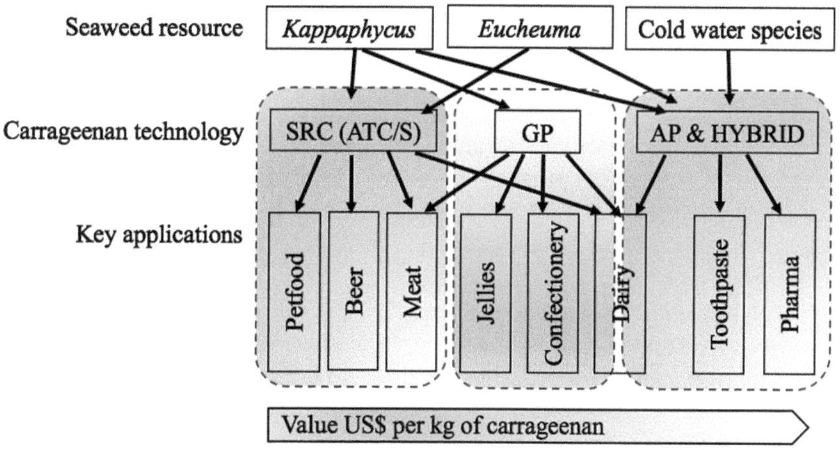

Fig. 13.10 Schematic of the carrageenan industry at present, showing major segmentation by seaweed resource, production technology and key applications

2017). With usage currently reportedly at 50–70% (or lower), this investment appeared to be somewhat unnecessary. However, the roadmap was also driving a restructuring of the industry, with the intent to shift away from export of high volume, cheaper grades, towards the development of higher value products and blends for domestic markets (from discussions with Indonesian processors).

Changes to the Indonesian situation will likely have most impact on China as the primary importer of their seaweed. However, the number one Chinese company: Shanghai Brilliant has located a facility to produce 10,000 MT year^{-1} of ATC/SRC in Indonesia, could this be indicative of future changes? With the Chinese industry being dominated by only a few large players, such a change could result in smaller companies being squeezed out. We have already heard reports that smaller, Chinese producers were toll-manufacturing in Indonesia and the Philippines.

Given the above however, over-capacity still seems likely, as does the reality that processors will have to diversify and add value if they are to remain competitive and prevent closures. If this is the case, then it is possible that it will be in the SRC and GP sectors of the industry that we will see the most innovation going forward. Multi-stream processing or biorefinery approaches are not new phenomena for the seaweed industry and are indeed discussed in detail elsewhere in this book. The MUZE (**m**ulti-stream **z**ero **e**ffluent) concept was proposed several years ago as a straightforward means to cost effectively and sustainably produce SRC/ATC grades, in addition to recovering other value-added products, such as extracts for the agri-horticulture sector (Neish 2012). Since then, it has been evaluated as a means to produce carrageenan and biostimulant extracts from *Kappaphycus* in India (Neish 2013) and to add value to the carrageenan/seaweed value-chain in Zanzibar (Neish and Msuya 2013). From discussions with processors, we are aware that there does appear to be a general interest to add value through diversification and multi-stream processing, particularly in Indonesia. According to an analysis of the commercial seaweed market (Grandview Research 2016), the category of liquid seaweed extracts is one of the predicted growth areas going forward, in particular to service the food, cosmetics and agri-horticulture sectors.

On a more sophisticated level, the seaweedbiorefinery concept has become a key focus for "Blue Economy" research strategies. Various models are proposed that deliver a diverse range of product streams and generally end up with a hydrocolloid (fiber) stream once all other high value components have been removed (Note: unless intended for generation of biofuels). Seaweed in general, has become a hot topic for research with national and regional marine strategies being implemented to promote sustainable exploitation of marine resources and to re-build and re-vitalise coastal, marine economies. As such, funding and opportunities for new cross-disciplinary research have probably never been better. Research into fields that have not traditionally been associated with the traditional seaweed industry, such as biotechnology and bio-pharma, are now well and truly on the radar and some products which exploit the bioactive properties of carrageenan (e.g. anti-microbial and anti-viral activity) are already commercially available.

The hydrocolloids industry, being based on the extraction of target polymers, is inherently wasteful and generates effluent streams through processing. In a world where there is a global push towards achieving greater resource efficiency, minimization of waste and food security, there is certainly ample room for the industry to look for more innovative and smarter 100% usage of its raw materials.

Seaweed in general is also enjoying renewed interest from the food industry. Consumer and industry media has been reporting a "surge" in the use of seaweed and microalgal ingredients and has gone as far as to say that seaweed and microalgae are driving NPD in many sectors (Food Ingredients First 2017). In 2016, seaweed was considered to be a top food industry trend with 30% growth in the US snack market alone (worth of $500 million) being reported (FoodBevMedia 2015; NNB 2016). However, the situation with regard to carrageenan was fundamentally different and the "Carrageenan Controversy" (Bixler 2017) as it is increasingly referred to, was undoubtedly having an impact on consumer attitudes in some markets. The media-hype surrounding the misguided portrayal of carrageenan as being unsafe for human consumption has been around for a number of years, but was more recently ramped up. There is now widespread misinformation and confusion being circulated in the social media and consumer news.

Changes in consumer attitudes towards the use of carrageenan in food and pet food markets in North America, in particular, were very apparent, as of early 2017. The decision by the NOSB (**N**ational **O**rganic **S**tandards **B**oard) which advises the US Department of Agriculture, to ban carrageenan from organically registered products would undoubtedly have further negative impacts in the North American markets in particular. Indeed carrageenan was being replaced by some companies, particularly in premium brands such as ice cream. In 2016, Nestlé publically rolled out a range of new, "simpler" ice cream formulations, in which carrageenan and xanthan gum were replaced with other hydrocolloids, with a more "natural" image, i.e. pectin. We have received word of an estimated 30–40% reduction (globally) in carrageenan intelligence usage in 2016, however, these figures were not supported by the data which suggested that imports of carrageenan to the US were down by only 4–5% (CyberColloids own data).

The US pet food industry was also reporting similar consumer driven changes. In the US, carrageenan is declared on pet food labels, unlike in Europe, where it falls under a natural ingredient classification. Manufacturers are seeking "clean label" ingredients and some premium brands (especially cat food) started to replace carrageenan with agar (also a red seaweed extract; from discussions with industry contacts). The decision to replace carrageenan very much depended on the cost differential. The pet food industry is typically cost-driven and is based on smaller margins for large volume production. This market segment would therefore be particularly sensitive to increased ingredient costs. There did not appear to be a significant push, as yet, for carrageenan to be replaced by non-US pet food manufacturers (from discussions with industry contacts).

As with pet food, the decision to replace carrageenan in food also depends on the cost differential and at present, replacement is mostly being seen as a brand protection strategy for premium products only. Carrageenan is a relatively cheap, reliable ingredient and is typically only used at quite low dose rates. Even so, to replace carrageenan with agar or pectin could cost around three times more and to be replaced with gellan could cost up to ten times more. However, cost is not the only barrier, as some functionality and organoleptic issues may also arise from the replacement of carrageenans with other texturizers. It remains to be seen whether, and how far, this trend would spill-over into non-premium brands and applications, or even into products where higher dose rates were required to achieve rheological functionality.

It will indeed be interesting to see how the North American situation plays out if such trends continue. Will the corporate, western companies invest, or take a stand against the media buzz in order to chase the North American market or is it too risky? Or, will they look to the rapidly growing Chinese and other Asian markets where consumer attitudes and market demands are different? Outside of North America, we are seeing a response to the wider global trends for "label friendly" ingredients and healthier foods and of course, carrageenans (like all other hydrocolloid additives) have been targeted in this respect, particularly in Europe where carrageenan carries an E number (i.e. E407 or E407a). We are aware that manufacturers in some sectors are addressing the issue of carrageenan replacement and undoubtedly would consider a cost-effective alternative if presented with one.

Replacement of hydrocolloids with label friendly alternatives is not a new theme, hydrocolloids are being substituted with ingredients that have a more natural perception. One exciting and growing area, is the use of functional food fibres. Functional food fibres are typically holistically processed vegetable, fruit and cereal materials which contain both soluble and insoluble fibre fractions, with different textural and often physiological functionalities. These fibres differ in functionality from standard fibre equivalents in that processing has been applied in order to activate, or modify, certain characteristics which provide improved water-binding, viscosity, gelling behaviour and mouth-feel properties. Some fibres even have fat mimetic properties on account of their superior water-binding capacity (Hotchkiss and Trius 2016a, b). Products made from oat, citrus, vegetable and cellulose-rich sources are currently in the market and are finding increased applications in healthy formulations. One seaweed fibre derived from *Gracilaria verrucosa* was launched into the market relatively recently (Phytafiber ® from Java Biocolloid). It is marketed on the basis of its natural image, water-binding and gelling functionality and as a source of dietary fibre. The soluble fibre component here is agar. The question arises as to why similar products are not available from the carrageenan-bearing, red seaweeds or indeed, for that matter, from alginate-bearing, brown seaweeds. There seem to be many opportunities to be realised in this regard.

Given our recent observations and returning to the general overview of the industry as depicted in Fig. 13.10 above, it was clearly evident to us that a further polarization of the industry between Asia and the west was beginning to occur. Both in terms of supply and markets. The carrageenan controversy was undoubtedly having an effect in the west (as of March 2017) but it was not a global issue and was unlikely to impact Asian markets, or suppliers in the near future. There had been no reported investment in the industry, by key players, in the Americas and Europe in recent years and some appeared to be simply boosting their efforts elsewhere in the hydrocolloids space, e.g. CP Kelco making further significant financial commitments to pectin in 2013. Is this indicative of further stagnation and possibly the start of the decline of the carrageenan industry in the west? This remains to be seen and will no doubt be followed by key commentators. With production and markets growing rapidly in Asia, we predict that this is the area where there will be the most innovation going forward.

References

Bixler HJ (2017) The carrageenan controversy. J Appl Phycol. doi: 10.1007/s10811-017-1132-4

Bixler HJ, Porse H (2011) A decade of change in the seaweed hydrocolloids industry. J Appl Phycol 23(3):321–335

CyberColloids (2012) Carrageenan industry report. July 2012

CyberColloids (2017) Carrageenan market report. March 2017

Dakay (2010) The future of the Carrageenan Industry. In: Proceedings SEABEXIII, Surabaya, Indonesia, July 15, 2010

Food Ingredients First (2017) Special report: seaweed and microalgae driving new product development. Food ingredients first website. Accessed March 2017

FoodBevMedia (2015) Article accessed online http://www.foodbev.com/news/seaweed-could-be-next-big-snacking-opportunity-firm-says. September 2015

Future Market Insights (2017) Carrageenan gum market: food and beverages application segment projected to record highest revenue growth throughout the forecast period: global industry analysis and opportunity assessment, 2016–2024

Grandview Research (2016) Commercial seaweed market analysis by product (Brown Seaweed, Red Seaweed, Green Seaweed), by form (Liquid, Powdered, Flakes), by application (Agriculture, Animal Feed, Human Consumption) and segment forecasts to 2024

Hotchkiss S, Trius A (2016a) Functional food fibres and their use in healthier fat reduced formulations. Monographic special issue: food: free from. Agro FOOD Industry Hi Tech – vol 27(3) – May/June 2016

Hotchkiss S, Trius A (2016b) Food fibers: creating healthy texture systems. Hydrocolloid Innovation. The World of Food Ingredients. March 2016

Hotchkiss S, Brooks M, Campbell R, Philp K, Trius A (2016) The use of carrageenan in food. In: Pereira L (ed) Carrageenans: sources and extraction methods, molecular structure, bioactive properties and health effects, 1st edn. Nova Science Publications Inc., New York, pp 1–293

Hotchkiss S, Campbell R, Hepburn C (2016b) Carrageenan: sources and extraction methods. In: Pereira L (ed) Carrageenans: sources and extraction methods, molecular structure, bioactive properties and health effects, 1st edn. Nova Science Publications Inc., New York, pp 1–293

Hurtado AQ, Neish IA, Critchley AT (2015) Developments in production technology of *Kappaphycus* in the Philippines: more than four decades of farming. J Appl Phycol 27(5):1945–1961

Markets and Markets (2016) Hydrocolloids market global forecast by type (Gelatin, Xanthan, Carrageenan, Alginate, Agar, Pectin, Guar, Locust Bean, Gum Arabic, and CMC), Function (Thickener, Stabilizer, Gelling, Fat Replacer, and Coating), Source, Application, & by Region – Global Forecast to 2020. Markets and Markets, February 2016

Neish IC (2012) MUZE. Maximise seaweed crop value using multi-stream, zero-effluent (MUZE) processing. Satoumi.co Monograph. MUZEV214OCT12.PDF

Neish IC (2013) Innovation as a driver for Indonesian seaweed trade and market development. Seaweeds aquaculture, processing and trade: prospects and developments. Jakarta, Indonesia, October 2013

Neish IC (2015) A diagnostic analysis of seaweed value chains in Sumenep Regency, Madura Indonesia. Report submitted for UNIDO Project no 140140 "preparatory assistance to develop a seaweed value chain development project for Madura"

Neish IC, Msuya FE (2013) Seaweed value chain assessment of Zanzibar. Report submitted for UNIDO Project No. 13083 "Building Seaweed Processing Capacities in Zanzibar and Pemba: creating value for the poor"

New Nutrition Business (2016) The next big opportunity in snacking: five steps to creating a success with seaweed. NNB Report 985

Porse H, Ladenburg S (2015) Seaweed value chain final report. Report submitted to Smart-Fish Indonesia under the UNIDO Seaweed Value Chain Programme

Seisun D (2016) Food hydrocolloids. IMR International hydrocolloid Conference, Barcelona, Spain, April 2016

Index

A.Q. Hurtado et al. (eds.), *Tropical Seaweed Farming Trends, Problems and Opportunities*, Developments in Applied Phycology 9, DOI 10.1007/978-3-319-63498-2

F

G

T

MIX
Papier aus verantwortungsvollen Quellen
Paper from responsible sources
FSC® C105338

If you have any concerns about our products,
you can contact us on
ProductSafety@springernature.com

In case Publisher is established outside the EU,
the EU authorized representative is:
Springer Nature Customer Service Center GmbH
Europaplatz 3, 69115 Heidelberg, Germany

Printed by Libri Plureos GmbH
in Hamburg, Germany